Technology and Sustainable Development

Technological change is at the core of all major disruptions in human history, and revolutions, wars, and general development are regularly connected to some sort of technological change. However, not all development is beneficial. While technology has fueled great innovations and rapid development, the notion of sustainable development has gained prominence as we now experience serious social, economic, and environmental challenges.

This book examines whether technology can be used to fix the very problems caused by technology, as the various chapters examine different aspects related to how technology has brought us where we are today (which some will say is the best place humanity's been at according to a range of metrics), and whether technology helps or hinders us in our efforts to solve the challenges we currently face. The issues discussed cover the three sustainability dimensions and include topics such as the materiality of AI, technology in education, AI for gender equality, innovation and the digital divide, and how technology relates to power, the political system, and capitalism. The chapters all build on the theoretical backdrop of technological change, sustainable development, and the UN's Sustainable Development Goals are actively used throughout this book, both to examine how these goals capture or overlook central elements of sustainable development, and also to facilitate and create a common framework of engagement between the chapters.

This book provides a novel combination of traditional theories that are explored through different case studies, providing the ground for a better understanding of how and when technology can – and cannot – be the enabler of sustainable development. It is thus an important resource for students of all disciplines, technologists, and those developing and applying new technologies. It is also a valuable resource for politicians and regulators attempting to harness the power of technology for good, while limiting its negative potential.

ABOUT THE EDITOR

Henrik Skaug Sætra is a political scientist with a broad and interdisciplinary background and approach, mainly focusing on the political, ethical, and social implications of technology. He focuses specifically on the sustainability-related impacts of AI and has previously published a book and several articles on AI and the UN's Sustainable Development Goals.

Technology and Sustainable Development

The Promise and Pitfalls of Techno-Solutionism

Edited by
Henrik Skaug Sætra

Routledge
Taylor & Francis Group
Milton Park and New York

Designed cover image: Getty Images

First published 2023
by Routledge
605 Third Avenue, New York, NY 10158

and by Routledge
4 Park Square, Milton Park, Abingdon, Oxon, OX14 4RN

Routledge is an imprint of the Taylor & Francis Group, an informa business

ISBN: 978-1-032-35059-2 (hbk)
ISBN: 978-1-032-35056-1 (pbk)
ISBN: 978-1-003-32508-6 (ebk)

DOI: 10.1201/9781003325086

Typeset in Minion Pro
by Apex CoVantage, LLC

Contents

Contributors

Harald Borgebund, Østfold University College, Norway. Harald Borgebund holds a PhD in political philosophy from the University of York (UK). His research and teaching interests are liberal democracy, democratic theory, political communication, and the role of capitalism and technology in contemporary society.

Benedetta Brevini, University of Sydney, Australia. Benedetta Brevini is Associate Professor of Communication at the University of Sydney and Senior Visiting Fellow at the London School of Economics. Her latest volumes are *Carbon Capitalism and Communication: Confronting Climate Crisis* (2017) and *Climate Change and the Media* (2018). *Is AI good for the Planet* (2021) is her newest work.

Marianna Capasso, Sant'Anna School of Advanced Studies, Italy. Marianna Capasso is Postdoctoral Research Fellow at the Sant'Anna School of Advanced Studies, Pisa, Italy. Her main research interests are philosophy of technology, applied ethics, and political theory. In particular, her research focuses on Neorepublicanism, Meaningful Human Control, Social Dimensions of Responsibility with AI-driven systems, and Value Sensitive Design.

Ricardo Colomo-Palacios, Universidad Politécnica de Madrid, Spain. Ricardo Colomo-Palacios is Full Professor at Universidad Politécnica de Madrid, Spain. Formerly, he worked at Østfold University College, Norway, and Universidad Carlos III de Madrid, Spain. His research interests include software governance, management information systems, software project management, people in software projects, software and services process improvement, and management information systems.

Anders Dechsling, Østfold University College, Norway. Anders Dechsling is researcher working within special education and with a focus on projects related to neurodevelopmental conditions and virtual reality. Additional research interests are in the domains of social media, substance use disorders, dementia, stimulus control, and behavior analysis.

Hadassah Drukarch, Leiden University, the Netherlands. Hadassah Drukarch is currently a student at the Advanced LL.M. in Law and Digital Technologies at Leiden University (NL). She also works as a paralegal at Considerati and is the founder and host of The Law of Tech Podcast which revolves around the interaction between law and digital technologies.

Jo Ese, Østfold University College, Norway. Jo Ese was a sociologist with a particular interest in the cooperation between academia and society. He was interested in public health partnerships between universities and local/regional governmental bodies on social stratification and social sustainability. He also worked on the digitalization of working life, organizations, and the public sphere.

Eduard Fosch-Villaronga, Leiden University, the Netherlands. Dr. Fosch-Villaronga, PhD, LLM, MA, is Assistant Professor and the Director of Research at the eLaw Center for Law and Digital Technologies at Leiden University (NL), where he investigates legal and regulatory aspects of robot and AI technologies.

Marco Giraudo, University of Turin, Italy. Marco Giraudo is currently a postdoc researcher at the University of Turin. He holds a master's degree in Comparative Law and a PhD in Law and Economics from the same university. His research focuses on the economics of the legal foundations of markets. Law, uncertainty, and legal instability are the issues at the core of his research agenda.

Imad Antoine Ibrahim, University of Twente, the Netherlands. Imad Antoine Ibrahim is Assistant Research Professor at the College of Law, Qatar University, Doha, Qatar. He has been working on global issues for the last decade. He is currently involved in several projects in places such as Europe, MENA region, Central Asia, and China.

Ivar Jonsson, Østfold University College, Norway. Ivar Jonsson, D.Phil., is Professor of Innovation and Entrepreneurship at Østfold University College, Norway. Previous affiliations include the universities of Sussex, Iceland, Greenland, Akureyri, Bifröst, Gothenburg, and Luleå University of Technology, Sweden. He has participated in international research projects on Arctic societies, political economy, digitalization, and sustainable development.

Petter Kvalvik, Institute for Energy Technology, Norway. Petter Kvalvik is Head of business development digitalization at Institute for Energy Technology and program manager for My Digital City. He has a background in safety and security-critical systems, strategic program and portfolio management, and applied computer science within data ecosystems, and is passionate about the human-centered digitalization of society.

Stuart Mills, University of Leeds, Great Britain. Dr Stuart Mills is Lecturer (Assistant Professor) in Department of Economics at the University of Leeds. His research focuses on behavioral economics, digital economy, and political economy.

Lilja Mósesdóttir, Østfold University College, Norway. Lilja Mósesdóttir is Professor of Business Administration at Østfold University College, Norway. Her previous affiliations include University of Manchester, University of Iceland, and Luleå University of Technology, Sweden. She has participated in European and Nordic research projects on gender and employment, innovation in the digital age, and sustainable development.

Anders Nordahl-Hansen, Østfold University College, Norway. Anders Nordahl-Hansen is Professor of Special Education at Østfold University College, Institute of Education, ICT and Learning, Norway. His research interests involve autism and neurodevelopmental conditions. He is head of the research group DeveLeP and associate editor of the journals Research in Developmental Disabilities and Frontiers in Digital health.

Erlend Ingridsønn Nordrum, Østfold University College, Norway. As a part of the research initiative The Digital Society at Østfold University College, Erlend investigates how digital technology affects the life chances of various groups. He is Doctoral Research Fellow at Østfold University College and holds a master's degree in Sociology from the University of Oslo.

Hanan Salam, New York University Abu Dhabi, UAE. Hanan Salam is Assistant Professor in Computer Science at New York University Abu Dhabi, where she is directing the Social Machines & Robotics Lab. She is also the co-founder of Women in AI, an international nonprofit whose mission is to close the gender gap in the field of AI.

Mary Sánchez-Gordón, Østfold University College, Norway. Mary Sánchez-Gordón is Associate Professor at the Computer Science and Communication Department of the Østfold University College, Norway. She holds a PhD degree in Computer Science from the Universidad Carlos III de Madrid. She has been working as Software Engineer, Project Manager, and Software Engineering Consultant in several companies.

Faridun Sattarov, University of World Economy and Diplomacy, Uzbekistan. Faridun Sattarov holds a PhD degree in Philosophy from the University of Twente and is Chair of the Political Science Department of the University of World Economy and Diplomacy, Uzbekistan. Previously, he worked as a researcher at the University of Twente, the University of Liverpool, the Technological University of Eindhoven, and the UNESCO.

Neil Selwyn, Monash University, Australia. Neil Selwyn has been researching and writing about digital education for the past 25 years. He is currently Distinguished Professor at Monash University, Melbourne. Recent books include *Should Robots Replace Teachers? AI and the Future of Education* (Polity, 2019) and the third edition of *Education and Technology: Key Issues & Debates* (Bloomsbury, 2021).

Henrik Skaug Sætra, Østfold University College, Norway. Henrik Skaug Sætra is a political scientist with a broad and interdisciplinary background and approach, mainly focusing on the political, ethical, and social implications of technology. He focuses specifically on the sustainability-related impacts of AI, and he has published *AI for the Sustainable Development Goals* (CRC Press, 2022) and several articles on AI and the UN Sustainable Development Goals.

Marisa Tschopp, scip AG, Women in AI, Switzerland. Marisa Tschopp is a corporate researcher at scip AG and associated researcher at IWM, where she conducts research about

Artificial Intelligence from a psychological perspective, with a focus on ethical implications. She is a frequent speaker at international events, including TEDx, and board member of the nonprofit Women in AI.

Steven Umbrello, Delft University of Technology, the Netherlands. Steven Umbrello is a postdoctoral research fellow at the Delft University of Technology. He is also Managing Director at the Institute for Ethics and Emerging Technologies. His research is on how to design emerging and transformative technologies for human values.

Richard Whittle, University College London, United Kingdom. Dr Whittle is a CAPE-funded fellow in economic policy, hosted at the Institute of Innovation and Public Purpose at UCL. He has a background in "critical behavioral public policy" and is Co-Investigator and Co-director of the £7 million Yorkshire Policy Engagement and Research Network.

CHAPTER 1

Introduction

The Promise and Pitfalls of Techno-solutionism

Henrik Skaug Sætra

CONTENTS

1.1 INTRODUCTION

Rarely does a day pass without us being reminded of the social, economic, and environmental challenges we now face. During 2 years of living with a pandemic that seemed to never pass, a series of social and economic issues emerged. Businesses struggled with restrictions and citizens' cautiousness, while regulators struggled to balance economic and business needs against uncertain – but serious – public health considerations. People also experienced pandemic fatigue generated through the numerous minor respites followed by the next and potentially more threatening Greek letter variety of the coronavirus. During this period, the social ramifications of lockdown and restrictions became abundantly clear, and so did the recognition that people were unevenly affected by the pandemic. In addition to the effects on our local communities and states, the pandemic highlighted challenges related to radical inequalities with regard to, for example, capacities for producing and procuring vaccines, but also to use national resources to support and maintain citizens and businesses.

Simultaneously, we can no longer avoid being exposed to the reality of various environmental challenges threatening to drastically alter the trajectory of our future as a species. Climate change and the loss of biological diversity are two key issues, and while some are concerned for the natural world because they consider it to be valuable in itself, others are mainly concerned because these environmental changes are having major social and economic repercussions – for humans.

DOI: 10.1201/9781003325086-1

Meanwhile, all these challenges are continually being assessed and attempted tackled by researchers, developers, and businesses. More often than not, technology is heralded as the cure for our ills. In the face of COVID-19, for example, vast amounts of research were conducted, and companies launched new AI- and Big Data-based solutions aimed at getting us out of the predicament. Apps for tracking infections, Big Data analysis for pandemic control, and not least AI-based solutions aimed at diagnosing and understanding COVID-19 in a variety of ways. As the tidal wave of research and solutions waned, we found ourselves with meager successes stemming from AI and Big Data (Chakravorti, 2022). The technology that enables vaccines, however, stood out as a very effective technological remedy for the most severe effects of the virus. But vaccines also amply demonstrate how technology relates to inequality. While developed nations with high coverage contemplated a third and fourth booster dose for their population, developing and least developed nations had hardly gotten started. Inequity aside, the unrestricted spread of the virus in some regions can also be a cause of ever-new strands of the virus which threaten to undermine the efforts of the developed and developing nations alike. These are issues where technology and sustainability interlink, and where we see that enabling universal, equitable, and fair access to technology on a global scale is often the only way toward effectively facing challenges (United Nations, 2015).

Climate change is a particularly interesting example of how technology relates to efforts to develop new modes of operation and new solutions which might allow us to escape the direst consequences of humanity's industrial activity. Some argue that the only true solution is to radically change our ways – consume less, produce less, and let go of the notion of growth as we take the notion of *limits* seriously (Farley & Smith, 2020; Latouche, 2009). Others, however, place their faith in *green growth* and human ingenuity (Jacobs, 2013). For the latter group, technology is essential, as it allows for the development of new products with reduced environmental impact, for finding substitutes for resources exploited beyond repair, and not least for manipulating – and even fixing and restoring – our environment (Cao, 2011). Geoengineering and technologies such as carbon capture and storage (CCS) are illustrative examples of the search for technological solutions for mitigating and adapting to climate change (Stuart et al., 2020). By developing new technologies and solutions, we can – or might – add to existing technologies to solve the challenges created by more primitive technologies. The result being a socio-technical system in which the technical elements are ever more complicated (Winner, 1977), and its effective operation is increasingly important for the future prospects of humankind, our environment, and all other species that happen to coinhabit this world of ours. This gives rise to questions regarding humanity's fundamental relationship with technology. Is it a curse, a cure, or even both (Müller, 2016)?

1.2 TECHNO-SOLUTIONISM AND HUMANITY'S SYMBIOTIC RELATIONSHIP WITH TECHNOLOGY

The subtitle of this book refers to *techno-solutionism*, which is in simple terms the idea that we can and should use technology to solve the challenges we happen to face. It relates to the notion that everything is a nail to a person with a hammer, as most problems can be fixed with a more comprehensive application of technology for the most ardent techno-solutionists. The notion of technological "fixes" is another way to describe this attitude (Drengson, 1984). Morozov (2013) describes how *solutionism* entails a faith in technology, but also a tendency to fundamentally change how we perceive and analyze social phenomena. The notion of faith also suggests a tight link to a closely related term, namely, techno-*optimism*. While techno-optimism most obviously points toward an openness to the possibility that technology can provide a better world (Danaher, 2022), techno-solutionism is more comprehensive. It also entails the step that we can, but need not, act on such optimism in the active pursuit of an agenda in which we organize our societies in ways that make them amenable to the technological solutions we perceive. Exploring how technology changes both our approach to and understanding of fundamental sustainability-related challenges is the central objective of this book. Technology is not just a tool to be used instrumentally for human purposes, as it has profound effects on how we think, the solutions and opportunities we perceive, and even how we encounter and relate to each other (Müller, 2016).

A different concept related to the techno-solutionist approach, particularly as technology relates to our relationship with the environment, is *prometheanism* (Müller, 2016), which originates in the ancient Greek myth of the titan Prometheus. For example, Farley and Smith (2020) argue that the belief that technology can ultimately replace natural ecosystem services suggests a *promethean* perspective, and the terms prometheanism and techno-optimism are widely used in environmental ethics and environmentalism. Popular targets of criticism by the prometheans are the skeptics that have been proven wrong, and few are mentioned more often than Thomas Malthus. In 1798, he released *An Essay on the Principle of Population* (Malthus, 1798), in which he argued that the earth's limitations would inevitably also limit the human population. Long after Malthus, the Club of Rome released *The Limits to Growth* (Meadows et al., 1972), in which they similarly argued that we were running into hard limits that would eventually limit human development. Malthus was certainly proven wrong by technological developments in, for example, agriculture, and most now also seem to argue that technological development has revealed clear limitations in the modeling which is the basis of *The Limits to Growth*.

Prometheus was a titan who stole fire from the Gods and gave it to the mortals, much to Zeus' and the other God's frustration. This fire is usually interpreted as *techne* – rationality, art, and knowledge – which is the cause of human development in terms of increased technological power and capabilities for domesticating and harnessing the natural world (Aeschylus, 2012; Müller, 2016). The price Prometheus paid for this theft was twofold: He was bound and eternally tormented by an eagle eating his ever-regenerating liver, but he was also left concerned with whether he improved or worsened the situation of the mortals. The story is timeless, and the story of the garden of Eden and the tree of knowledge is an early example, while there are countless more recent stories related to the challenges caused by taking our technologies too far. One particularly famous example is Mary Shelley's (2012) *Frankenstein*, aptly subtitled "The Modern Prometheus".

Techno-solutionism is often referred to as prometheanism because it entails a faith in the notion that we can control and take charge of the world in which we live, and that we have sufficient knowledge and technology to do this in a way that will *improve* our situation. This book examines whether such an approach is well founded, and in particular whether it can help us face and effectively solve the challenges related to environmental, social, and economic sustainability.

Technological change is at the core of all major disruptions in human history, and revolutions, wars, and general development are often connected to and explained by some sort of technological change (Barley, 2020). When I say revolutions, I mainly refer to the industrial kind, which are inextricably linked to the emergence of both new technologies and, more importantly, socio-technical systems. By emphasizing the latter, I allude to a recurring topic in the following chapters, namely that focusing on technologies in isolation will rarely allow us to grasp the full potential or all of the pitfalls that accompany them.

Engineers, developers, and analysts of technology arguably tend to focus on how a particular technology can be used to achieve certain beneficial effects – how it allows us to solve a specific challenge. However, developing and applying technology entails consequences far beyond those intended by their progenitors, and any approach not factoring in such consequences is referred to as an *isolationist* approach to technology and technological change (Barley, 2020).

To really understand the implications of new technology, this book advocates for a broader approach to the analysis of techno-solutionism – one that takes account of the interdependence of different technologies and processes in what Barley (2020) refers to as *stacks*, and also the indirect ripple effects technology has across different social, economic, and environmental domains (Farley & Smith, 2020; Sætra, 2022b). The notion of unintended effects is a well-known term for parts of what must be accounted for in such an approach, but even more so is the notion that technologies are interlinked in complex socio-technical systems (Winner, 1977). In such systems, changes in particular technologies entail changes in the system as a whole. Of crucial importance is the realization that individuals and our societies are integral parts of these systems and thus are also affected by and in a position to influence technological change (Morozov, 2013).

1.3 SUSTAINABILITY AND TECHNOLOGY

Technology is also linked to how humans interact, what sort of traits lead to success for both individuals and groups, and what sort of political arrangements make sense. Technology is arguably what has allowed humans to develop into what we are today (Müller, 2016); our use of various primitive and advanced tools, constructing buildings as shelters, ways of farming land, and medicines and science are crucial components of what make life as we know it possible. Saying that all technology is bad is consequently close to absurd. However, this book is premised on the idea that not all *development* is beneficial (Næss, 1999). While technology has fueled great innovations and rapid development, the notion of sustainable development has now gained prominence because we experience great social, economic, and environmental challenges due to the very growth technology has enabled.

Sustainability is often equated with the notion of sustainable *development*, with its mainstream definition originating with the UN report *Our Common Future* written by the Brundtland commission (Brundtland et al., 1987). The commission emphasized how *sustainability* encompasses more than just the environmental dimension, and that in order to solve environmental challenges we must also focus on social challenges such as inequality and poverty, and economic issues related to, for example, innovation and inclusive economic growth.

As the evidence of environmental threats continues to amass, the use of the term sustainability has proliferated and become close to ubiquitous (Farley & Smith, 2020). The need to face these threats is perceived as increasingly obvious, and politicians and businesses alike scramble to find the best path toward sustainable development. Everyone agrees that we need to change our ways, and everyone agrees on ambitious targets for the future, with a particular focus on the years 2030 and 2050 (Guterres, 2020; UNFCCC, 2022). But far fewer agree on what actually needs to be done. And all the while, emissions are still rising, biological diversity is reduced, and our oceans are filled with plastics; humanity faces a number of health-related challenges we do not fully understand, but which we assume to be associated with, for example, biological diversity, our use of chemicals, and the food we eat (IPBES, 2022). In short, the world community's ambitions are high, but we are arguably not yet on a path to solving our problems.

These ambitions have been codified in a number of international frameworks, which will be explored in more detail in the next chapter. The most important framework today is arguably the UN's Sustainable Development Goals (SDGs) (United Nations, 2015), which consist of 17 goals and 169 targets, which constitute the *Agenda 2030*. These goals relate to social, economic, and environmental sustainability, and examples of goals are the elimination of poverty, improved health, ending discrimination, reducing inequality, decent work and economic growth, and combatting climate change.

As the SDGs have become the *lingua franca* of political and corporate sustainability, they will also be given extended attention throughout this book. Both because they are widely used and consequently important in the public discourse and because we will argue that reducing sustainability to the SDGs entails real dangers of not achieving the kinds of changes really needed. First, the set of goals does not cover all important aspects related to sustainability, and as such the framework must be complemented by other approaches (Sætra, 2022b). Second, the goals themselves are interrelated and even partially contradictory (Farley & Smith, 2020), and this necessitates a deeper analysis of the nature of such contradictions and how to prioritize our goals. Third, the SDGs and even the original concept of sustainable development are based on certain foundational assumptions that are potentially deeply problematic. These relate to whether or not there are meaningful limits to growth, and whether, for example, *strong* "green growth" might allow us to simultaneously achieve economic growth and improved environmental conditions (Jacobs, 2013). Connected to this is the core assumption questioned in this book, namely that technology can be used to solve and overcome all these challenges, which also makes technology an integral component of green growth.

A key question asked by the contributors is whether technology can be used to fix the very problems caused by technology, as the various chapters examine different aspects related to how technology has brought us where we are today, and whether technology helps or hinders us in our efforts to solve the challenges we currently face. We do take seriously the fact that technological change has played a vital role in allowing us to, for example, be more numerous, live longer, and be healthier. Some even say that we are at the historical peak of humanity's development, at least according to a large number of metrics (Pinker, 2011). However, this book emphasizes that these successes are also accompanied by a number of fundamental challenges related to social and environmental integrity.

The better part of this volume consists of cases where technology is used to overcome sustainability-related challenges. The cases are different, and so are the theories used and the technologies discussed. In unison, however, they provide the foundation for answering the core question asked: Does technology provide us with the means to solve sustainable development? As will become clear, the answer to this question is far more complicated than it might at first appear to be. This is partly because we show that technology can be both an enabler and inhibitor of sustainable development and the SDGs. However, it is also because the notion of sustainability must be analyzed in some detail before an answer is sought, and I thus support Farley and Smith (2020), who argue that mainstream understandings of sustainability are deeply flawed and in need of a corrective. This is partly because of the numerous assumptions often encompassed in people's use of the term, and to strengthen the shaky theoretical foundations of sustainability as a concept this book will at times return to a more fundamental question: What are the core values we seek to reach through sustainable development and the use of technology? By returning to the technology of philosophy of, for example, Winner (1977), combined with Næss's (1999) deep ecological critique of technology, this book provides a more nuanced conclusion on the proper role of technology in reaching sustainable development.

1.4 WHO IS THIS BOOK FOR?

Anyone interested in both the general and more specific implications of technology should find this book a valuable resource. It will allow engineers of all kinds, including developers and computer scientists, to better account for the impacts of what they develop, which is seen as increasingly important as various ethics of technology gain prominence (Dotan, 2021; Sætra & Danaher, 2022).

It will also be an important resource for managers and others involved in building technology-related businesses. As sustainability is gaining traction in financial markets, where it is often referred to as Environment, Social, Governance (ESG), investors, regulators, and business partners are increasingly interested in learning about the sustainability-related impacts of and on all businesses (Sætra, 2021, 2022a). Without a proper understanding of how technology is related to sustainable development and the SDGs, reporting on mandatory and voluntary frameworks will be exceedingly difficult. It is already evident that poor understanding of and reporting on such implications are punished in the market, and this book can help generate a fundamental understanding of the key relationships between technology and sustainable development.

Another primary audience for this book is politicians and regulators. Controlling technology is notoriously difficult (Collingridge, 1980), and what is referred to as the *pacing problem* describes how technology tends to outpace regulation and regulatory frameworks (Downes, 2009), something that might easily lead to situations in which techno-solutionism is allowed to shape our societies in ways unrooted in democratic processes and democratic will. A fundamental understanding of how technology enables or prevents sustainable development is essential for regulators to effectively shape technological development in the interest of our societies and pre-empt undesirable consequences as much as possible.

Penultimately, anyone working on promoting the UN's SDGs, or sustainable development in general, will find this book's specific focus on this framework useful for understanding both the potential and limitations of using technological fixes for sustainability-related challenges. It will also challenge often unstated assumptions related to the desirability of growth and the absence of hard limits to our societies' – or economies' – growth. Regardless of what position the reader ends up taking, it will be based on a consideration of a varied set of arguments and thus stronger.

Finally, students of any stripe, at all levels, will find something useful in this book. Regardless of what discipline one studies, technology is of some relevance, and it is arguably becoming increasingly important – or at least attracting more attention. Likewise, tomorrow's and today's students will most certainly have to continue the quest for solutions to the sustainability-related challenges generated through generations, and this will require all students to have some knowledge of what sustainability is and how technology relates to it.

1.5 STRUCTURE OF THIS BOOK

In this book, the reader will find a wide array of approaches to the question of how technology relates to sustainable development. To establish the basis on which to build, it begins with an exploration of the key concepts and the theoretical backdrop required to answer the core questions posed in this introduction. The UN's SDGs are actively used throughout this book, both to examine how these goals capture or overlook central elements of sustainable development, and to facilitate and create a common framework for engagement between the chapters. These goals are presented in Chapter 2 alongside other key concepts, with a particular emphasis on sustainability, sustainable development, technology, and technological change. Following this are chapters dealing with specific cases and examples of how technology hinders or helps the achievement of various areas related to sustainable development.

First, two chapters focusing on environmental sustainability follow. In Chapter 3, Benedetta Brevini discusses artificial intelligence and artificial solutions – focusing on the materiality of technology and how AI relates to the climate emergency. Marianna Capasso and Steven Umbrello then discuss the potential and limitations of geoengineering and technologies of planetary control in Chapter 4.

Next are five chapters dealing with social sustainability. In Chapter 5, I and Jo Ese discuss how trans people are using social media filtering to create a less hostile online environment. Then, Neil Selwyn establishes the key aspects to consider regarding the use of technology in education in Chapter 6. Anders Dechsling and Anders Nordahl-Hansen's Chapter 7 proceeds to focus on the specific case of how virtual reality technology is used in Autism Spectrum Disorder interventions. Issues of inequality are discussed in both Chapters 6 and 7, and Erlend Ingridsønn Nordrum drills down on inequality and the digital divide in Chapter 8. This is followed by Marisa Tschopp and Hanan Salam's Chapter 9, which focuses on SDG 5 and the question of gender inequality in – and within – AI.

Moving on to the economic, political, and legal aspects of technology and sustainable development, Eduard Fosch-Villaronga, Hadassah Drukarch, and Marco Giraudo present a legal sustainability approach to digital innovation in Chapter 10. This is followed up in Chapter 11, where Lilja Mosedottir and Ivar Jonsson explore issues of governance toward sustainable development. Politics is also central as Harald Borgebund proceeds to analyze the relationships between capitalism, sustainability, and democracy in Chapter 12. How politics is effected is also the topic of Chapter 13, where Stuart Mills and Richard Whittle deal with how nudges can be understood as ideational technologies – emphasizing *green* nudges.

The importance of politics is also stressed by Imad Antoine Ibrahim in Chapter 14, as he discusses what he refers to as the "fallacy of disruptive technologies". Proceeding to provide a possible answer to the question of why politics is seen as so important for understanding the potential of technology, Faridun Sattarov discusses how technology relates to the distribution of power in Chapter 15. Knowledge is power, and one might consequently argue that so is data. Proper governance of data is thus a vital issue, dealt with by Petter Kvalvik, Sánchez-Gordón, and Ricardo Colomo-Palacios in Chapter 16.

The final part of this book moves into the domain of radical approaches to sustainable development, starting with Ivar Jonsson and Lilja Mosedottir's Chapter 17 on Techno-solutionism facing post-liberal oligarcy. In Chapter 18, I discuss the role of technology in the radical approach to a more sustainable future, drawing on the work of, amongst others, Arne Næss.

In sum, this book provides a novel combination of traditional theories that are explored through different case studies, providing the ground for a better understanding of how and when technology can – and cannot – be the enabler of sustainable development. In the concluding Chapter 19, the overall findings presented by the different contributors are synthetized, with a particular focus on identifying the generally valid takeaways regarding the positive potential for using technology to achieve sustainable development, but also the pitfalls to avoid if such development is to be realized.

1.6 REFERENCES

Aeschylus. (2012). *Prometheus bound* (D. H. Roberts, Trans.). Hackett Publishing Company.

Barley, S. R. (2020). *Work and technological change*. Oxford University Press.

Brundtland, G. H., Khalid, M., Agnelli, S., Al-Athel, S., & Chidzero, B. (1987). *Our Common Future*: Report of the World Commission on Environment and Development. United Nations General Assembly document A/42/427.

Cao, S. (2011). Impact of China's large-scale ecological restoration program on the environment and society in arid and semiarid areas of China: Achievements, problems, synthesis, and applications. *Critical Reviews in Environmental Science and Technology*, *41*(4), 317–335.

Chakravorti, B. (2022, March 17). Why AI failed to live up to its potential during the pandemic. *Harvard Business Review*. https://hbr.org/2022/03/why-ai-failed-to-live-up-to-its-potential-during-the-pandemic

Collingridge, D. (1980). *The social control of technology*. Frances Pinter.

Danaher, J. (2022). Techno-optimism: An analysis, an evaluation and a modest defence. *Philosophy & Technology*, *35*(54). https://doi.org/10.1007/s13347-022-00550-2
Dotan, R. (2021). *The proliferation of AI ethics principles: What's next?* https://montrealethics.ai/the-proliferation-of-ai-ethics-principles-whats-next/
Downes, L. (2009). *The laws of disruption: Harnessing the new forces that govern life and business in the digital age*. Basic Books.
Drengson, A. R. (1984). The sacred and the limits of the technological fix. *Zygon*, *19*(3), 259–275. https://doi.org/10.1111/j.1467-9744.1984.tb00929.x
Farley, H. M., & Smith, Z. A. (2020). *Sustainability: If it's everything, is it nothing?* Routledge.
Guterres, A. (2020, December 11). *Carbon neutrality by 2050: The world's most urgent mission*. www.un.org/sg/en/content/sg/articles/2020-12-11/carbon-neutrality-2050-the-world's-most-urgent-mission
IPBES. (2022). *Nexus assessment: Thematic assessment of the interlinkages among biodiversity, water, food and health*. https://ipbes.net/nexus
Jacobs, M. (2013). Green growth. In R. Falkner (Ed.), *The handbook of global climate and environment policy* (pp. 197–214). Wiley Blackwell.
Latouche, S. (2009). *Farewell to growth*. Polity.
Malthus, T. R. (1798). *An essay on the principle of population*. J. Johnson.
Meadows, D. H., Meadows, D. H., Randers, J., & Behrens III, W. W. (1972). *The limits to growth: Report for the club of Rome's project on the predicament of mankind*. Universe Books.
Morozov, E. (2013). *To save everything, click here: The folly of technological solutionism*. Public Affairs.
Müller, C. J. (2016). *Prometheanism: Technology, digital culture and human obsolescence*. Rowman & Littlefield.
Næss, A. (1999). *Økologi, Samfunn, Livisstil*. Bokklubben dagens bøker.
Pinker, S. (2011). *The better angels of our nature: Why violence has declined*. Viking.
Sætra, H. S. (2021). A framework for evaluating and disclosing the ESG related impacts of AI with the SDGs. *Sustainability*, *13*(15). https://doi.org/10.3390/su13158503
Sætra, H. S. (2022a). The AI ESG protocol: A tool for evaluating and disclosing impacts and risks of AI and data capabilities, assets, and activities. *Sustainable Development*. https://onlinelibrary.wiley.com/doi/epdf/10.1002/sd.2438
Sætra, H. S. (2022b). *AI for the sustainable development goals*. CRC Press.
Sætra, H. S., & Danaher, J. (2022). To each technology its own ethics: The problem of ethical proliferation. *Philosophy & Technology*, *35*(93). https://link.springer.com/content/pdf/10.1007/s13347-022-00591-7.pdf?pdf=button)
Shelley, M. W. (2012). *Frankenstein: Or, the modern Prometheus*. Broadview Press.
Stuart, D., Gunderson, R., & Petersen, B. (2020). Carbon geoengineering and the metabolic rift: Solution or social reproduction? *Critical Sociology*, *46*(7–8), 1233–1249. https://doi.org/10.1177/0896920520905074
UNFCCC. (2022). *Paris agreement*. https://unfccc.int/process-and-meetings/the-paris-agreement/the-paris-agreement
United Nations. (2015). *Transforming our world: The 2030 agenda for sustainable development*. Division for Sustainable Development Goals.
Winner, L. (1977). *Autonomous technology: Technics-out-of-control as a theme in political thought*. MIT Press.

CHAPTER 2

Key Concepts

Technology and Sustainable Development

Henrik Skaug Sætra

CONTENTS

2.1 INTRODUCTION

Some concepts feel so familiar that a definition hardly seems necessary. Sustainability and artificial intelligence (AI), for example, are ubiquitous terms that most of us hear and even use daily. It might be tempting, then, to delve straight into the specifics of how technology relates to, for example, inequality, environmental degradation, and political stability and democracy. A central premise of this book, however, is that the key concepts discussed in the introduction are deeply ambiguous and potentially problematic. A crucial first step is consequently to explain how we use – and don't use – concepts such as technology, solutionism, and sustainable development – the concepts also referred to in the title of this book.

The goal of this chapter is, however, not to arrive at the one true meaning of the concepts in use. It is rather to explore and discuss various mainstream understandings, and partly to highlight which understandings are used throughout this book. Monism is not the goal, and in the chapters that follow, the authors pick up on particular threads and concepts established in this chapter, in order to further develop and analyze common understandings and definitions. *Sustainable development*, in particular, will be subject to sustained scrutiny, and it is only in the final chapters that we gather these conceptual threads to reach a more conclusive position on how sustainability-related concepts should be understood in relation to technology.

DOI: 10.1201/9781003325086-2

This chapter contains brief explanations of how we use the terms technology and technological change, which are central in the analyses in the following chapters. Also, key concepts related to sustainability are detailed. Sustainability, sustainable development, and the UN's Sustainable Development Goals (SDGs) take the main stage, along with the various explicit and implicit assumptions associated with these.

2.2 TECHNOLOGY AND TECHNOLOGICAL CHANGE

Technology is all around us, and while we all use and rely on technology every day, properly defining it is a challenging endeavor. The title mentions technology and sustainable development in general, and the natural question, then, is: What do we mean by *technology*? The introduction has already established that our approach to technology is broad and what we refer to as holistic. Much recent research focuses on the most obvious recent innovations in technology, such as AI, virtual and augmented reality, social robots, etc. These are indeed encompassed by what we refer to as technology, but while some chapters focus on such technologies, the overall approach of this book is to treat technology in a more fundamental sense.

2.2.1 What is Technology?

> *Technology is a bastard child of uncertain parentage, the result of a twisted genealogy cutting across multiple discourses. No scholarly discipline owns this term.*
>
> *(Schatzberg, 2018)*

The history of the philosophy of technology reveals how challenging it is to pin down what it is and how best to study it (Coeckelbergh, 2020). A central approach in Western thought has been an instrumental account of technology which sees technological objects as *neutral* tools that can be used for either good or evil purposes (Müller, 2016; Schatzberg, 2018). For example, guns can be used to repress or liberate, and AI can be used to empower and assist people or to manipulate and exploit them. Such an account shifts the entire focus to those who wield technology and neglects consideration of how technology changes – often in very subtle ways – *what* people can do and *how* they can do them. It entails mainly focusing on technology as instruments, such as various tools and machines.

Others, such as Ellul (1964) in *The Technological Society*, consider technology to permeate *everything*, including how we organize human activity. He describes how technology relates to mechanization and machines, but more importantly how these aspects of technology are inextricably linked to human organization and human activity in general. This approach is sometimes referred to as the *cultural* approach to technology, and Lewis Mumford is another important representative of this approach (Schatzberg, 2018). This book will not primarily focus on human organization, but we acknowledge that the impacts of technology are much broader than what can be understood through an analysis of mechanical technology alone.

Langdon Winner (1977) argues that Ellul's broad understanding of the concept corresponds well with the ubiquity of technology in modern society, and that it encompasses not just what we *make* but also what we *do*. He does, however, agree that technology as a concept is riddled with ambiguity and imprecision, which is obviously problematic in terms of fostering some sort of common understanding of what the term entails. When a term has come to mean just about anything, it runs the risk of becoming nothing. Winner (1977) states that this has been the case for technology, while Farley and Smith (2020) argue that the same is now true for sustainability.

One goal of this book is to ensure that the concepts used are defined to such a degree that the reader will be able to understand what is referred to in the discussions of technology and sustainable development, laying the grounds for what Arne Næss (2016) referred to as reasonable debate. Such debates require that the participants define the terms they use and ensure that all participants in the debate approximate a joint understanding in order to avoid "skin-deep disagreements" (Næss, 2016).

Conceptual tidiness is important as the definition of technology is sometimes presented as "a mess" (Schatzberg, 2018). The perspectives adopted in the coming chapters will vary with regard to the choice of a specific or foundational approach to technology, as different approaches enable us to highlight different implications of technology. This will, however, subsequently require some effort to generate a complete picture of what is actually implied by the different chapters, and this will entail taking it all back to a cultural understanding of technology in Chapter 18 and the conclusion in Chapter 19. This enables us to explain and analyze how technology will have specific beneficial effects – for example, on innovation related to new energy sources – while simultaneously having consequences for how we organize our societies and perceive both ourselves as individuals and as parts of a society. The latter elements take us all the way to the constitutive aspect of technology and how it shapes our perceptions of possibilities, ourselves, and others (Müller, 2016).

2.2.2 Technological Change

Technology is, however, never static, and the study of how technology relates to sustainable development requires us to understand how technological *change* leads to changes in the sustainability-related impacts of technology. We do so in line with the approach developed by the organizational theorist and industrial sociologist Stephen R. Barley (2020).

A central aspect of his approach is to distinguish between *substitutional* and *infrastructural* technological change. When technological change mainly result in existing tasks being performed more effectively, without changing the broader socio-technical context, this is *substitutional* change (Barley, 2020). For example, robots that help nurses with a limited set of physically taxing tasks, such as lifting patients, could in theory be seen as mainly substitutional, as it makes workers a bit more effective, without necessarily changing the structure of work, institutions, educations, etc. (Sætra & Fosch-Villaronga, 2021).

However, technology will often also entail deeper changes, and these tend not to be immediately obvious. If technology enables new forms of education provision, for example, this could have a wide range of effects on aspects such as students' need to relocate for studies or not and the number, kind, and education of teachers and professors. This might in turn change policies related to both employment and education, and influence the viability of smaller communities, etc. When technology has these broader effects, this is *infrastructural* change (Barley, 2020). Following the example of a robot in healthcare, we can imagine a situation in which robots have more fully replaced human caretakers, as described by, for example, Sharkey and Sharkey (2012), Coeckelbergh (2012), and Sparrow (2016). If this leads to situations in which traditional healthcare workers become obsolete, the nature of eldercare facilities radically changes, and our perceptions of what such facilities are, and when the elderly should be in such facilities, the ground for infrastructural change is prepared (Sætra & Fosch-Villaronga, 2021).

As a starting point, I posit that pure substitutional change is relatively rare and that most technological change will inevitably involve wider societal effects. This makes technological change a more serious matter than it would have been otherwise, because infrastructural change necessarily entails change in power relations and shifts in social structures. Technology can, to a certain degree, be argued to empower workers and individuals in general, but the effects on structural power and the broader distribution of power in society are arguably more important as it carries the potential for more radical social changes (Sattarov, 2019). This links technological change directly to fundamental political processes and sustainable development.

A classic example is how the introduction of snowmobiles changed the lives of Skolt society in Skolt Lapland (Pelto, 1987). As a traditional reindeer herding community, the snowmobile was initially seen quite simply as a tool for more effectively herding and gathering reindeer. However, it ended up leading to fundamental changes in all aspects of their society, including societal institutions, social relations, the economy, and the distribution of wealth and work. This occurred because the snowmobiles allowed for new scales and modes of herding reindeer, and the full implications of this development were hard to anticipate and understand. As noted by Collingridge (1980), it is often the case that when technologies are new it is easy to regulate them, but exceedingly difficult to foresee their consequences. On the other hand, mature technologies have known implications, but regulating or limiting them often seems close to impossible. This is known as Collingridge's dilemma. In the domain of sustainable development, to which we soon turn, the notion of the *precautionary principle* relates quite directly to such challenges (Brundtland et al., 1987).

2.2.3 The Materiality of Technology

Changes in social power structures and social organization are not the end of technological change, however. To understand the environmental implications of technology, we must also recognize the material basis of technology (see Brevini (2021) and Chapter 3). Whenever we discuss technologies in the form of physical devices and machines, they are quite obviously tied to the material world, and this also applies to new and smart digital technologies. Computer systems, for example, are composed of a wide range of materials, and they consume power and consequently have direct environmental impacts (Brevini, 2021). In addition to this, the use and application of such machines will also have impacts "downstream", as they might both exacerbate and remediate, for example, climate change and the loss of biological diversity. One way to account for such effects is through what Barley (2020) refers to as *embedded* analysis of technology, which entails including processes of extraction, refinement, components, subsystems, etc., in our analysis of technological change (Figure 2.1).

In order to make use of such an approach, each technology can be presented as a *stack* of technology and materials (Barley, 2020), which show how all technologies link back to basic material processes such as iron mining and petrochemical extraction (Figure 2.2). Such analyses not only highlight the links between technology and materiality but also allow us to consider how the technical and material bases of our object of analyses have

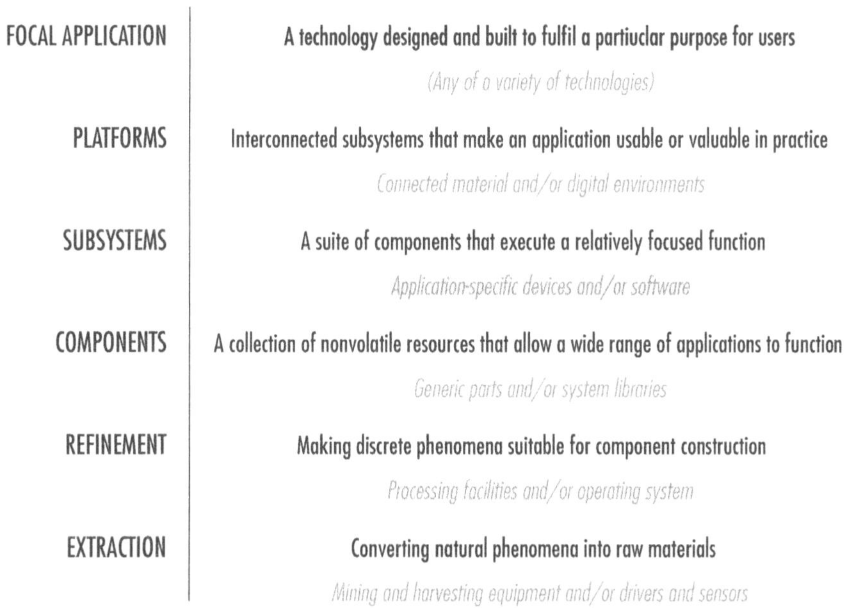

FIGURE 2.1 Embedded analysis of technological change. (After Barley, 2020.)

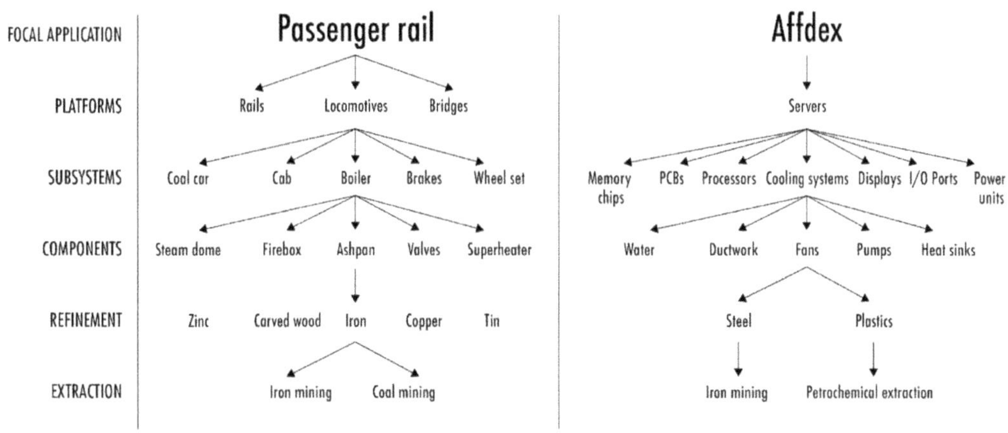

FIGURE 2.2 Two different technology “stacks”. (After Barley, 2020.)

broad *social* implications. New technologies might, for example, change or create a dependence on or new demand for materials that lead to shifts in entire industries. Once we start to account for raw materials and whole value chains, issues related to how minerals are mined and exploited, often in situations not particularly compatible with the goals of Agenda 2030 and human rights, can more easily be factored in.

2.2.4 Levels of Impacts – Micro, Meso, and Macro

Finally, when analyzing the implications of technology, it is often useful to distinguish between implications on the micro-, meso-, and macro-levels (Sætra, 2022). The micro-level refers to individuals, and how, for example, a single reindeer herder's situation was affected by getting access to a snowmobile. The meso-level relate to groups of various sizes, and implications at this level will often revolve around how different technologies are not equally accessible or useful to all. For example, one group of reindeer herders might have had more resources and an organizational and social structure that allowed them to more quickly adopt the snowmobile in their operations than another group. This might provide the resource-rich group key competitive advantages over the other. On the macro-level, we consider higher social and political levels and the economy as a production system (Jonsson, 2016). The macro-level could be national, regional, or global, depending on the scope of the analysis. Here, for example, we might find that the snowmobile led to rapid economic growth, new demand for educational offerings, shifts in the structure of businesses and demand for goods, and the need for social innovation to deal with unemployment.

The distinction between the different levels will be particularly important for highlighting how technology will often have quite different effects on different levels (Sætra, 2021a). For example, Gellers and Gunkel (2022) use this framework of analysis to explore how emerging technologies impact international human rights on the micro-, meso-, and macro-levels. What might be good for certain individuals, for example, might lead to increased inequalities between groups, while simultaneously producing beneficial effects on the macro-level related to innovation and the scientific foundations required to reach the SDGs. Distinguishing between the levels will then also allow us to account for how effects on the meso-level might in turn impact both individuals and the macro-levels, as shown in Figure 2.3.

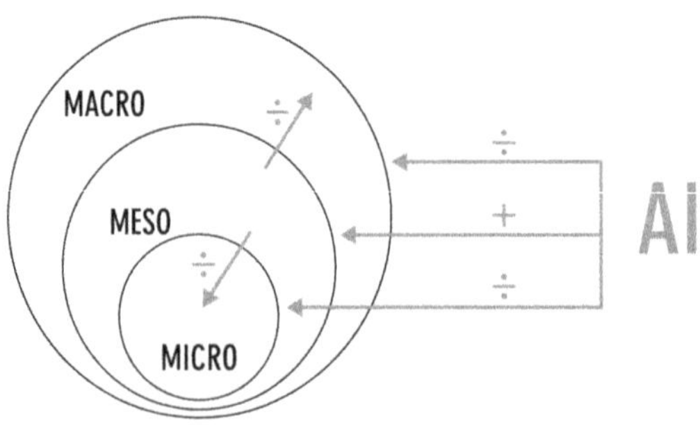

FIGURE 2.3 Three levels of analysis. (Courtesy of Satra, 2021a.)

2.3 SUSTAINING THE DEVELOPMENT OF WHAT AND HOW?

The concept of sustainability might have an intuitive core – related to matching inputs with outputs – and as long as we're talking about an isolated resource it is relatively easy to evaluate whether a particular practice is sustainable or not for a given time period. In scientific forestry, for example, the notion of sustained yield represents the idea that humans can, in theory, use a particular resource forever if they do so in a specific way (Guha, 2014). When Guha (2014) compares three varieties of environmentalism, *scientific conservationism* is presented as an approach focused on using science and technology to control the excesses of human activity. The other main approaches which he refers to as *back-to-the-land* and *the wilderness idea* are far less concerned with figuring out how to engender effective and lasting exploitation of natural resources, and focus instead on what might be *lost* with the advent of technology and the industrial society (Guha, 2014).

One problem is that the term "sustainability" is now used by different people to describe radically different approaches to the relationship between humans and the environment and also the relations *between* humans. The title of this book refers to sustainable *development*, and this allows us to narrow our focus slightly, as this connects more directly to a set definition and historical framework for understanding what is inferred by the sustainability component of the concept. Sustainable development is, as we will see, arguably most easily aligned with the foundational approach referred to as scientific conservationism than with the more radical approaches. However, the more radical approaches will also get their time in the spotlight. This will particularly be the case when the notion of growth-based sustainable development is critiqued in Chapters 17 and 18. However, many chapters contain critiques of the underlying assumptions and mainstream understanding of sustainable development and/or the SDGs.

Sustainable development is now inextricably linked to the work of the United Nation's Brundtland commission who published *Our Common Future* in 1987 (Brundtland et al., 1987). Here, they argue that it entails:

> [T]he ability to make development sustainable to ensure that it meets the needs of the present without compromising the ability of future generations to meet their own needs.

This definition indicates an instrumental approach to sustainability, as it implies that we must restrain our activities to fulfil certain human needs. We do not restrain ourselves because the environment, or animals, have value in themselves, but because humans need both resources and biological diversity to thrive. Such an instrumental approach is quite different from an approach based on recognizing the intrinsic value (Næss, 1999), and perhaps even rights (Nolt, 2014), of nature. The definition provided above is not sufficient to understand what lies in this concept, however, and the commission's full report clearly shows that while there are clear acknowledgments of certain *limits* to growth, growth is seen as necessary for human development and thus desirable (Brundtland et al., 1987). Sustainable development is consequently mainly about promoting *human* development and is at times referred to as utilitarian in how it mainly focuses on the consequences for human well-being (Farley & Smith, 2020).

However, to achieve human flourishing, our activity must be conducted within certain limits. These limits are partly environmental, but they are also technological, social, and political. This gives rise to the important distinction between environmental sustainability, social sustainability, and economic sustainability as three core *dimensions* of sustainable development (Brundtland et al., 1987). While the commission's approach may be anthropocentric and instrumental in how it sees nature as important because it fulfills human needs, their work was nevertheless important for highlighting how these dimensions are interrelated. The environment must be managed so that we maintain the potential for human life and well-being, but this cannot occur without us also addressing *social* issues related to poverty and equality, which again requires us to focus on *economic* issues related to economic systems, mechanisms of distribution, innovation, etc. (Brundtland et al., 1987). Sustainable development consequently requires a holistic approach to planning the meeting of future human needs, and this implies that the analysis of sustainability-related impacts requires the broad approach presented in Chapter 1.

While current understandings of sustainability tend to build on the notion of sustainable development, a goal-related approach to sustainable development has gained traction over the past decades. The UN's Millennium Development Goals were 8 goals introduced in 2000 with a 15-year goal period (Sachs, 2012), and they were consequently succeeded by the *Sustainable Development Goals* (SDGs) at the end of this period, in 2015. The SDGs were launched with the document *Transforming our world: The 2030 agenda for sustainable development* (United Nations, 2015), and they are explicitly based on the UN definition of sustainable development discussed earlier. The 17 top-level goals are shown in Figure 2.4.

FIGURE 2.4 The Sustainable Development Goals. (Courtesy of the United Nations, 2015.)

A number of these goals are covered directly in the following chapters, while some are mainly indirectly treated through the exploration of how technology relates to the three dimensions in which they can be categorized, until they are all discussed in the concluding chapters. It must also be noted that the SDGs are in a sense somewhat broader than these three dimensions (Sætra, 2022). The goals related to politics and partnerships (particularly 16 and 17) are perhaps most difficult to categorize, but they are here considered mainly as part of the social dimension of sustainable development. Another way to categorize the SDGs and the impact of technology is the concept of "five Ps" discussed in *Agenda 2030*. These highlight the different areas where action is required to reach the goals, and they are *people, planet, prosperity, peace*, and *partnership* (United Nations, 2015). But in this typology as well, the political goals are somewhat difficult to place, and they are arguably relevant for all the Ps. The place of politics in the SDGs is consequently a topic worthy of attention, which it will receive in this book. We might also note at the outset that while democracy might be assumed to be central to the SDGs, the word is mentioned only once in *Agenda 2030*, and not at all in the description of the goals and various targets (United Nations, 2015). SDG 16 does, however, discuss representative and responsible institutions, for example, and while this can be assumed to promote democracy, it need not be so.

While the focus of the 17 goals may seem quite obvious from the titles as shown in Figure 2.4, it is important to note that the intent of the goals are often quite specific and somewhat narrow. This is described both in the introduction to *Agenda 2030* and more directly in the numerous *targets* that make up each goal. Each goal has a number of targets, and there are 169 in total. While the targets are relatively specific, the framework is also built on the idea that the targets must be contextualized in order to make sense in different settings. For example, SDG 2 titled "Zero hunger" is by the Norwegian government interpreted widely in order to highlight challenges related to obesity, unhealthy eating habits, and malnutrition for the elderly (United Nations Association of Norway, 2022).

As becomes obvious from a quick glance at the goal headlines, the SDGs are both highly ambitious and broad in scope (Pekmezovic, 2019). While some might argue that goals such as "No poverty" are so ambitious that it becomes more discouraging than motivating, the targets help operationalize and temper the first impression. Still, the SDGs are arguably developed as *stretch goals* (Gabriel & Gauri, 2019), which imply that they are intentionally ambitious in order to provide something to continue to strive for throughout the goal period.

A key aspect of both sustainable development and the SDGs is that with three dimensions – and many goals – we must understand and explore the interlinkages between the three dimensions and also between the different goals *within* one dimension (Sætra, 2022). For example, economic growth could in general lead to increased resource use, emissions, etc., and innovation will often benefit some more than others and could consequently lead to increased inequality. Part of this is handled through how the SDGs, for example, qualify economic growth and state that it must be sustainable and inclusive, and that innovations and infrastructure should be accessible and affordable (United Nations, 2015). Despite this, trade-offs will often have to be made, and another example is how combatting climate change (SDG 13) might at times clash with the desire to preserve natural habitats and biodiversity (SDG 15) and life in the water (SDG 14). In fact, Nerini et al. (2019) show how

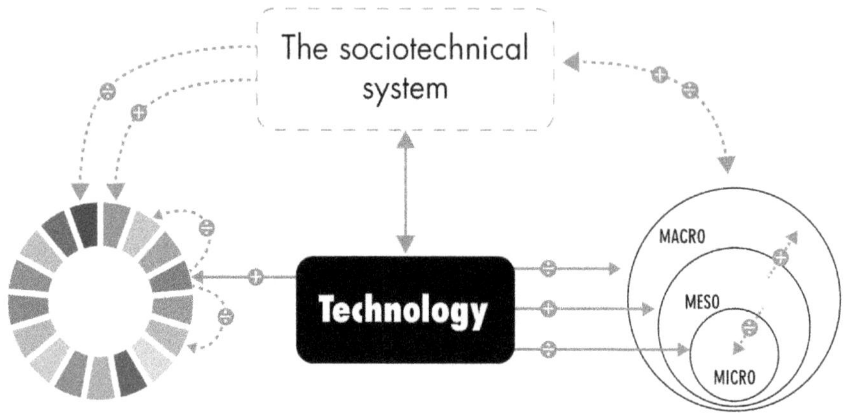

FIGURE 2.5 The main analytical framework. (Inspired by Sætra, 2022.)

achieving SDG 13 might undermine 12 other SDGs. These interlinkages will be explored in the chapters that follow, and a key point in this book is that we will not be able to understand the potential for technology to promote or hinder sustainable development unless we accept the importance of these interlinkages (Le Blanc, 2015; Nilsson et al., 2016).

2.4 SUMMARY: ANALYZING HOW TECHNOLOGY RELATES TO SUSTAINABLE DEVELOPMENT

With a more comprehensive understanding of technology at hand, the following chapters will go into more detail on what sort of technology is referred to when we consider the promise and pitfalls of techno-solutionism. Techno-solutionism, and techno-optimism, might very well be appropriate approaches to take, but it is important to understand that we refer to these terms as indicating a foundational belief in the potential for technology in general to contribute to good outcomes. Some chapters will focus on relatively isolated technologies and phenomena, and this will in turn be taken into account in the subsequent chapters and analyses to complete the picture.

Regarding good outcomes, this will here be linked to the notion of sustainable development. However, as should have become clear through this chapter, this is not an unproblematic term. Even more challenging are the SDGs, and while a basic understanding of what these concepts entail, and how they relate to each other, the real work of critically assessing how technology relates to these goals and what sort of assumptions are embedded in the goals starts now.

In a broad sense, the framework in Figure 2.5 shows the overall approach presented in this chapter, which is used in varying degrees in the chapters that follow. It shows how technology relates to the broader socio-technical system which in turn indirectly generates sustainability-related effects related to the SDGs and different levels of analysis (Sætra, 2021a). Technology is also shown to potentially have different effects on the micro-, meso-, and macro-levels (Sætra & Fosch-Villaronga, 2021). Finally, impacts on one SDG entail ripple effects on other SDGs that will also have to be considered (Sætra, 2021b).

2.5 REFERENCES

Barley, S. R. (2020). *Work and technological change*. Oxford University Press.

Brevini, B. (2021). *Is AI good for the planet?* Polity.

Brundtland, G. H., Khalid, M., Agnelli, S., Al-Athel, S., & Chidzero, B. (1987). *Our Common Future*: Report of the World Commission on Environment and Development. United Nations General Assembly document A/42/427.

Coeckelbergh, M. (2012). "How I learned to love the robot": Capabilities, information technologies, and elderly care. In *The capability approach, technology and design* (pp. 77–86). Springer.

Coeckelbergh, M. (2020). *Introduction to philosophy of technology*. Oxford University Press.

Collingridge, D. (1980). *The social control of technology*. Frances Pinter.

Ellul, J. (1964). *The technological society*. Vintage Books.

Farley, H. M., & Smith, Z. A. (2020). *Sustainability: If it's everything, is it nothing?* Routledge.

Gabriel, I., & Gauri, V. (2019). Towards a new global narrative for the sustainable development goals. in J. Walker, A. Pekmezovic, & G. Walker (Eds.), *Sustainable development goals: Harnessing business to achieve the SDGs through finance, technology, and law reform*. John Wiley & Sons Ltd.

Gellers, J. C., & Gunkel, D. J. (2022). Artificial intelligence and international human rights law: Implications for humans and technology in the 21st century and beyond. In A. Zwitter & O. J. Gstrein (Eds.), *Handbook on the politics and governance of big data and artificial intelligence*. Cheltenham.

Guha, R. (2014). *Environmentalism: A global history*. Penguin UK.

Jonsson, I. (2016). *The political economy of innovation and entrepreneurship: From theories to practice*. Routledge.

Le Blanc, D. (2015). Towards integration at last? The sustainable development goals as a network of targets. *Sustainable Development*, *23*(3), 176–187.

Müller, C. J. (2016). *Prometheanism: Technology, digital culture and human obsolescence*. Rowman & Littlefield.

Næss, A. (1999). *Økologi, Samfunn, Livisstil*. Bokklubben dagens bøker.

Næss, A. (2016). *En del elementære logiske emner*. Universitetsforlaget.

Nerini, F. F., Sovacool, B., Hughes, N., Cozzi, L., Cosgrave, E., Howells, M., . . . Milligan, B. (2019). Connecting climate action with other Sustainable Development Goals. *Nature Sustainability*, *2*(8), 674–680.

Nilsson, M., Griggs, D., & Visbeck, M. (2016). Policy: Map the interactions between Sustainable Development Goals. *Nature*, *534*(7607), 320–322.

Nolt, J. (2014). *Environmental ethics for the long term: An introduction*. Routledge.

Pekmezovic, A. (2019). The UN and goal setting: From the MDGs to the SDGs. In J. Walker, A. Pekmezovic, & G. Walker (Eds.), *Sustainable development goals: Harnessing business to achieve the SDGs through finance, technology, and law reform*. John Wiley & Sons Ltd.

Pelto, P. J. (1987). *The snowmobile revolution: Technology and social change in the Arctic*. Waveland Pr Inc.

Sachs, J. D. (2012). From millennium development goals to sustainable development goals. *The Lancet*, *379*(9832), 2206–2211. https://doi.org/10.1016/S0140-6736(12)60685-0

Sætra, H. S. (2021a). AI in context and the sustainable development goals: Factoring in the unsustainability of the sociotechnical system. *Sustainability*, *13*(4). https://doi.org/10.3390/su13041738

Sætra, H. S. (2021b). A framework for evaluating and disclosing the ESG related impacts of AI with the SDGs. *Sustainability*, *13*(15). https://doi.org/10.3390/su13158503

Sætra, H. S. (2022). *AI for the sustainable development goals*. CRC Press.

Sætra, H. S., & Fosch-Villaronga, E. (2021). *Healthcare digitalisation and the changing nature of work and society. Healthcare*, *9*(8), 1007. https://doi.org/https://doi.org/10.3390/healthcare9081007

Sattarov, F. (2019). *Power and technology: A philosophical and ethical analysis*. Rowman & Littlefield.

Schatzberg, E. (2018). *Technology: Critical history of a concept*. University of Chicago Press.

Sharkey, N., & Sharkey, A. (2012). The eldercare factory. *Gerontology*, *58*(3), 282–288.

Sparrow, R. (2016). Robots in aged care: A dystopian future? *AI & Society*, *31*(4), 445–454.

United Nations. (2015). *Transforming our world: The 2030 agenda for sustainable development*. Division for Sustainable Development Goals.

United Nations Association of Norway. (2022, February 17). *Utrydde sult*. www.fn.no/om-fn/fns-baerekraftsmaal/utrydde-sult

Winner, L. (1977). *Autonomous technology: Technics-out-of-control as a theme in political thought*. MIT Press.

CHAPTER 3

Artificial Intelligence, Artificial Solutions

Placing the Climate Emergency at the Center of AI Developments

Benedetta Brevini

CONTENTS

3.1 INTRODUCTION

The COVID-19 global pandemic has caused the worst economic contraction since the Great Depression. It has underscored the need to rethink what type of economy and society we want to build as we face the worsening climate emergency. Europe is leading the way in developing strategies for a Green Recovery. Technological innovations and digital services are at the core of recovery with the potential to create millions of jobs and boost economies devastated by the pandemic. The European Commission proposed a major recovery plan for Europe on May 26, 2020, approved by the European Council on July 21, 2020. Alongside the recovery package, EU leaders agreed on a €1,074.3 billion long-term EU budget for 2021–2027. Among others, the budget will support investment in the digital and green transitions and resilience.

The newly published Communication by the European Commission titled "Strategic Foresight Report 2022" on "Twinning the green and digital transition in the new geopolitical

DOI: 10.1201/9781003325086-3

context", published on June 29, stresses once again the crucial role of the "twin transition", green and digital, both at the top of the EU's political agenda. What is crucial about this Communication (European Commission, 2022) is that for the first time, the European Commission is explicit about the fact that digital technologies will also bring additional environmental burdens to them.

In particular, it explains

> Unless digital technologies are made more energy-efficient, their widespread use will increase energy consumption. Information and communications technology (ICT) are responsible for 5–9% of global electricity use and around 3% of greenhouse gas emissions. . . . However, studies show that ICT power consumption will continue to grow, driven by increasing use and production of consumer devices, demand from networks, data centres, and crypto assets.
>
> (European Commission, 2022, p. 2)

It further acknowledges that "further tensions will emerge in relation to electronic waste and environmental footprints of digital technologies" (ibid., p. 3).

However, despite growing attention to the environmental costs of ICT systems, Artificial Intelligence (AI) gets principally heralded as the key technology to solve contemporary challenges, including the environmental crisis, which is one of the goals of sustainable development. As explained in the introduction to this book, *sustainability* comprehends much more than the environmental challenges we are facing, as every environmental concern is a social, economic, and political concern.

Unfortunately, debates on Green Recovery plans and AI developments continue to avoid a crucial question: How green is Artificial Intelligence? And, considering that the most important international framework to achieve sustainability is enshrined in the UN's Sustainable Development Goals (SDGs) (United Nations, 2015; Sætra, 2022), is AI enabling or hindering SDGs specifically related to sustainable environmental development?

This chapter builds on the agenda of inquiry established in the collection Carbon Capitalism and Communication (Brevini & Murdock, 2017; Murdock & Brevini, 2019) in which communication systems are approached as assemblages of material devices and infrastructures, capable of depleting scarce resources in their manufacturing, usage, and disposal. It will also build on the volume *Is AI good for the Planet?* (Brevini, 2021) where AI applications were investigated as technologies, machines, and infrastructures that demand excessive amounts of energy to compute, analyze, and categorize; they use limited resources in their production, consumption, and disposal, potentially exacerbating problems of waste and pollution. After reflecting on a definition of AI that considers its materiality (Brevini, 2021) away from mainstream hypes, this chapter explores the multifaceted ways in which AI is impacting the climate emergency, thus impacting sustainable environmental development (and specifically, for example, SDG 13 (climate action), SDG 14 (life below water), SDG 15 (life on land).

It concludes by offering a set of solutions to limit the direct challenges that AI poses to SDGs.

3.2 PANDEMIC, CLIMATE CRISIS, AND ENERGY CONSUMPTION

The pandemic has hastened our reliance on technology and the massive acceleration of the adoption of AI, Big Data, cloud computing, and video technologies. We eat, socialize, work, study, exercise online, and plug in the cloud. New research from Milkround (2021) in the United Kingdom reveals that video conferencing has surpassed e-mail as the most widely used form of business communication during the lockdown.

So, we are reliant on communication systems as never before, while the planet is facing the biggest crisis ever faced. We now know that unless emissions fall by 7.6% each year between 2020 and 2030, the world will miss the opportunity to get on track toward the 1.5°C goal. We also know that we are currently on a trajectory for a temperature rise of over 3°C (United Nations Environment Programme, 2019). Yet, for almost 2 years we have been constantly bombarded by media reports that the pandemic has been incredibly good for the climate crisis, reducing climate emissions, taming transport, flights, and movements (Gössling & Humpe, 2020, p. 2).

On the contrary, even despite the lockdowns of 2020, greenhouse gas emissions have remained stubbornly high. Daily global carbon dioxide emissions fell by as much as 17% in early April 2020. But, as the world's economy started to recover, emissions rebounded; and the UN showed that 2020 only saw a 4–7% decline in carbon dioxide relative to 2019 (UN News, 2020). While transportation and industrial activity declined from January 2020, electricity consumption remained constant, which partly explains the minimal drop in emissions (IEA, 2020). How, you may ask? According to the World Energy Outlook 2019, globally 64% of the global electricity energy mix comes from fossil fuels (coal 38%, gas 23%, oil 3% (IEA, 2019)). Since fossil fuels are the largest source of greenhouse gas emissions, without fundamental shifts to renewable resources in the global energy production we shall not be able to prevent incalculable loss of life.

The book *Carbon Capitalism and Communication* has focused specifically on developing a type of communication scholarship that focuses on the materiality of communication systems: Communication systems run on machines and infrastructures that deplete scarce resources in their production, consumption, and disposal, thus increasing the amounts of energy in their use, and exacerbating problems the climate crisis (Brevini & Murdock, 2017). Researchers Lotfi Belkhir and Ahmed Elmeligi estimate that the tech industry's carbon footprint could increase to 14% by 2040, "accounting for more than half of the current relative contribution of the whole transportation sector" (Belkhir & Elmeligi, 2018, p. 448). Data centers will make up 45% of this footprint (up from 33% in 2010) and network infrastructure 24% (ibid., p. 457).

3.3 UNDERSTANDING ARTIFICIAL INTELLIGENCE AND ITS ENVIRONMENTAL TOLL

While more information is being collected on the environmental toll of data centers, little is being discussed about the impact of communication technologies, specifically Artificial Intelligence (AI). If we are to understand AI as an emerging communication technology, one deeply reliant on data to power its machine learning capabilities, more research needs to be done to understand what resources will be needed, as well as the ensuing environmental costs and damages, to operate it.

In mainstream debates, AI has been defined as the ability of machines to mimic and perform human cognitive functions. These include reasoning, learning, problem-solving, decision-making, and even the attempt to match elements of human behavior such as creativity.

In recent scholarship within communication studies, for example, within Human–Machine Communication (HMC), an emerging area of communication research defined AI as the study of the "creation of meaning among humans and machines" (Guzman & Lewis, 2019, p. 71). Others instead focused on refinement and theory related to people's interactions with technologies such as agents and robots (Spence, 2019).

In *Is AI Good for the Planet* (Brevini, 2021, p. 40), I argued that definition adopted by the latest *White Paper on Artificial Intelligence* issued by the European Commission serves as a good starting point to regain an understanding of the *materiality* of AI highlighting the connection between AI, data, and algorithms: "AI is a collection of technologies that combine data, algorithms and computing power. Advances in computing and the increasing availability of data are therefore key drivers of the current upsurge of AI" (ibid.).

Embracing the tradition of critical political economy of communication, in which communication systems are approached as assemblages of material devices and infrastructures (Brevini & Murdock, 2017), AI then can be better understood as technologies, machines, and infrastructures that demand amounts of energy to compute, analyze, and categorize. As a consequence, these communication technologies use scarce resources in their production, consumption, and disposal, exacerbating problems of waste and pollution.

3.4 THE POTENTIALS OF ARTIFICIAL INTELLIGENCE: AI FOR THE CLIMATE

Artificial Intelligence – so we are told – is helping to solve some of the world's biggest challenges, from treating chronic diseases and reducing fatality rates in traffic accidents to fighting Climate Change and anticipating cybersecurity threats (Brevini, 2020, p. 2).

Hence, it's not surprising that it also promises to tackle the most urgent emergency: The Climate Crisis that the earth is facing. A famous report titled *Harnessing Artificial Intelligence for the Earth*, published in January 2018 by the World Economic Forum, reiterated that the solution to the world's most pressing environmental challenges is to employ technological innovations and more specifically AI (World Economic Forum, 2018).

"We have a unique opportunity to harness this Fourth Industrial Revolution, and the societal shifts it triggers, to help address environmental issues and redesign how we manage our shared global environment" (World Economic Forum, 2018, p. 3). "The intelligence and productivity gains that AI will deliver can unlock new solutions to society's most pressing environmental challenges: climate change, biodiversity, ocean health, water management, air pollution, and resilience, among others" (ibid., 19).

Beyond these glorified claims, AI applications that enhance environmental management are growing at a rapid rate and there are increasing numbers of scientists committed to employ AI tools to forecast adverse effects of future climate change (Rolnick et al., 2022; Donti, 2020). For example, Treeswift, a spin-off from Penn Engineering, provides an

AI-powered forest-monitoring system that uses autonomous drones and machine learning to capture data, images, and inventory in order to map forest biomass. Treeswift can provide carbon capture data, deforestation monitoring, growth forecasting, and support forest management with targeted applications across preservation, the timber industry, and wildfire control (Lopez, 2020) all in principle aligned with SDGs 13 and 15. AI is also predicted to assist in the integration and spread of renewable energy through ductile price mechanisms and efficient energy storage and load operation (SDG 13). By enhancing the productivity of the agriculture industry, AI is said to play a key role in resource management, to minimize the environmental impact of farming, and to increase global resilience to extreme climate through various applications focused on data, on informed decision-making, and on augmented responses to changes in supply and demand (Mann, 2021). This will be supported in part by the budding field of climate informatics, in which AI and deep learning networks are leveraged to revolutionize our understanding of weather and climate change. AI is also progressively applied in water management (SDG 15). For example, in analyzing the conditions of a mountainous watershed in Northern China, AI methods identified climatological–hydrological relationships and projected future temperature, precipitation, and streamflow along with annual hydrological responses to these variables (Zhu et al., 2019). Other relevant applications are explored by Umbrello and Capasso in Chapter 4 of this book.

3.5 TECHNO-SOLUTIONISM, TECH OPTIMISM, AND ECOMODERNISM

Technology has long been considered a fix-all solution to the inequalities of capitalism. As the introduction to this chapter has succinctly explained, *Techno-solutionism* can be easily connected to the concept of *Techno-optimism* (Danaher, 2022, p. 1), with its clear view "that technology, when combined with human passion and ingenuity, is the key to unlocking a better world". As Mosco eloquently argued, "one generation after another has renewed the belief that, whatever was said about earlier technologies, the latest one will fulfil a radical and revolutionary promise" (Mosco, 2004, p. 8). Embedded in this neoliberal, techno-determinist discourse is a belief digital technology can disrupt inequalities and power asymmetries, without the need to challenge the status quo.

Linked to this concept, but specifically addressing the environmental problem is the credo of Ecomodernism (Asafu-Adjaye et al., 2015). Against those who place the unequal capitalist power relations at the center of the climate emergency (Brevini & Murdock, 2017; Foster, 2002), the Ecomodernist Manifesto (Asafu-Adjaye et al., 2015) cites technology as our answer to the ecological crisis, evading the need to confront the inherent environmental destructiveness of capitalism.

Authored by a group of sustainability figures from the Breakthrough Institute, *An Ecomodernist Manifesto* argues that "meaningful climate mitigation is fundamentally a technological challenge" (Asafu-Adjaye et al., 2015). For Ecomodernists, limitless economic growth is not disputed but encouraged.

Ecomodernism is also being adopted in leftist circles (Isenhour, 2016), among scholars who claim "the idea that the answer to Climate Change is consuming less energy – that a shift to renewables will necessarily mean a downsizing in life – feels wrong" (Bastani, 2017).

For Bastani, a proponent of Fully automated green communism (ibid.), "rather than consuming less energy, developments in wind and solar (and within just a few decades) should mean distributed energy of such abundance that we won't know what to do with it" (ibid.).

Despite its discussions around limiting greenhouse gas emissions, the International Kyoto Protocol also did little to dissuade an Ecomodernist agenda, instead encouraging environmental advocates in the United States (see Al Gore's presidential campaign) to push for technological improvement in energy efficiency as a way of averting environmental disaster (Foster, 2001, 2002).

This view, which we similarly find in cybertarian Silicon Valley circles, turns into a powerful apology for the status quo and is embraced by the same corporate giants that traditionally opposed action on Climate Change. Unfortunately, "a fundamental faith in growth" and a "foundational techno-optimism" (Sætra, 2022, p. 103, see also Chapter 18) are also very engrained in the framework of SDGs.

3.6 INEQUALITY AND EXPLOITATION: UNDERSTANDING THE ENVIRONMENTAL COSTS OF AI AS COMMUNICATION TECHNOLOGIES

After re-establishing the focus on the material basis of Artificial Intelligence, completely overlooked in green recovery debates and SD frameworks, I want to focus specifically on the multiple environmental costs of AI.

The starting point of every discussion should be an analysis of global supply chains of Artificial Intelligence, starting with the extractivism and neglect of social and environmental justice that AI currently require to produce, transport, train, and dispose (Brevini, 2021), certainly at odds with SDG 12 (responsible consumption), SDG 13 (climate action), SDG 15 (life on land), SDG 7 (clean energy) but also with more generic sustainability goals like with SDG 12 (responsible consumption).

In order to produce the material devices needed for AI to run, we need to start exploring its planetary costs by considering the extraction of rare metals and mineral sources that are needed happens following logics of colonialism.

In her work on digital developments with humanitarian structures, Mirca Madianou (2019) has developed the notion of "technocolonialism" in order to analyze how "the convergence of digital developments with humanitarian structures and market forces reinvigorate and rework colonial legacies" (2019, p. 2). The same "tenacity of colonial genealogies and inequalities" (Madianou, 2020, p. 1) characterize the global supply chains of Artificial Intelligence, as the extractive nature of technocolonialism resides in the minerals that need to be mined to make the hardware for AI applications. So, for example, the demand for mineral resources is growing exponentially, because of the AI uptake, thus compromising several SDGs (13, 15, 12 to list a few). The European Communication has stressed, for example, that of lithium in the EU, mainly in batteries, which is projected to raise by 3500% by 2050 (European Commission, 2022). This of course stresses the contradictions highlighted by authors such as Sætra (2022) and in Chapter 18 of this book between the drive to "growth" and preservations of land and see SDGs 14–15.

Moving to the second phase of the global supply chain, the production of AI model also shows high environmental costs, thus challenging SDGs.

A study published in 2019 by the College of Information and Computer Sciences at University of Massachusetts Amherst (Strubell et al., 2019) quantifies the energy consumed by running artificial intelligence programs. In the case examined by the study, a common AI training model in Linguistics can emit more than 284 tons of carbon dioxide equivalent. This is comparable to five times the lifetime emissions of the average American car. It is also comparable to roughly 100 return flights from London to NYC (Brevini, 2021, p. 68). Meanwhile, the converged communication systems upon which AI relies generate a plethora of environmental problems of their own, most notably energy consumption and emissions, material toxicity, and electronic waste (Brevini & Murdock, 2017). According to the International Energy Agency, if the energy demand continues to accelerate at the current pace, the residential electricity needed to power electronics will rise to 30% of global consumption by 2022 and 45% by 2030 (Maxwell, 2015).

Artificial Intelligence relies on data to work. At present, cloud computing eats up energy at a rate somewhere between the national consumption of Japan and that of India (Brevini, 2021; Murdock & Brevini, 2019). Today, data centers' energy usage averages 200 terawatt hours (TWh) each year (Jones, 2018; IEA, 2017) more than the national energy consumption of some countries, including Iran. Moreover, the information and communications technology (ICT) sector that includes mobile phone networks, digital devices, and television amounts to 2% of global emissions (Jones, 2018). Greenhouse gas emissions from the Information and Communication Industry (ICT) could grow from roughly 1–1.6% in 2007 to exceed 14% worldwide by 2040, accounting for more than half of the current relative contribution of the whole transportation sector, thus raising serious challenges to SDG 7 and SDG 13, for example.

Moreover, data centers require large, continuous supplies of water for their cooling systems, raising serious policy issues in places like the United States and Australia where years of drought have ravaged communities (Mosco, 2017), again compromising SDG 15. As the website of Google's Deepmind website explains (Evans & Gao, 2016),

> One of the primary sources of energy use in the data centre environment is cooling . . . . Our data centres – which contain servers powering Google Search, Gmail, YouTube, etc. – also generate a lot of heat that must be removed to keep the servers running. This cooling is typically accomplished via large industrial equipment such as pumps, chillers and cooling towers.

According to Deepmind, the solution to this problem is of course Machine Learning, which is also extremely energy consuming and generative of carbon emissions.

At the end of the global supply chain, we should also consider the problem of disposal of the devices employed in AI.

When communication machines are discarded, they become electronic waste or e-waste, saddling local municipalities with the challenge of safe disposal. This task is so burdensome that it is frequently offshored, and many countries with developing economies have become digital dumping grounds for more privileged nations (Brevini & Murdock, 2017).

Finally, while promising to solve the climate emergency, AI companies are marketing their offers and services to coal, oil, and gas companies, thus compromising efforts to reduce the emissions and divest from fossil fuels. A new report on the future of AI in oil and gas market published by Zion Market Research (Zion Market Research, 2019) found that the sector of AI in oil and gas is expected to reach around USD 4.01 billion globally by 2025 from 1.75 billion in 2018. AI companies around the world are pushing their capabilities to the oil and gas sectors to increase their efficiencies, optimize their operations, and increase productivity: In other words, they are selling their services to increase the pace and productivity of excavation and drilling. Exxon Mobil, for example, signed a partnership in February this year with Microsoft to deploy AI programs, while oil and gas exploration in the fragile ecosystem of Brazil has seen recent employment of AI technology by state oil giant Petrobras; similarly, European oil major Royal Dutch Shell has signed a partnership with AI company C3 (Joppa & Herweijer, 2018).

3.7 PLACING THE CLIMATE EMERGENCY AT THE CENTER OF SCHOLARSHIP

New developments of Artificial Intelligence place escalating demands on energy, water, and resources in their production, transportation, and use, reinforce a culture of hyper-consumerism, and add to the accumulating amounts of waste and pollution already generated by accelerating rates of digital obsolescence and disposal (see Brevini, 2021; Gabrys, 2013). Instead of embracing new developments in Communication technologies and AI as a new utopia that will fix the world and capitalism problems, we should start quantifying and considering the environmental costs and damages of the current acceleration of algorithm-powered data communication that can too easily compromise SDGs (Sætra, 2021; Sætra, 2022).

We need to ask who should own and control the essential infrastructures that power data communication and Artificial Intelligence and make sure to place the climate emergency at the center of the debate on sustainable development. How can we shape the future of Artificial intelligence to be one of collective well-being and minimized climate impact?

Progress is being made at global fora and national levels as international agreements, legislative frameworks, position papers and guidelines are being drawn up by the European Union and Council of Europe, and UNESCO is in the midst of developing a Recommendation on the Ethics of Artificial Intelligence.

Despite this, however, it seems that global discussions the climate emergency – for example, in the context of UN COP – are yet to connect environmental with AI policy discussions, and more research is needed to ascertain the environmental damage caused by Artificial Intelligence.

As this chapter showed, if we consider the material basis of AI and look at its techno-colonialist character, we should consider all its environmental costs. They start with mineral extractions, water, energy, and natural resources necessary for hardware and machine production (generating huge challenges to SDGs 6, 7, 13, 14, and 15); it then generates additional resource depletion for distribution, transportation, and post-consumption of material technology (challenging SDGs 7, 13, 14, and 15) to end with major e-waste disposal needs (SDGs 6, 7, 13, 14 and 15). Added to this is the major environmental cost of data extraction, computing, and analysis (SDGs 7 and 13).

We know many corporations now audit the production conditions of sub-contractors' factories, but there is still an urgent need to demand accountability for those who own clouds and data centers. One crucial intervention could be government-mandated *Green Certification* for server farms and centers to achieve zero emissions. Given AI's increasing computing capabilities, the disclosure of its carbon footprint could be a first step in the right direction. This could take the form of a *Tech Carbon Footprint* Label, which would provide information about the raw materials used, the carbon costs involved, and what recycling options are available, resulting in stronger public awareness about the implications of adopting a piece of smart technology.

Making transparent the energy used to produce, transport, assemble, and deliver the technology we use daily would enable policymakers to make more informed decisions and to the public to make more informed choices. Added to this could be policy intervention which requests manufacturers to lengthen the lifespan of smart devices and provide spare parts to replace faulty components.

Global policymaking should encourage educational programs to enhance *green tech literacy* and raise awareness of the costs of hyperconsumerism, as well as the importance of responsible energy consumption as crucially linked to SDGs 3 and 4.

In line with SDG 4, *green tech literacy* programs should also entail interventions to ban production of products that are too data demanding and deplete too much energy.

As Artificial Intelligence, like all technologies, is always, in "a full sense social" (Williams, 1981, p. 227), the choice to develop the kind of "green AI" that can enhance environmental sustainable goals rests on us. Unfortunately, the current development of AI does not display the kind of environmental ethos that is needed to address the climate emergency we are facing.

3.8 REFERENCES

Asafu-Adjaye, J. et al. (2015). *An ecomodernist manifesto*. Breakthrough Institute. www.ecomodernism.org

Bastani, A. (2017). *Fully automated green communism*. Novara Media. https://novaramedia.com/2017/11/19/fully-automated-green-communism.

Belkhir, L., & Elmeligi, A. (2018). Assessing ICT global emissions footprint: Trends to 2040 & recommendations. *Journal of Cleaner Production, 177*, 448–463. https://doi.org/10.1016/j.jclepro.2017.12.239

Brevini, B. (2020). Black boxes, not Green: Mythologizing artificial intelligence and omitting the environment. *Big Data & Society, 7*(2), 205395172093514. https://doi.org/10.1177/2053951720935141.

Brevini, B. (2021). *Is AI good for the planet*. Polity.

Brevini, B., & Murdock, G. (2017). *Carbon capitalism and communication*. Palgrave Macmillan.

Danaher, J. (2019). Automation and utopia. In *Automation and utopia*. Harvard University Press.

Donti, P. (2020). How machine learning can help tackle climate change. XRDS: Crossroads. *The ACM Magazine for Students, 27*(2), 58–61.

European Commission. (2020). *White paper on artificial intelligence: A European approach to excellence and trust.* European Commission.

European Commission. (2022). *Communication from the Commission to the European Parliament and the Council "Strategic Foresight Report Twinning the green and digital transitions in the new geopolitical context".* https://op.europa.eu/en/publication-detail/-/publication/d04b2f02-f86b-11ec-b94a-01aa75ed71a1/language-en

Evans, R., & Gao, J. (2016). DeepMind AI reduces Google data centre cooling bill by 40%. *Deepmind.* https://deepmind.com/blog/article/deepmind-ai-reduces-google-data-centre-cooling-bill-40

Foster, J. B. (2001), *Ecology against capitalism. Monthly Review, 53*(5), p. 1.

Foster, J. B. (2002). Capitalism and ecology: The nature of the contradiction. *Monthly Review, 54*(4).

Gabrys, J. (2013). Plastic and the work of the biodegradable. In J. Gabrys, G. Hawkins, & M. Michael (Eds.), *Accumulation: The material politics of plastic* (pp. 208–227). Routledge.

Gössling, S., & Humpe, A. (2020). The global scale, distribution and growth of aviation: Implications for climate change. *Global Environmental Change, 65,* 1–12. https://doi.org/10.1016/j.gloenvcha.2020.102194

Guzman, A. L., & Lewis, S. C. (2019). Artificial intelligence and communication: A human–machine communication research agenda. *New Media & Society, 22*(1), 70–86. https://doi.org/10.1177/1461444819858691

IEA. (2017). *Digitalisation and energy.* IEA. www.iea.org/reports/digitalisation-and-energy.

IEA. (2019). *World energy outlook 2019.* IEA. www.iea.org/reports/world-energy-outlook-2019.

IEA. (2020). *Global energy review 2020.* IEA. www.iea.org/reports/global-energy-review-2020

Isenhour, C. (2016). Unearthing human progress? Ecomodernism and contrasting definitions of technological progress in the Anthropocene. *Economic Anthropology, 3*(2), 315–328.

Jones, N. (2018, September 12). How to stop data centres from gobbling up the world's electricity. *Nature.* www.nature.com/articles/d41586-018-06610-y

Joppa, L., & Herweijer, C. (2018). *How AI can enable a sustainable future.* PricewaterhouseCoopers LLP and Microsoft. https://www.pwc.co.uk/sustainability-climate-change/assets/pdf/how-ai-can-enable-a-sustainable-future.pdf

Lopez, I. (2020). *Treeswift's autonomous robots take flight to save forests.* Phys.Org

Madianou, M. (2019). Technocolonialism: Digital innovation and data practices in the humanitarian response to the refugee crisis. *Social Media and Society, 5*(3). https://doi.org/10.1177/2056305119863146

Madianou, M. (2020). *Reproducing colonial legacies: Technocolonialism in humanitarian biometric practices.* https://www.ohchr.org/sites/default/files/Documents/Issues/Racism/SR/RaceBordersDigitalTechnologies/Mirca_Madianou.pdf

Mann, A. (2021). *Food in a changing climate.* Emerald Group Publishing.

Maxwell, R., & Miller, T. (2015, September 11). High-tech consumerism, a global catastrophe happening on our watch. *The Conversation.* https://theconversation.com/high-tech-consumerism-a-global-catastrophe-happening-on-our-watch-43476

Milkround. (2021). *Gen Z lead the way through lockdown with tech skills that boost productivity.* Retrieved February 8, 2022, from www.milkround.com/advice/gen-z-lead-the-way-through-lockdon-with-tech-skills-that-boost-productivity

Mosco, V. (2004). *The digital sublime: Myth, power, and cyberspace.* MIT Press.

Mosco, V. (2017). The next internet. In *Carbon capitalism and communication* (1st ed.). Palgrave Macmillan.

Murdock, G., & Brevini, B. (2019). Communications and the Capitalocene: Disputed ecologies, contested economies, competing futures. *The Political Economy of Communication, 7*(1), 51–82.

Rolnick, D., Donti, P. L., Kaack, L. H., Kochanski, K., Lacoste, A., Sankaran, K., . . . Bengio, Y. (2022). Tackling climate change with machine learning. *ACM Computing Surveys (CSUR), 55*(2), 1–96.

Sætra, H. S. (2021). AI in context and the sustainable development goals: Factoring in the unsustainability of the sociotechnical system. *Sustainability, 13*(4). https://doi.org/10.3390/su13041738

Sætra, H. S. (2022). *AI for the sustainable development goals.* CRC Press.

Spence, P. R. (2019). Searching for questions, original thoughts, or advancing theory: Human-machine communication. *Computers in Human Behavior, 90*, 285–287.

Strubell, E., Ganesh, A., & McCallum, A. (2019). *Energy and policy considerations for deep learning in NLP.* Cornell University, arXiv:1906.02243.

UN News. (2020). *Carbon dioxide levels hit new record; COVID impact 'A tiny blip', WMO says.* United Nations. https://news.un.org/en/story/2020/11/1078322

United Nations. (2015). *Transforming our world: The 2030 agenda for sustainable development.* Division for Sustainable Development Goals.

United Nations Environment Programme. (2019). *Emissions gap report 2019.* UNEP.

Williams, R. (1981). Communication technologies and social institutions. In R. Williams (Ed.), *Contact: Human communication and its history.* Thames & Hudson.

World Economic Forum. (2018). *Harnessing artificial intelligence for the earth.* WE Forum. https://www3.weforum.org/docs/Harnessing_Artificial_Intelligence_for_the_Earth_report_2018.pdf

Zhu, R., Yang, L., Liu, T., Wen, X., Zhang, L., & Chang, Y. (2019). Hydrological responses to the future climate change in a data scarce region, Northwest China: Application of machine learning models. *Water, 11*(8), 1588. https://doi.org/10.3390/w11081588

Zion Market Research. (2019, July 18). Global AI in oil and gas market. *Intrado GlobeNewsare.* www.globenewswire.com/news-release/2019/07/18/1884499/0/en/Global-AI-In-Oil-and-Gas-Market-Will-Reach-to-USD-4–01-Billion-By-2025-Zion-Market-Research.html

CHAPTER 4

Sustainable Climate Engineering Innovation and the Need for Accountability

Marianna Capasso and Steven Umbrello

CONTENTS

4.1 INTRODUCTION

Although still highly controversial, the idea that we can use technology to radically alter our environment to mitigate the challenges we now face is becoming an ever more discussed approach. The potential for cloud brightening, solar radiation management, and carbon capture technologies, among others, have been debated for a long time. Still, it was not long ago that research on such topics was largely suppressed. Much of this historical aversion to this research can be primarily laid at the feet of the idea being that there is a moral hazard involved in even exploring the *potential* for fixing our problems, not through a radical change in individual behavior, consumption, and the systems of production but through improving the symptoms. Moral hazard arguments are ubiquitous in the public debate and the academic literature on climate engineering, seeing it as a "techno-fix" compromise instead of addressing systemic and broader moral and institutional reforms (Wagner & Zizzamia, 2021). However, we are now seeing increasing acceptance of such technologies, and carbon capture and storage, in particular, is relatively close to mainstream. Many promoters of climate engineering argue that it is necessary to counteract climate change, with the need to serve the moral imperative of mitigation and provide

DOI: 10.1201/9781003325086-4

adaptation for vulnerable people across the globe (Horton & Keith, 2016). However, scholars recently recognized that these arguments often lack an in-depth analysis informed by moral and political theory since they neglect the power dynamics inherent in climate engineering research and implementation (Gardiner & McKinnon, 2020; Hourdequin, 2021; Smith, 2018).

This chapter highlights how both climate engineering innovation and SDGs framework should be seen not as policy-neutral and objective sites, but as sites for politics, sites for ongoing debate and deliberation on their normative ends and governance. Our aim is to show how a more nuanced, multidimensional definition of accountability is needed in order to permit responsible innovation of climate technologies that align with the ideal of sustainable development. This chapter is divided as follows. First, it starts by describing what climate engineering is and uses one particular form, carbon capture, usage, and storage (CCUS), as a use case. Second, it explores how the synergy between the responsible deployment of climate engineering innovation and the achievement of the SDGs targets should unpack the socio-political significance of both frameworks, since they are both depending on political preferences and social acceptability, and on how normative justifications and decisions about innovation and sustainable strategies and constraints are managed, taken, and communicated.

Then, this chapter concentrates on what accountability is, how it has been traditionally understood in the literature, and why a more expansive and polysemic definition of accountability is required if climate engineering technologies like CCUS are actually to support sustainable development. Specifically, this chapter discusses possible strategies to theorize and implement accountable and sustainable frameworks for climate engineering innovation, starting from the creation of shared standards to matters of responsibility among social actors and of answerability, which requires that conduct and information are reported, explained, and reasonably justified in the context of these climate models. Finally, the conclusions recap the main arguments sustained in this chapter and explore their connection to the key topics of the volume.

4.2 CLIMATE ENGINEERING

Climate Engineering technologies are a class or family of technologies proposed to ameliorate or mitigate climate change's causes and/or effects on both local and global scales. Although the term has been appropriated in the past as a theoretical application to terraforming another planet, like Mars (e.g., see Jakosky & Edwards, 2018), to be habitable, in this context, we are referring to the technology family that aims to act on the Earth's climate system to reduce atmospheric greenhouse gases or, more radically by transforming physical and/or chemical biosphere mechanisms to achieve direct climate control (Buchinger et al., 2022).

There are various member technologies of this technology family, including but not limited to carbon capture, usage, and storage (CCUS) and solar radiation management (SRM). The former refers to technologies that can remove existing CO_2 from the atmosphere, which, consequently, can feasibly ameliorate existing emissions, thus impacting temperature regulation (Bui et al., 2018; Hanssen et al., 2020). SRM, on the other hand, are technologies that are designed to transform how the biosphere interacts with solar radiation (Ming et al., 2014). One of the ways that this has been proposed to function on the global scale is by creating a dense cloud of particles in the stratosphere, which are designed to reflect part of the solar radiation, thus reducing global temperatures. However, there are more local approaches to SRM, such as employing heat reflection systems to protect and restore snow or glaciers (Applegate & Keller, 2015). The *time-to-market* of this technology family is considered "Short to medium for small and regional scale deployment, medium to long term for large-scale and global deployment, and most advanced applications" (Buchinger et al., 2022, p. 38). Given the relative urgency underlying the development of this technology family, as well as the high research and industrial relevance, it merits considering the various ethical concerns that emerge when considering CCUS and SRM, such as those concerning who will be impacted both directly and indirectly by them, who can or will have access to these technologies, who will decide how and where these systems will be implemented, as well as the various concerns surrounding the value of sustainability.

Naturally, there are various arguments in favor and against the design, deployment, and use of these climate engineering technologies (Brooks et al., 2022). For example, those in favor often levy arguments that since global climate warming is anthropogenic, it is likewise humans' moral imperative to take action to ameliorate such change. Likewise, arguments are made concerning our collective responsibility to future generations and their well-being, as well as the argument of delaying the inevitable consequence of warming, which is made for both CCUS and SRM (Stilgoe, 2016). In the latter case, proponents argue that SRM techniques would help deflect some proportion of the warming effect until atmospheric emissions are effectively reduced. At the same time, CCUS would feasibly permit more short-term warming, namely emissions which would then be ameliorated with later CCUS techniques.

However, some arguments against these technologies are usually political in their orientation, arguing that many of these approaches require crossing national and geospatial boundaries, thus implicating notions of the sovereignty of those countries wishing to use/not use such technologies (Proelss & Güssow, 2011). Similarly, given that the effects of such technologies across time are neither immediate nor certain, this questions whether and how we can intervene in a complex system like the climate with positive effects. In the event of adverse effects, can we have a reasonable certainty of the ability to reverse such impacts (Raza et al., 2019)? The findings of a review on geoengineering carried out by the UK Royal Society in 2009 revealed major uncertainties and potential risks concerning effectiveness, social, and environmental impacts of geoengineering projects (Royal Society, 2009). At the beginning of 2022, a coalition of scientists and governance scholars launched an initiative calling for a ban on research and deployment of SRM, claiming that the current global governance system is unfit to maintain a fair political control of it (Biermann et al., 2022). These are some of the arguments discussed within the discourse on climate engineering technologies like SRM and carbon capture, usage, and storage. The following subsection will take up CCUS as the case we will be looking at for this chapter.

4.2.1 Carbon Capture, Usage, and Storage (CCUS)

Spurred primarily by the United Nations Economic Commission for Europe's (UNECE's) objective of achieving net-zero emissions, carbon capture, usage, and storage (CCUS) systems have been proposed and sustained as one of the most conceptually effective ways of achieving this goal of removing large volumes of CO_2 from the atmosphere. CCUS systems are understood as technologies that capture CO_2 emissions from power generation sources that use fossil fuels and industrial processes for storage deep underground or re-use (Figure 4.1). This re-use is often for producing synthetic materials such as other fuels, chemicals, building materials, etc.

There are two general routes for CCUS: carbon usage and carbon storage. Concerning the latter, carbon is removed directly from either the air or facilities and industrial processes, stored in the compressed form, and then transported to sequestration areas to be stored permanently underground in geological formations like saline, oil, and gas reservoirs (Metz et al., 2005). Concerning carbon usage, the captured and compressed carbon is re-used in other processes such as being pumped into greenhouses to make them more efficient, in the synthesis of materials, chemicals, and fuels, as well as in essential commercial products like carbonated soft drinks (Ho et al., 2019; Psarras et al., 2017). Using captured carbon as fuels and in other industrial and manufacturing processes increases net efficiency while simultaneously reducing net waste, thus contributing to the infrastructure underlying the circular economy (Budzianowski, 2017). Still, sequestration could feasibly permit augmented usage of existing emission sources, given the ability to directly capture emissions from the atmosphere and these emission facilities (Tcvetkov et al., 2019).

Still, there are some barriers to both carbon capture and storage and carbon capture and usage. Concerning storage, many projects are currently in operation on a global scale; however, the technical equipment necessary for this process to be undertaken is exceptionally costly and serves as an obstacle for many sources of emissions, particularly in the Global South (Rubin & Zhai, 2012; Román, 2011). This goes hand in hand with other barriers, such as the lack of technical expertise necessary to run and maintain such systems and uncertain return on investment (Roussanaly et al., 2021). Unlike the more commercialized

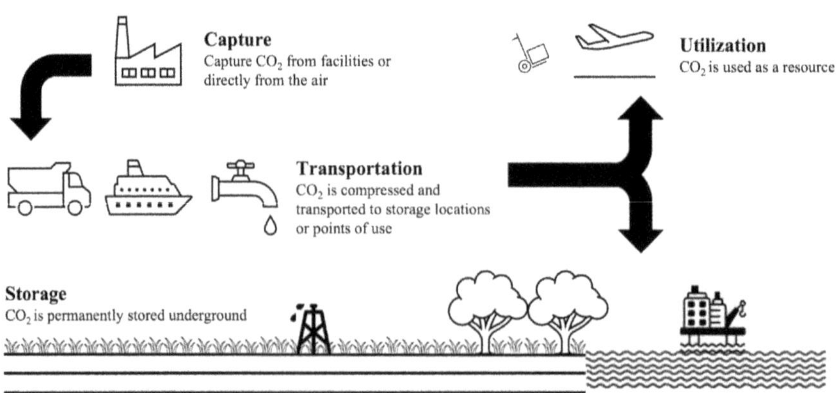

FIGURE 4.1 Carbon capture, use, and storage schema.

storage technologies, carbon utilization technologies are more novel. Likewise, to ensure that both the ecological and economic boons are achieved, thus ensuring long-term and ubiquitous adaptability of carbon utilization technologies, what is required is low-carbon hydrogen and vast volumes of renewable energy, all at affordable costs (Yu et al., 2021; Brändle et al., 2021).

4.3 A SITE FOR POLITICS

CCUS has entered the discourse on climate models to counteract or delay climate change. However, its long-term consequences are still unknown, as are its impacts as a broader paradigm shift that is different from adaptation and mitigation measures. Technologies such as CCUS have been said to be morally problematic "techno-fix" compromises to climate change, in the sense that they alone are inadequate solutions that address merely the setting of behaviors and not how behavioral failures come into being, that is, the failure of people to behave in an appropriate and climate-friendly way, and the underlying social, political, and economic dynamics (Scott, 2012; Borgmann, 2012). Moreover, CCUS is considered by many unjust and incompatible with the ideal of sustainable development, since they would have several detrimental effects, including the displacement and marginalization of local communities, the undermining of food rights and land rights, and, finally, the infringement of biosphere and natural ecosystems' integrity, leading to the creation of new vast-scale infrastructures and industries that can reproduce the emissions problem instead of ameliorating it (Schneider, 2019). For example, an SDG that is potentially impacted by CCUS is the SDG 6 on clean water, since such technologies can create significant land and water trade-offs, and adverse impacts on local water quality (IPCC, 2022, Chaps. 6, 12). Also, the SDG 7 on affordable and clean energy can be impacted due to the high energy demand of some of CCUS methods (IPCC, 2022, Chap. 12).

Widespread claims suggest that technologies like CCUS are intrinsically troubling: they are often embedded in undemocratic systems of innovation and knowledge that disregard the underlying causes and patterns of climate change and increase the dependence of developing countries and vulnerable groups while strengthening the power and control of developed countries and technocratic, corporate elites (Gardiner & McKinnon, 2020). In particular, in the range of potential injustices raised from climate engineering technologies, the most debated one is the exacerbation of power asymmetries and the fact that those tech-mediated climate models can generate profound and global relations of domination (Smith, 2018, 2021). Narratives or claims on climate engineering proposals might be portrayed as objective, unbiased, and policy-neutral; hence, they might de-politicize the climate change discourse, obscuring the political motivations behind their reasoning and legitimizing structures of power that perpetuate oppression and exploitation (Sikka, 2021; O'Lear et al., 2021).

However, even if the climate engineering literature tends to recognize equity concerns, often, no normative political dimension is adopted for evaluating the monitoring and control mechanisms for the assessment, development, and policy dimensions surrounding those technologies (McLaren, 2018). The governance frameworks and democratic processes needed to develop and sustain technologies such as CCUS responsibly remain largely neglected by policymakers and the academic research community at large (Bellamy et al., 2021). Similarly, scholars have noted how Responsible Research and Innovation activities often remain separate and self-referential, without appropriate processes for citizens' engagement (Stahl et al., 2021), by failing to be a "site for politics", that is, a site for ongoing debate and deliberation about the normative ends of innovation and its governance (Owen et al., 2021).

Also in the sustainable development literature, it is widely accepted that the achievement of the SDGs depends on democratic and effective governance mechanisms, to the point that governance has been considered the "fourth pillar of sustainable development" (Kanie et al., 2014, p. 6). Nonetheless, there is no consensus or clear conceptualization on the theoretical foundation of governance for sustainable development and its different aspects (Glass & Newig, 2019). Moreover, empirical studies have found how policies for the achievement of SDGs paradoxically obscure the trade-offs and political assumptions upon which sustainable development rests, leading to a situation of "anti-politics" that does not account for a space where incoherencies from dominant private, market-based organizations can be discussed and contested (Yunita et al., 2022). Detractors of SDGs have conceived this set of normative principles as a political framework or ideology that can compromise public decision-making mechanisms and privilege commercial interests, leading to unjust and exclusionary policies instead of promoting just structural change (Weber, 2017).

Therefore, a critical political question arises, asking to whom, by whom, and to what ends the sustainable development trajectories should be designed and deployed. At the same time, the central question for CCUS technologies is no longer whether but how, to what extent, by whom, and to whom they should be pursued (Bellamy & Geden, 2019). This means that the choice of CCUS technologies will depend on the evolution of political preferences and social acceptability, and on how sustainability constraints are managed by governments (IPCC, 2022, Chap. 12, p. 62).

Rather than being a purely technical matter, climate engineering innovation processes are political in the sense that they are strictly entangled with the same broader socio-political contexts and power structures in which are embedded (on the normative political dimensions of technologies see the recent Coeckelbergh, 2022; Waelen, 2022). Moreover, those processes cannot avoid confronting the theoretical underpinnings of sustainable development. Synergies between the responsible deployment of such climate models and the achievement of SDGs targets should unpack the political rationale in the transformative potential of the UN 2030 Agenda, and should encompass governance methods for inclusion and empowerment.

4.4 REVISITING ACCOUNTABILITY

Among the few scholarly studies on SDGs politics, a recent thesis that has been advanced is that sustainable development goal setting and fulfillment are particularly adapted to study long-term political decisions, interactions, and structures and are in urgent need of political normative frameworks that scrutinize normative qualities of governance such as legitimacy, responsibility, and accountability (Bexell & Jönsson, 2021). Leaving aside the questions of legitimacy and responsibility, these studies define accountability as the "retrospective mirror of political responsibility" and connect it to monitoring and sanctioning mechanisms: social actors that deal with sustainable development should be liable for how they exercise power and how they make strategic socio-political choices about goals (Bexell & Jönsson, 2017, pp. 17–18, 2021, p. 3).

Also, in the philosophy of technology literature, accountability has been identified as a form of retrospective, backward-looking (van de Poel, 2011) or passive (Pesch, 2015) responsibility, namely as a form of *ex-post* scrutiny that requires justification for a state of affairs and constitutes the basis for blameworthiness. Only in these last few years have some scholars recognized that accountability also has a preventive and anticipatory role since it engages with a relation between an actor and a forum, in which conducts are exposed, justified, and debated in a back-and-forth exchange (Verdiesen et al., 2021, based on Bovens, 2007; Bovens et al., 2014; Santoni de Sio & Mecacci, 2021).

This definition is more aligned with debates on accountability in normative political theory, where accountability has been the object of various discussions but usually refers to the self-determination of citizens that keep/hold their representatives accountable and responsive (Palumbo & Bellamy, 2010). In political studies, responsiveness has been identified as a "potential readiness to respond" (Pitkin, 1967, p. 233) to citizens with whom ultimate responsibility for the actions and decisions should rest (Urbinati & Warren, 2008). However, citizens need "meaningful" forms of participation, understood as opportunities for real influence in the polity (Pateman, 1970, pp. 70–71). This generates a whole range of problems, as responsiveness might be at odds with political equality and influence in civic life, especially when economic standing or socio-political resources and powers might make some individuals or groups more likely to voice concerns and influence policy strategies and outcomes (Papadopoulos & Warin, 2007). Thus, the establishment of meaningful forms of accountability and responsiveness implies not only the likelihood of substantive forms of representation but also, more importantly, a contribution to equality in policy outcomes and long-term fair distribution of public goods (Grimes & Esaiasson, 2014).

Therefore, accountability is not merely retrospective and connected to sanctioning measures but involves an ex-ante account of governance that involves mutual deliberation on public goods, the creation of shared standards, and monitoring and scrutiny mechanisms. As a normative concept, it consists of the respect of various dimensions in the accountability relation: to whom (accountees); by whom (accounters); for what and by which shared standards this relation is assessed; answerability, that is, through what process and in which modalities conduct and information are reported, explained, and reasonably justified and accountees informed; and enforceability, that is, what effects or consequences arise when someone is held accountable and violates the conditions necessary for a meaningful relationship with the accountees (on the multidimensional nature of accountability see also Mashaw, 2006; Buchanan & Keohane, 2006, p. 426; Callies, 2018; Villalona, 2021, p. 19).

Accountability has been explored to some extent in the UN 2030 Agenda, with an explicit reference to "effective, accountable and inclusive institutions at all levels" in SDG #16.[1] The UN 2030 Agenda envisages a follow-up and review framework to promote accountability to citizens and leaves this task to the institution of the High-Level Political Forum (HLPF) and to voluntary national review systems, which may have multiple different modalities in their national policy choices for SDGs implementation (United Nations, 2015, para 72–91; Karlsson-Vinkhuyzen et al., 2018, p. 1380-ff). In SDGs literature, accountability is depicted as an indispensable factor. Still, surprisingly there is no clear understanding of its nature and how it can facilitate the strategy design for SDGs implementation at the national level and social value creation (Abhayawansa et al., 2021). The most significant challenges to accountability in the Global SDG Accountability Report are the lack of institutional coordination across governments and the low public awareness of SDGs among citizens and stakeholders (Villalona, 2021, pp. 29–33, 36). Thus, the definition of accountable relations is not clear and settled in the SDGs literature. In the following pages, this discourse on the polysemic nature of accountability provides some interesting theoretical implications for the question of sustainable development and climate engineering innovation.

4.5 ACCOUNTABLE AND SUSTAINABLE CLIMATE ENGINEERING

Scholars involved in the normative discussion on climate engineering tend to focus on institutional legitimacy as a criterion to guide responsible climate engineering and climate engineering experiments (Callies, 2018; Bellamy et al., 2017). However, accountability might be an equally relevant normative criterion that both the sustainable development framework and climate engineering innovation should confront. Indeed, accountability as a criterion might provide a guide for complex processes by which parameters for sustainable development come to be defined, as well as an approach to responsibly conducting climate engineering innovation. SDGs have been considered a starting point for the development of criteria for climate engineering (Stelzer, 2020). However, as mentioned, even if intended to provide an inclusive approach to societal stakeholders, the SDGs framework still needs approximation and reflection on how to realize this global effort. Hence, the polysemic nature of accountability above delineated and its articulations in multiple dimensions might form a basis for philosophical reflection on how to responsibly implement climate engineering innovation, in modalities that also align with the ideal of sustainable development.

First, the dimensions of accountability require identifying accounteers and accountees, the need for shared standards upon which conduct and relations are assessed, and, consequently, a dimension of enforceability in scenarios of violations. Naturally, these shared standards could take the form of international law, given the global impacts of climate engineering technologies. No global roles, obligations, or rights exist concerning these technologies. However, existing ancillary international and regional frameworks do provide the foundations for such international treaties to be formed. Human rights law, State responsibility, Environmental law, Climate change law, Space law, and Maritime law provide starts for how law between nations governing international geographies can be approached concerning climate engineering technology innovation and deployment. Taking human rights law as an example, we can already see how framing the multidimensional understanding of accountability for climate engineering can take place. Procedural rights, for example, would implicate the need for citizens to have access to information, participate in public affairs, and, of course, have access to legal remedies. Substantive rights provide the grounding on which such procedural rights take place concerning climate engineering, particularly an individual's right to life, healthy environment, health, food, and water. More abstractly, however, there are also rights concerning the scientific research into climate engineering innovation, in particular, the freedom to conduct said research, the right to benefit from scientific progress, and, of course, the related moral and material interests derived from such research. Although there are no current international statutes delineating this concerning climate engineering, projects are undergoing aiming at providing shared standards for both the design of these technologies and their eventual implementation.[2]

However, some scholars argue that just formal or informal governance of climate engineering is impossible, since it would require novel international organizations with unprecedented enforcement powers (Biermann et al., 2022). Others have emphasized how, even if global climate change mitigation is recognized as a global public good (i.e., the benefits of which are available to everyone and nobody can be excluded) requiring aggregate efforts, the cooperation of some or most nations in this case may fail because it is vulnerable to cases of free riding and relies on unbalanced premises, since countries with the largest number of poor people tend to be those who have contributed least to the problem of climate change and to be less prone to be involved in a carbon-free development path (Barrett, 2007). Still, this does not mean that what restrains climate engineering from being an object of political governance and accountability in the context of climate change mitigation should be ignored. Instead, this point and the related issues deserve further attention, also to avoid ungoverned spaces, or situations of "de facto governance" on the part of industrialized, developed countries and private sector lobbies, in ways that do not involve the consideration of other countries or vulnerable groups (Gupta & Möller, 2019; Biermann & Möller, 2019).

An ideally "just" governance should be aware of the interlinkages between different dimensions (institutional, socio-technical, technical) in climate engineering innovation, and promote separate regulatory strategies and adaptive and progressive approaches toward risk allocation, in ways that are not unilateral and recognize common but differentiated responsibilities among social actors, who have different capabilities to adapt, different institutions, and different incentives to promote climate-friendly policies in the collective action problem of climate change (Barrett, 2008, 2014).

To avoid the spread of narratives on climate engineering proposals that pretend to be policy-neutral and objective, a societal reflection that evaluates what is "sustainable" in possible guiding governance principles should be put forward. For example, in the sustainable development literature, many have criticized the increasing "countability" as a guiding principle for sustainable proposals, which relies on quantitative indicators of outcomes that are depicted as value-neutral (Bexell & Jönsson, 2017, 2021). The same has been done in the climate engineering literature, where many have claimed how poorly might be a "portfolio" approach in the context of technologies like CCUS since rather than foster a coherent vision, it just adds and combines CCUS as an option within idealized and coordinated scenarios or portfolios, and so it does not consider the competing relations and trade-offs with other resources (land, energy, water) and with policy and institutional layers (Sovacool et al., 2022). Thus, in policy decisions regarding climate engineering, the implementation and justification of decisions should go beyond a mere quantitative assessment of risks and sustainable indicators and instead involve better-informed investigations dealing with the various normative uncertainties related to those climate proposals (see, e.g., Taebi et al., 2020). For example, an empirical study has recently demonstrated how a slow, robust, and bottom-up governance intervention for novel carbon-removal options might positively impact other dimensions, such as mitigating social backlash and improving technical and environmental design (Sovacool et al., 2022).

Regarding the modalities for implementing and monitoring shared standards or governance principles, one solution might be the promotion of forms of meaningful horizontal accountability, which works in contexts where there are no clear hierarchies but peer relations with various stakeholders (Schillemans, 2008). This kind of accountability might be the most decisive in the SDGs context, where different national and voluntary accountability mechanisms for implementation present competing powers, such as audit institutions, courts, and parliaments (Breuer & Leininger, 2021). Although the SDGs are not legally binding, national governments are expected to improve their governmental and intergovernmental mobilization efforts and develop specific indicators for climate engineering options. However, even if the inclusion of CCUS into mitigation portfolios has received an increasing consideration, few countries are pursuing a reliable implementation of carbon dioxide removal strategies into long-term national mitigation portfolios so far (IPCC, 2022, Chap. 12, pp. 39, 62).

At the international level, the UN Framework Convention on Climate Change (UNFCCC) and its Paris Agreement (PA) do not explicitly mention climate engineering technologies. Still, PA procedural mechanisms and nationally determined contributions might provide a basis for future deliberations on climate engineering proposals, promoting collective cooperation and transparency (Craik & Burns, 2019). The latest report from the United Nations' Intergovernmental Panel on Climate Change (IPCC) states that the governance of carbon dioxide removal methods can draw on a "political commitment" to formal integration into existing climate policy frameworks, and that a crucial governance challenge would be to establish reliable systems for monitoring, reporting, and verification (MRV) of the carbon flow and mitigation outcomes (IPCC, 2022, Chap. 12, p. 6). The report also affirms that the SDGs framework serves as a "template" to evaluate the long-term implications of mitigation on sustainable development and vice versa (IPCC, 2022, Technical Summary, p. 133). In this sense, the IPCC report suggests that coordinated and cross-sectoral policies integrating mitigation with SDGs on other sectoral policy actions (health, nutrition, equity, and biodiversity) should be adopted to alleviate or avoid many trade-offs of carbon dioxide removal methods (IPCC, 2022, Chap. 12). The creation and maintenance of shared standards on technologies like CCUS would thus require interaction and integration of different actions in the context of the SDGs to enable just transition pathways[3] and accountable infrastructures. As stated in the volume's introductory chapter, trade-offs between SDGs may emerge, and one crucial aspect of the governance of technologies is to acknowledge the interlinkages between different dimensions of sustainable development (Sætra, 2022).

Finally, the answerability dimension requires the practice of holding accounters as appropriate actors of justificatory challenge and thus susceptible to response about their conduct (Smith, 2012). Defining accountability as mere transparency concerning outcomes is a partial way to view it (Andersson & Wikström, 2014). The way carbon dioxide removal strategies are communicated is likely to influence their use and the way people conceptualize them; hence not only transparency ex-post is needed but also the framing of information presented to the public needs considerable scrutiny (Spence et al., 2021). Institutional commercial or scientific actors might misrepresent adverse information and frame climate engineering interventions as societal camouflages, reflecting how social actors prefer to instrumentally or implicitly describe technologies in ways that avert opposition or debate (Low et al., 2022). Public awareness of technologies like CCUS is still very low, but the engagement of public and civil society organizations is very relevant to shape equitable carbon-removal and storage projects that consider human health, energy needs, ecological integrity, and local community engagement (IPCC, 2022, Chap. 12, p. 65).

In this scenario, accountability may also require space for bottom-up and community strategies or for contestation (Heidelberg, 2017). Recent empirical studies on climate engineering models have reported the positive role of controversy and opposition from ENGOs, social groups, media, and delegates at the international conventions; in addition, they have also motivated the growing need for additional forms of societal appraisal, co-benefits methods, and citizen, indigenous and entrepreneurial involvement, which are still not settled for carbon-removal experimentation or are too vague for providing concrete public engagement (Low et al., 2022). Accountability as a normative criterion involves relations of responsiveness that aim to promote a dynamic co-variation of people's interests and policies (Morales, 2014). Thus, accountability for climate engineering innovation should deal with this co-variation, even if, due to the early research stage of these technologies, it is not clear how participatory RRI approaches and their emphasis on inclusivity can guide toward sustainable solutions, instead of introducing conflict-prone diversity perspectives that can also hamper or set-back research (Stelzer, 2020). Thus, "No one will be left behind" (United Nations General Assembly, 2018) is still a work in progress. A civil space that seeks to promote the participation of different views is necessary and valuable but still requires novel solutions and continued scrutiny to foster meaningful accountability relations for the governance of emerging technologies like those of climate engineering.

4.6 CONCLUSIONS

Climate engineering technologies are a technology family whose goal is to change the Earth's temperature such that we can readily combat climate change and remediate the damage that has already been done. This chapter took up a specific climate engineering technology, namely carbon capture, usage, and storage (CCUS) and showed how these technologies pose unique, global socio-political issues. This chapter has explored at how climate engineering innovation can be supplemented with a polysemic and multidimensional account of accountability. This has provided a theoretically informed basis for reflection on how to implement not only the responsible innovation of climate engineering technologies but also a dynamic landscape in which the innovation of climate engineering technologies can be built to support sustainable development more broadly.

Climate engineering innovation should avoid the risk of adopting an apolitical façade, which treats governance arrangements as neutral sites and fosters an illusory techno-optimism over the management of such a complex tech-mediated climate model. We have highlighted how the consideration of these models as mere techno-fixes does not go far enough. Indeed, techno-fix solutions can be included in the general vision of techno-solutionism and optimism, as the belief that technologies can contribute to good outcomes (see Chapters 1 and 2). But too much reliance on techno-fixes can lead to the progressive depoliticization of planetary environmental issues, and can foster a distorted binary vision in which the climate crisis is resolved by either withdrawing from technology (i.e., rejection) or accelerating it (i.e., solutionism) (Dillet & Hatzisavvidou, 2022). Instead, more balanced approaches that expand and deepen the understanding of socio-political responses, fundamental and complex social changes to the climate crisis, and the governance of technologies like CCUS are needed.

We have shown how climate engineering innovation should deal with analyzing power asymmetries and their problematic dimensions, in line with considerations on infrastructural technological change as sustained in the introductory chapters. Infrastructural technological change means that technologies may involve wide societal effects and relevant shifts in social structures (Barley, 2020). Therefore, our aim in this chapter has been that of highlighting how climate engineering innovation can be properly considered object of socio-political theorizing, since its core implications (e.g., the possibility of generating power asymmetries, and inequality more generally) can generate examples and paradigms of injustice, as well as require regulatory strategies, enforcements, and normative justifications on how decisions about innovation and sustainable strategies are taken and communicated. A reliable implementation of carbon dioxide removal and storage strategies into long-term national mitigation portfolios and public awareness of such strategies are still very low. And at the same time further work is needed to assess what responsible climate engineering innovation means, in modalities that also align with the ideal of sustainable development. In examining how and to what extent the concept of accountability is polysemic and multidimensional, our aim was to show how climate engineering innovation involves broad socio-political processes, and, more fundamentally, requires holistic approaches that take into consideration the responsibility of the actors involved, mechanisms of distribution and participation, and democratic governance on its sustainability-related impacts.

NOTES

1 UN 2015, target 16.6, but accountability is also present in SDG #17 in "Data for monitoring and accountability" and SDG #5 and #10, on gender inequality and inequality between countries, respectively.

2 For example, the TechEthos (EU Horizon 2020 Grant Agreement no. 101006249) project aims to provide "ethics by design" guidelines as well as legal recommendations for climate engineering technologies (among others), see TechEthos Project (2022) and Porcari et al. (2021).

3 In those recent years, "just transition" as a concept emerged from labor unions, environmental justice groups and the EU policy environment, encompassing the equitable shift toward

a regenerative economy in which principles and processes can respect and promote environmental and climate justice, see for example, Morena et al., 2020; *European Commission, Just Transition Platform, available at* https://ec.europa.eu/regional_policy/en/funding/jtf/just-transition-platform *(Last Access 7 Oct 2022).* The same SDGs framework that is based on the "leave no one behind" principle requires among its goals the pursuing of a just transition, as an energy transition that is shared widely and supports fair distribution (United Nations General Assembly 2015: Preamble). In this chapter, we do not devote much space to the "just transition" concept, since we are not exclusively interested in inclusiveness and matters of distributive justice in climate engineering innovation, that is, in the principles and processes that distribute benefits and burdens across members of society. However, we concentrate on the dimension of accountability, which is linked to matters of responsibility among members in society, and shared standards and normative justifications on actions. Justice issues related to energy or environment have not only components related to distributive justice but most importantly to responsibility, see Pellegrini-Masini et al., 2020. On the interdependence of different types of justice in energy justice, see the recent Astola et al., 2022.

4.7 REFERENCES

Abhayawansa, S., Adams, C. A., & Neesham, C. (2021). Accountability and governance in pursuit of Sustainable Development Goals: Conceptualising how governments create value. *Accounting, Auditing & Accountability Journal, 34*(4), 923–945. https://doi.org/10.1108/AAAJ-07-2020-4667

Andersson, J., & Wikström, E. (2014). Constructing accountability in inter-organizational collaborations. The implications of a narrow performance-based focus. *Journal of Health Organization and Management, 28*(5), 619–634. https://doi.org/10.1108/JHOM-10-2013-0220

Applegate, P. J., & Keller, K. (2015). How effective is albedo modification (solar radiation management geoengineering) in preventing sea-level rise from the Greenland Ice Sheet? *Environmental Research Letters, 10*(8), 084018. https://doi.org/10.1088/1748-9326/10/8/084018

Astola, M., Laes, E., Bombaerts, G. et al. (2022). Community heroes and sleeping members: Interdependency of the tenets of energy justice. *Science and Engineering Ethics, 28*(45). https://doi.org/10.1007/s11948-022-00384-3

Barley, S. R. (2020). *Work and technological change.* Oxford University Press.

Barrett, S. (2007). *Why cooperate? The incentive to supply global public goods.* Oxford University Press.

Barrett, S. (2008). Climate treaties and the imperative of enforcement. *Oxford Review of Economic Policy, 24*(2), 239–258.

Barrett, S. (2014). Solar geoengineering's brave new world: Thoughts on the governance of an unprecedented technology. *Review of Environmental Economics and Policy, Association of Environmental and Resource Economists, 8*(2), 249–269.

Bellamy, R., & Geden, O. (2019). Govern CO_2 removal from the ground up. *Nature Geoscience, 12*, 874–876. https://doi.org/10.1038/s41561-019-0475-7

Bellamy, R., Geden, O., Fridahl, M., Cox, E., & sPalmer, J. (2021). Editorial: Governing carbon dioxide removal. *Frontiers in Climate, 3*. https://doi.org/10.3389/fclim.2021.816346

Bellamy, R., Lezaun, J., & Palmer, J. (2017). Public perceptions of geoengineering research governance: An experimental deliberative approach. *Global Environmental Change, 45*, 194–202. https://doi.org/10.1016/j.gloenvcha.2017.06.004

Bexell, M., & Jönsson, K. (2017). Responsibility and the United Nations' sustainable development goals. *Forum for Development Studies, 44*(1), 13–29.

Bexell, M., & Jönsson, K. (2021). *The politics of the sustainable development goals: Legitimacy, responsibility, and accountability.* Routledge.

Biermann, F., & Möller, I. (2019). Rich man's solution? Climate engineering discourses and the marginalization of the Global South. *International Environmental Agreements: Politics, Law and Economics, 19*(2), 151–167. https://doi.org/10.1007/s10784-019-09431-0

Biermann, F., Oomen, J., Gupta, A., Ali, S. H., Conca, K., Hajer, M. A., Kashwan, P., Kotzé, L. J., Leach, M., Messner, D., Okereke, C., Persson, Å., Potočnik, J., Schlosberg, D., Scobie, M., & VanDeveer, S. D. (2022). Solar geoengineering: The case for an international non-use agreement. *WIREs Climate Change, 13*(3), e754. https://doi.org/10.1002/wcc.754

Borgmann, A. (2012). The setting of the scene. Technological fixes and the design of the good life. In C. J. Preston (Ed.), *Engineering the climate: The ethics of solar ration management* (pp. 189–199). Lexington Books.

Bovens, M. (2007). Analysing and assessing accountability: A conceptual framework 1. *European Law Journal, 13*(4), 447–468. https://doi.org/10.1111/j.1468-0386.2007.00378.x

Bovens, M., Goodin, R. E., & Schillemans, T. (2014). Chapter 1. Public accountability. In M. Bovens, R. E. Goodin, & T. Schillemans (Eds.), *The Oxford handbook of public accountability*. Oxford University Press.

Brändle, G., Schönfisch, M., & Schulte, S. (2021). Estimating long-term global supply costs for low-carbon hydrogen. *Applied Energy, 302*, 117481. https://doi.org/10.1016/j.apenergy.2021.117481

Breuer, A., & Leininger, J. (2021). Horizontal accountability for SDG implementation: A comparative cross-national analysis of emerging national accountability regimes. *Sustainability, 13*. https://doi.org/10.3390/su13137002

Brooks, L., Cannizzaro, S., Umbrello, S., Bernstein, M. J., & Richardson, K. (2022). Ethics of climate engineering: Don't forget technology has an ethical aspect too. *International Journal of Information Management, 63*, 102449. https://doi.org/10.1016/j.ijinfomgt.2021.102449

Buchanan, A., & Keohane, R. O. (2006). The legitimacy of global governance institutions. *Ethics & International Affairs, 20*(4), 405–437. https://doi.org/10.1111/j.1747-7093.2006.00043.x

Buchinger, E., Kinegger, M., Zahradnik, G., Bernstein, M. J., Porcari, A., Gonzalez, G., Pimponi, D., & Buceti, G. (2022). *TechEthos technology portfolio: Assessment and final selection of economically and ethically high impact technologies*. Deliverable 1.2 to the European Commission. TechEthos Project Deliverable. www.techethos.eu.

Budzianowski, W. M. (2017). Implementing carbon capture, utilisation and storage in the circular economy. *International Journal of Global Warming, 12*(2), 272–296. https://doi.org/10.1504/IJGW.2017.084510

Bui, M., Adjiman, C. S., Bardow, A., Anthony, E. J., Boston, A., Brown, S., . . . Mac Dowell, N. (2018). Carbon capture and storage (CCS): The way forward. *Energy & Environmental Science, 11*(5), 1062–1176. https://doi.org/10.1039/C7EE02342A

Callies, D. E. (2018). Institutional legitimacy and geoengineering governance. *Ethics, Policy & Environment, 21*(3), 324–340. https://doi.org/10.1080/21550085.2018.1562523

Coeckelbergh, M. (2022). *The political philosophy of AI*. Polity Press.

Craik, N., & Burns, W. C. (2019). Climate engineering under the Paris agreement. *Environmental Law Reporter News & Analysis, 49*, 11113.

Dillet, B., & Hatzisavvidou, S. (2022). Beyond technofix: Thinking with Epimetheus in the Anthropocene. *Contemporary Political Theory, 21*, 351–372. https://doi.org/10.1057/s41296-021-00521-w

Gardiner, S., & McKinnon, C. (2020). The justice and legitimacy of geoengineering. *Critical Review of International Social and Political Philosophy, 23*(5), 557–563. https://doi.org/10.1080/13698230.2019.1693157

Glass, L. M., & Newig, J. (2019). Governance for achieving the sustainable development goals: How important are participation, policy coherence, reflexivity, adaptation and democratic institutions? *Earth System Governance, 2*, 100031.

Grimes, M., & Esaiasson, P. (2014). Government responsiveness: A democratic value with negative externalities? *Political Research Quarterly, 67*(4), 758–768. https://doi.org/10.1177%2F1065912914543193

Gupta, A., & Möller, I. (2019). De facto governance: How authoritative assessments construct climate engineering as an object of governance. *Environmental Politics, 28*(3), 480–501. https://doi.org/10.1080/09644016.2018.1452373

Hanssen, S. V., Daioglou, V., Steinmann, Z. J. N., Doelman, J. C., Van Vuuren, D. P., & Huijbregts, M. A. J. (2020). The climate change mitigation potential of bioenergy with carbon capture and storage. *Nature Climate Change, 10*(11), 1023–1029. https://doi.org/10.1038/s41558-020-0885-y

Heidelberg, R. L. (2017). Political accountability and spaces of contestation. *Administration & Society, 49*(10), 1379–1402.

Ho, H. J., Iizuka, A., & Shibata, E. (2019). Carbon capture and utilization technology without carbon dioxide purification and pressurization: A review on its necessity and available technologies. *Industrial & Engineering Chemistry Research, 58*(21), 8941–8954. https://doi.org/10.1021/acs.iecr.9b01213

Horton, J., & Keith, D. (2016). Solar geoengineering and obligations to the global poor. In C. J. Preston (Ed.), *Climate justice and geoengineering: Ethics and policy in the atmospheric Anthropocene* (pp. 79–92). Rowman and Littlefield International.

Hourdequin, M. (2021). Environmental ethics: The state of the question. *Southern Journal of Philosophy, 59*, 270–308. https://doi.org/10.1111/sjp.12436

IPCC. (2022). *Climate change 2022: Mitigation of climate change. contribution of working group III to the sixth assessment report of the intergovernmental panel on climate change* (P. R. Shukla, J. Skea, R. Slade, A. Al Khourdajie, R. van Diemen, D. McCollum, M. Pathak, S. Some, P. Vyas, R. Fradera, M. Belkacemi, A. Hasija, G. Lisboa, S. Luz, & J. Malley, Eds.). Cambridge University Press. https://doi.org/10.1017/9781009157926.

Jakosky, B. M., & Edwards, C. S. (2018). Inventory of CO 2 available for terraforming Mars. *Nature Astronomy, 2*(8), 634–639. https://doi.org/10.1038/s41550-018-0529-6

Kanie, N., Zondervan, C., & Stevens, C. (2014). *Ideas on governance 'of' and 'for' sustainable development goals: UNI-IAS/POST2015 conference report.* United Nations University Institute for the Advanced Study of Sustainability.

Karlsson-Vinkhuyzen, S., Dahl, A. L., & Persson, Å. (2018). The emerging accountability regimes for the Sustainable Development Goals and policy integration: Friend or foe? *Environment and Planning C: Politics and Space, 36*(8), 1371–1390. https://doi.org/10.1177/2399654418779995

Low, S., Baum, C. M., & Sovacool, B. K. (2022). Taking it outside: Exploring social opposition to 21 early-stage experiments in radical climate interventions. *Energy Research & Social Science, 90*, 102594. https://doi.org/10.1016/j.erss.2022.102594

Mashaw, J. L. (2006). Accountability and institutional design: Some thoughts on the grammar of governance. In M. Dowdle (Ed.), *Public accountability: Designs, dilemmas and experiences* (pp. 115–156). Cambridge University Press.

McLaren, D. (2018). Whose climate and whose ethics? Conceptions of justice in solar geoengineering modelling. *Energy Research & Social Science, 44*, 209–221. https://doi.org/10.1016/j.erss.2018.05.021

Metz, B., Davidson, O., De Coninck, H., Loos, M., & Meyer, L. (2005). *IPCC special report on carbon dioxide capture and storage.* Cambridge University Press.

Ming, T., Liu, W., & Caillol, S. (2014). Fighting global warming by climate engineering: Is the Earth radiation management and the solar radiation management any option for fighting climate change? *Renewable and Sustainable Energy Reviews, 31*, 792–834. https://doi.org/10.1016/j.rser.2013.12.032

Morales, L. (2014, September). *A conceptual and theoretical approach to governmental policy responsiveness between elections.* Paper presented at the 2014 ECPR Conference, University of Glasgow.

Morena, E., Dunja, K., & Dimitris, S. (2020). *Just transitions: Social justice in the shift towards a low-carbon world.* Pluto.

O'Lear, S., Hane, M. K., Neal, A. P., Stallings, L. L. M., Sierra, W., & Park, J. (2021). Environmental geopolitics of climate engineering proposals in the IPCC 5th assessment report. *Frontiers in Climate, 3*, 718553. https://doi.org/10.3389/fclim.2021.718553

Owen, R., von Schomberg, R., & Macnaghten, P. (2021). An unfinished journey? Reflections on a decade of responsible research and innovation. *Journal of Responsible Innovation*, *8*(2), 217–233. https://doi.org/10.1080/23299460.2021.1948789

Palumbo, A., & Bellamy, R. (Eds.). (2010). *Political accountability*. Routledge.

Papadopolous, Y., & Warin, P. (2007). Are innovative, participatory and deliberative procedures in policy making democratic and effective? *European Journal of Political Research*, *46*, 445–472. https://doi.org/10.1111/j.1475-6765.2007.00696.x

Pateman, C. (1970). *Participation and democratic theory*. Cambridge University Press.

Pellegrini-Masini, G., Pirni, A., & Maran, S. (2020). Energy justice revisited: A critical review on the philosophical and political origins of equality. *Energy Research and Social Science*, *59*, 101310.

Pesch, U. (2015). Engineers and active responsibility. *Science and Engineering Ethics*, *21*(4), 925–939. https://doi.org/10.1007/s11948-014-9571-7

Pitkin, H. (1967). *The concept of representation*. University of California Press.

Porcari, A., Pimponi, D., Gonzalez, G., Buceti, G., Buchinger, E., Kienegger, M., Bernstein, M., & Zahradnik, G. (2021). *Description of selected high socio-economic impact technologies*. Deliverable 1.1 for the European Commission. TechEthos Project Deliverable. www.techethos.eu

Proelss, A., & Güssow, K. (2011). Carbon capture and storage from the perspective of international law. In *European yearbook of international economic law 2011* (pp. 151–168). Springer. https://doi.org/10.1007/978-3-642-14432-5_7

Psarras, P. C., Comello, S., Bains, P., Charoensawadpong, P., Reichelstein, S., & Wilcox, J. (2017). Carbon capture and utilisation in the industrial sector. *Environmental Science & Technology*, *51*(19), 11440–11449. https://doi.org/10.1021/acs.est.7b01723

Raza, A., Gholami, R., Rezaee, R., Rasouli, V., & Rabiei, M. (2019). Significant aspects of carbon capture and storage–A review. *Petroleum*, *5*(4), 335–340. https://doi.org/10.1016/j.petlm.2018.12.007

Román, M. (2011). Carbon capture and storage in developing countries: A comparison of Brazil, South Africa and India. *Global Environmental Change*, *21*(2), 391–401. https://doi.org/10.1016/j.gloenvcha.2011.01.018

Roussanaly, S., Berghout, N., Fout, T., Garcia, M., Gardarsdottir, S., Nazir, S. M., . . . Rubin, E. S. (2021). Towards improved cost evaluation of Carbon Capture and Storage from industry. *International Journal of Greenhouse Gas Control*, *106*, 103263. https://doi.org/10.1016/j.ijggc.2021.103263

Royal Society. (2009). *Geoengineering the climate: Science, governance and uncertainty': Royal society policy document*. Retrieved October 7, 2022, from https://royalsociety.org/-/media/Royal_Society_Content/policy/publications/2009/8693.pdf

Rubin, E. S., & Zhai, H. (2012). The cost of carbon capture and storage for natural gas combined cycle power plants. *Environmental Science & Technology*, *46*(6), 3076–3084. https://doi.org/10.1021/es204514f

Sætra, H. S. (2022). *AI for the sustainable development goals*. CRC Press.

Santoni de Sio, F., & Mecacci, G. (2021). Four responsibility gaps with artificial intelligence: Why they matter and how to address them. *Philosophy & Technology*, *34*, 1057–1084. https://doi.org/10.1007/s13347-021-00450-x

Schillemans, T. (2008). Accountability in the shadow of hierarchy: The horizontal accountability of agencies. *Public Organization Review*, *8*, 175–194. https://doi.org/10.1007/s11115-008-0053-8

Schneider, L. (2019). Fixing the climate? How geoengineering threatens to undermine the SDGs and climate justice. *Development*, *62*, 29–36. https://doi.org/10.1057/s41301-019-00211-68

Scott, D. (2012). Insurance policy or technological fix: The ethical implications of framing solar radiation management. In C. J. Preston (Ed.), *Engineering the climate: The ethics of solar ration management* (pp. 113–131). Lexington Books.

Sikka, T. (2021). An intersectional analysis of geoengineering: Overlapping oppressions and the demand for ecological citizenship. In J. P. Sapinski, H. J. Buck, & A. Malm (Eds.), *Has it come to this? The promises and perils of geoengineering on the brink* (pp. 99–118). Rutgers University Press.

Smith, A. M. (2012). Attributability, answerability, and accountability: In defense of a unified account. *Ethics, 122*(3), 575–589. https://doi.org/10.1086/664752

Smith, P. T. (2018). Legitimacy and non-domination in solar radiation management research. *Ethics, Policy & Environment, 21*(3), 341–361. https://doi.org/10.1080/21550085.2018.1562528

Smith, P. T. (2021). Who may geoengineer: Global *Domination*, revolution, and solar radiation management. *Global Justice: Theory Practice Rhetoric, 13*(01), 138–165.

Sovacool, B. K., Baum, C. M., & Low, S. (2022). Risk–risk governance in a low-carbon future: Exploring institutional, technological, and behavioral tradeoffs in climate geoengineering pathways. *Risk Analysis*, 1– 22. https://doi.org/10.1111/risa.13932

Spence, E., Cox, E., & Pidgeon, N. (2021). Exploring cross-national public support for enhanced weathering as a land-based carbon dioxide removal strategy. *Climatic Change, 165*, 23. https://doi.org/10.1007/s10584-021-03050-y

Stahl, B. C. et al. (2021). From responsible research and innovation to responsibility by design, *Journal of Responsible Innovation, 8*(2), 175–198. https://doi.org/10.1080/23299460.2021.1955613

Stelzer, H. (2020). Responsible innovation and climate engineering. A step back to technology assessment. *Philosophy of Management, 19*, 297–316. https://doi.org/10.1007/s40926-020-00127-z

Stilgoe, J. (2016). Geoengineering as collective experimentation. *Science and Engineering Ethics, 22*(3), 851–869. https://doi.org/10.1007/s11948-015-9646-0

Taebi, B., Kwakkel, J. H., & Kermisch, C. (2020). Governing climate risks in the face of normative uncertainties. *WIREs Climate Change, 11*, e666. https://doi.org/10.1002/wcc.666

Tcvetkov, P., Cherepovitsyn, A., & Fedoseev, S. (2019). The changing role of CO_2 in the transition to a circular economy: Review of carbon sequestration projects. *Sustainability, 11*(20), 5834. https://doi.org/10.3390/su11205834

TechEthos Project. (2022). TechEthos in a nutshell. *TechEthos Project Factsheet*. www.techethos.eu

United Nations. (2015). *Transforming our world: The 2030 agenda for sustainable development*. Division for Sustainable Development Goals.

United Nations General Assembly. (2018). *No one will be left behind*. Geneva Academy Briefing No. 11. Geneva.

Urbinati, N., & Warren, M. E. (2008). The concept of representation in contemporary democratic theory. *Annual Review of Political Science, 11*(1), 387–412. https://doi.org/10.1146/annurev.polisci.11.053006.190533

van de Poel, I. (2011). The relation between forward-looking and backward-looking responsibility. In N. Vincent, I. van de Poel, & J. van den Hoven (Eds.), *Moral responsibility*. Library of Ethics and Applied Philosophy (Vol. 27). Springer. https://doi.org/10.1007/978-94-007-1878-4_3

Verdiesen, I., Santoni de Sio, F., & Dignum, V. (2021). Accountability and control over autonomous weapon systems: A framework for comprehensive human oversight. *Minds & Machines, 31*, 137–163. https://doi.org/10.1007/s11023-020-09532-9

Villalona, C. (2021). *Global SDG accountability report: A snapshot on the state of accountability for the 2030 Agenda*. United Nations. https://secureservercdn.net/166.62.112.219/9bz.99d.myftpupload.com/wp-content/uploads/2021/06/GlobalSDGAccountabilityReport_pages_hRes-1.pdf

Waelen, R. (2022). Why AI ethics is a critical theory. *Philosophy & Technology, 35*, 9. https://doi.org/10.1007/s13347-022-00507-5

Wagner, G., & Zizzamia, D. (2021). Green moral hazards. *Ethics, Policy & Environment*, 1–17. https://doi.org/10.1080/21550085.2021.1940449

Weber, H. (2017). Politics of 'leaving no one behind': Contesting the 2030 Sustainable Development Goals agenda. *Globalizations, 14*(3), 399–414. https://doi.org/10.1080/14747731.2016.1275404

Yu, M., Wang, K., & Vredenburg, H. (2021). Insights into low-carbon hydrogen production methods: Green, blue and aqua hydrogen. *International Journal of Hydrogen Energy*, *46*(41), 21261–21273. https://doi.org/10.1016/j.ijhydene.2021.04.016

Yunita, A., Biermann, F., Rakhyun, E. K., & Marjanneke, J. V. (2022). The (anti-)politics of policy coherence for sustainable development in the Netherlands: Logic, method, effects. *Geoforum*, *128*, 92–102. https://doi.org/10.1016/j.geoforum.2021.12.002

CHAPTER 5

Shinigami Eyes and Social Media Labeling as a Technology for Self-care

Henrik Skaug Sætra and Jo Ese

CONTENTS

5.1 INTRODUCTION

Since its inception, the role of social media as part of the public sphere (Habermas, 1991) has been debated. Ranging from the panegyric praise of the emancipating role of Twitter and Facebook during the Arab Spring to the condemnation of the corrosive effects of the very same platforms during the last two US presidential campaigns, social media as part of our infrastructure for a sustainable and sound public debate has been contested (Kruse et al., 2018; Vaidhyanathan, 2018). Furthermore, social media platforms are constantly developing, with changing algorithms nudging the public debate in different directions, potentially profoundly influencing interactions, the public debate, and eventually how our democracies and societies are structured.

However, if there ever existed such a thing as a Habermasian public sphere in present-day society, the concept is certainly challenged by recent developments. In this chapter,

DOI: 10.1201/9781003325086-5

we examine one particular challenge exemplified by the *Shinigami Eyes* web browser extension (Kiran, 2022). The extension is developed as a tool to aid trans people and their allies in identifying and avoiding transphobic users and content, while easily recognizing trans-friendly users. However, the add-on has received criticism due to its labeling of social media users without the users knowing about or agreeing to the labeling, and some feel that such a tool undermines the very idea of a public sphere and free and open public debate.

Although society's treatment and acceptance of trans individuals have gone through major developments in the last decades, marginalization, exclusion, abuse, and harassment of trans individuals are still widespread (Ciszek et al., 2021). As with other forms of harassment, a lot of the anti-trans harassment is found online. Social media is a prominent part of the lives of modern individuals in major parts of the world, and LGBTQ+ people are also actively using social media to navigate their existence and construct their identities (Jenzen, 2017; Lucero, 2017; Southerton et al., 2021). However, in this chapter, we discuss how these individuals, and trans people in particular, experience harassment and hostility on social media (Jenzen, 2017), which when seen in conjunction with other forms of marginalization poses clear threats relevant to a number of aspects related to sustainable development and the Sustainable Development Goals (SDGs) (Ongsupankul, 2019).

For example, "full and effective participation and equal opportunities for leadership at all levels of decision-making in political, economic and public life" is a target of SDG 5 (United Nations, 2015), and while this might seem to offer opportunities to promote the rights of LGBTQ+ people, we show how these individuals are erased and omitted from major frameworks such as the SDGs. Nevertheless, we might interpret the SDGs more broadly (Dorey & O'Connor, 2016), and if so point out how, for example, participation in public life requires an online presence and online participation. If these arenas are perceived as threatening and dangerous, it can be argued that these threats undermine sustainable development. A tool like Shinigami Eyes, which potentially helps trans individuals avoid content perceived as harmful, thus makes it easier to participate in this important domain of public life (Lucero, 2017).

However, public life is arguably also about the exchange of arguments and ideas, including facing, experiencing, arguing against, and sometimes learning from ideas and opinions that we disagree with. Although a technology like Shinigami Eyes protects individuals from content they find uncomfortable, some fear it could also insulate individuals from opinions that do not resemble their own. Without communication and exchanges of ideas, arguments, and experiences between different groups, conflict and polarization may arise. Furthermore, those who are labeled also have certain rights and interests. For example, SDG 16.10 specifically addresses fundamental freedoms like freedom of thought and freedom of expression, which align with arguments used by the Norwegian data protection authority to *ban* Shinigami Eyes from processing data in Norway (Datatilsynet, 2022a). Furthermore, the labeling of individuals challenges the Habermasian ideal of the public sphere as a place where the *content* of a message – and not the identity of the speaker – is what should be considered, an ideal also promoted by Merton (1942) in the domain of science.

In this chapter, we examine the sustainability-related implications of trans people's online lives and Shinigami Eyes labeling, and raise the following questions:

1. How do issues related to the rights of transgender people relate to sustainable development and the SDGs?
2. Is social media potentially so hateful and violent toward vulnerable groups that the kind of labeling described in this chapter is necessary and legitimate?
3. Which conflicts between the rights of different individuals and different SDGs are central to understanding the controversy surrounding Shinigami Eyes?

5.2 SUSTAINABILITY, LGBTQ+ RIGHTS, AND SOCIAL MEDIA

Sustainability, as discussed in Chapter 2, encompasses economic, social, and environmental issues. The broader notion of social sustainability, including the principles of inclusion, safety, and justice for all, is clearly amenable to include issues related to, for example, discrimination and hate-speech based on sexual orientation, gender identity and expressions, and sexual characteristics (SOGIESC).[1] One might, then, be inclined to assume that the SDGs also provide protection for people experiencing exclusion and discrimination based on SOGIESC. Such an assumption would, however, be unwarranted.

Quite the contrary, the SDGs do not mention SOGIESC or elements of it anywhere (Mills, 2015; Ongsupankul, 2019; United Nations, 2015), as proponents of LGBTQ+ rights and equality have long pointed out (Dorey & O'Connor, 2016; Ongsupankul, 2019). Some go so far as to state that, for example, gender-diverse and trans people are "completely omitted and erased from the scope" of key SDGs, such as SDG 5 (Matthyse, 2020). Others support the notion that LGBTQ+ people are erased in Agenda 2030, and also that their sexual rights are, problematically, omitted from the SDGs (Logie, 2021). Politics is a crucial part of the SDGs and similar frameworks (Majeedullah et al., 2016), and while efforts to include all or parts of the LGBTQ+ community in international frameworks such as the SDGs achieve intermittent success, the advances made in negotiations are often defeated before final publication. Ongsupankul (2019), for example, relates how LGBT rights were said to be "off the table" in the lead-up to Agenda 2030, as the need to compromise with countries where LGBTQ+ people and their concerns are not accepted or recognized thoroughly complicates the negotiation process.

Despite the lack of explicit mention, many are arguing that SOGIESC is encompassed by the core goal of the SDGs, namely to *leave no one behind* (United Nations, 2015). This concept is consequently highlighted by many of those exploring the rights of LGBTQ+ people from a sustainability perspective (Logie, 2021; Ongsupankul, 2019). The most optimistic accounts of SOGIESC and the SDGs emphasize this and argue that while SOGIESC issues are not explicitly mentioned, LGBTQ+ people are included through what is referred to as an "other" status in various reports from, for example, the United Nations, the Human Rights Council, and the World Health Organization (Divan et al., 2016; Logie, 2021; Mills, 2015).

Stonewall International has released a guide for LGBT inclusion and the SDGs, and while they state that the SDGs "could have gone further", this particular publication highlights how the SDGs can in fact be used to improve the situation of LGBTQ+ people through the focus on leaving no one behind and equality for *all* in Agenda 2030 (Dorey & O'Connor, 2016). Their analysis of the most relevant goals is summarized in the following.

SDG 1 aims to end poverty in all its forms everywhere and entails social protection systems for all (target 1.3), making sure that all have equal rights to economic resources, access to services, and control over land and property (target 1.4). As LGBTQ+ people are excluded and discriminated in a wide range of settings – in education and work – they are economically vulnerable, and there is a need for more research on economic discrimination, including SOGIESC considerations in development projects, and social assistance projects (Dorey & O'Connor, 2016).

SDG 3 is about the promotion of healthy lives and well-being for all, and *universal* access to sexual and reproductive healthcare services (target 3.7) and universal health coverage for all (target 3.8) are of particular relevance. LGBTQ+ people's access to good and non-discriminatory health services is key, and this entails education and prevention measures, proper help for trans people seeking to transition, and not least mental health services and support (Dorey & O'Connor, 2016).

Quality education for all is the aim of SDG 4, and LGBTQ+ people experience bullying and exclusion throughout the educational system, leading many to dropout and miss future opportunities (Majeedullah et al., 2016). Increased awareness and competence for teachers and counselors, active policies against homo-, bi-, and transphobic bullying, inclusion of LGBTQ+ inclusive and positive material in education, and general efforts to foster more inclusive cultures are among the key actions required for achieving this goal (Dorey & O'Connor, 2016).

SDG 5 is also mentioned, despite this goal explicitly and exclusively targeting *women* and *girls* (Ongsupankul, 2019). Dorey and O'Connor (2016) here emphasizes intersectionality and how lesbian, trans, and bi women are particularly vulnerable to discrimination and violence. In a critical analysis of SDGs and transgender equality, Matthyse (2020) argues that the omission of gender-diverse and trans people in SDG 5 has been achieved by the patriarchy to continue gendered oppression and inequality, and highlights the unfortunate impacts of such erasure in the global framework designed to include all. They proceed to argue that an "authentic version of gender equality to achieve freedom from oppression on the grounds of gender takes into account a multiplicity of gender identities and gender expressions" (Matthyse, 2020), something also supported by others (Majeedullah et al., 2016).

One of the goals often presented as the most obvious candidate for addressing SOGIESC challenges is SDG 10 (Majeedullah et al., 2016), which aims to reduce inequality within and among countries. Target 10.2 states that the goal is to empower and promote the inclusion of *all*, irrespective of a number of explicitly mentioned characteristics (age, race, sex, etc.) "or other status", which is taken to include LGBTQ+ people (Logie, 2021). Target 10.3 proceeds to state that equal opportunity should be promoted by eliminating discriminatory laws, policies, and practices. This goal is consequently crucial for addressing the discrimination of LGBTQ+ people through discriminatory laws and policies. Examples of such policies include requiring trans people to undergo sterilization in order to change gender, not providing opportunities for trans people to change legal gender, prohibiting same-sex or other relations, prohibiting the establishment of LGBTQ+ groups and organizations, and excluding LGBTQ+ people from social and health services. Dorey and O'Connor (2016) argue that generating awareness of how LGBTQ+ people are included in "other" status is key to achieving progress in this domain.

SDG 11 is also relevant, as it relates to the inclusiveness and safety of cities and human settlements. This could arguably be related to SOGIESC-related violence and exclusion, but Dorey and O'Connor (2016) focus on target 11.1 and access to affordable housing and basic services, as many LGBTQ+ people are forced to leave their families and homes and experience high prevalence of homelessness. Violence is dealt with under SDG 16, which is another goal with high thematic overlap with sexuality- and gender-related challenges (Majeedullah et al., 2016). SDG 16 has targets related to reducing all forms of violence and death threats (target 16.1) and ensuring access to justice (16.3). While all forms of violence are abhorrent, Dorey and O'Connor (2016) highlight the challenges related to how states are not taking violence against LBGTQ+ people seriously, and that these individuals in many areas consequently cannot rely on the state's protection. This makes these individuals more vulnerable to human rights abuse, and taking action to increase awareness of and the prioritization of hate crimes is crucial for reaching SDG 16 (Dorey & O'Connor, 2016).

It is important to be aware of country-to-country and even community-to-community differences regarding the situation of LGBTQ+ people. Even if some readers consider the situation for these individuals in their society to be relatively safe and that they have equal rights and protections as others, we are here considering the global context and how people from many different countries are coming together on platforms and social media that are often not clearly regulated through national legislation. On such platforms, those who do *not* respect LGBTQ+ rights cross paths with others who might *have* rights nationally. This gives rise to the conflict of rights and interests discussed later. Those who do not have rights in their homeland, such as people in Tunisia where homosexuality is criminalized and any offense is retained in a person's criminal record for 5 years (Ongsupankul, 2019), may be able to find support and the means to partly fight back through social media as they can connect with like-minded people and share and develop tools and systems (Haimson et al., 2020; Haimson et al., 2021).

Despite various efforts and initiatives to improve their situation, the reality is that LGBTQ+ people are amongst the most vulnerable individuals in many modern societies. And amongst this group, trans people are the most vulnerable of all (Jenzen, 2017; Ongsupankul, 2019). Trans people experience "extreme social exclusion" which result in increased vulnerability to physical and mental health conditions, discrimination in education and work, and a "general loss of opportunities for economic and social advancement" (Divan et al., 2016).

To add to this trans people who experience resentment, prejudice, and threatening environments often find themselves in criminalized contexts, discouraging them from seeking the aid of police and barring them from other avenues of justice (Divan et al., 2016). Trans people are often criminalized and misunderstood, deprived of basic services (Matthyse, 2020), and, for example, trans women have alarmingly low life expectancies – as low as 35 years old in the Americas (Inter-American Commission on Human Rights, 2018).

Remedies for these vulnerabilities require action at numerous levels, including legal reform and the criminalization of various forms of violence and hate speech against trans people, getting rid of laws that criminalize and expose trans people, and "general advocacy to sensitize the ill-informed about trans issues and concerns" (Divan et al., 2016). This involves community-level reforms of the societal attitudes and social practices that engender the marginalization of trans people and others (Majeedullah et al., 2016).

Summing up, trans people live in "extremely hostile contexts" (Divan et al., 2016) and live with a higher risk of experiencing various forms of violence and hate crimes (Matthyse, 2020). That a framework such as Agenda 2030 and the SDGs fail to mention and explicitly emphasize their rights highlights the potential need for *other*, and potentially untraditional, means to improve and safeguard the position of these individuals, both by the individuals themselves and by others who support their cause. While institutions, local communities, and families are often highlighted as arenas in which gender and sexuality norms are policed (Majeedullah et al., 2016), public spheres and online communities are potentially equally important arenas both of moral policing and crucial potential arenas for escape from the aforementioned hostile arenas (Haimson et al., 2020; Jenzen, 2017). Furthermore, as many sexual minorities live in contexts in which they cannot invoke legal rights or prefer not to report crimes committed against them out of fear of being arrested (Ongsupankul, 2019), self-care of various kinds becomes necessary (Edmond, 2022). One way for the trans community to protect themselves online is the marking and filtering of perceived harmful individuals in social media, to which we now turn.

5.3 THE CASE: SHINIGAMI EYES AND SOCIAL MEDIA LABELING

We focus on one specific technological response to the preceding challenges LGBTQ+ people experience online, namely a browser add-on named *Shinigami Eyes* (Kiran, 2022). The add-on is free and distributed on GitHub, and it works for Chrome and Firefox browsers. What it does is highlight "transphobic and trans-friendly social network pages and users with different colors" (Kiran, 2022). The publisher is an anonymous self-declared trans person publishing under the nickname *Kiran*. They describe making this extension to alleviate some of the uncertainty trans people experience when they face communities with members whose interests, views, and opinions might be hostile to trans people. This is particularly difficult, Kiran writes, when dealing with groups of intersectional interests, such as the feminist, lesbian, and atheist communities (Kiran, 2022).

> The purpose of this extension is to make transgender people feel more confident towards people, groups, and pages they can trust, and to highlight possible interactions with the trans-hostile ones.
>
> (Kiran, 2022)

Intersectionality refers to how minority characteristics seldom come alone, and how, for example, race and gender interact in shaping the experiences of marginalized people (Crenshaw, 1990; Matthyse, 2020). SOGIESC, and being trans as highlighted by Kiran, provides additional minority characteristics that interact with, for example, gender. While feminists and trans people share a broad experience of being marginalized, the interactions between trans people and sections of the feminist community are often quite hostile (Hines, 2019). Dealing with groups that are uniformly and openly hostile to trans people is described by Kiran as relatively easy, but dealing with groups with members who are seemingly progressive but at the same time hostile to trans people is difficult. This generates insecurity, as trans people are left guessing who their allies and enemies are (Kiran, 2022). In such a process, each individual is left to keep track of whom to trust and not which, needless to say, is a tall order on the supported platforms Facebook, Twitter, Reddit, Tumblr, Medium, YouTube, Wikipedia, and search engine results.

Edmond (2022) provides support for the usefulness of Shinigami Eyes in their account of how a user of the extension states that Shinigami Eyes does in fact aid in conducting the "vetting process" they were already undertaking (Edmond, 2022). As also emphasized by Haimson et al. (2021), tags and filtering "will be essential to any future trans technology", and Shinigami Eyes provides one way to achieve this on platforms not explicitly made for this group.

Taking one step back, the add-on has taken its name from the popular Japanese manga and anime show called *Death Note* (original Japanese title *Desu Nōto*) (Nakatani et al., 2006–2007; Ohba & Obata, 2003–2008). In this show, death gods – the *Shinigami* – enter the world of the humans and provide a select few with notebooks called Death Notes. These notebooks provide their human owners with the power to end the lives of others merely by writing their names in the notebook while imagining their faces. The Shinigami have special eyes that enable them to see the names and remaining life span of the humans they encounter. As the notebook requires someone's real name for them to die, this is an attractive ability for death note holders. The Shinigami consequently offer the human notebook owners a deal: They can have *Shinigami Eyes* in exchange for half of their remaining life span. This would then enable them to see the names and life span of their fellow humans in bright red and in real time. This brief background should suffice to show that the background of the add-on's name is quite sinister, and that its functionality is only superficially connected to the function of Shinigami Eyes in the anime. The publisher's nickname is also presumably derived from the series' protagonist's nickname – *Kira*.

The browser extension, when installed, colors people's usernames either red or green depending on whether they have been labeled trans-hostile or trans-friendly. Figure 5.1 shows how this appears on the authors' Twitter, where the usernames and other information are clearly colored and consequently stand out from other users.

The initial extension marked people based on the author's manual labeling mixed with "machine learning", and people can now contribute their own labels as well (Kiran, 2022). Figure 5.2 shows the user interface that allows for the labeling of users, over a post where the labeling of Margaret Atwood is being discussed.[2]

The main input for labeling seems to be manual, and the extension website offers guidelines for how to label. It urges a conservative attitude, with a set of examples of what is not enough – and what *is* enough – to label someone anti-trans or trans-friendly (Kiran, 2022). The criteria used are based on anti-trans and anti-nonbinary sentiment, and it is consequently targeted at a subset of the challenges related to SOGIESC more broadly.

FIGURE 5.1 Three labeled users on Twitter, the left being labeled trans-friendly while the center and right profiles are marked as trans-hostile.

FIGURE 5.2 Shinigami Eyes user interface.

If an existing label is thought to be wrong, or perhaps in need of updating as referred to in Figure 5.2, the users can simply add new labels, which "will help us [Kiran] improve this extension" (Kiran, 2022). Kiran also states that malicious and fake reports can be immediately overridden locally, while only changes that "pass some trustworthiness criteria (including human validation)" are included in the public dataset used to label accounts for all Shinigami Eyes users.

The browser extension provides an example of how technology enables marginalized and vulnerable groups to take action to protect themselves and create safer spaces online. Having seen the challenges faced by LGBTQ+ people, and trans people in particular, the perceived need to take action online is understandable. In fact, it might be the only way for these people to feel safe *enough* online to not have to fully withdraw from certain platforms and networks. Seeing how important these social arenas are in many peoples' lives today, this is no small matter (Jenzen, 2017).

However, the extension has also generated controversy related to how such tools potentially change social media as forms of public spheres and could have negative consequences for those being labeled, and we will shortly turn to such objections. First, however, we note that more technical issues are also threatening to derail Shinigami Eyes. For example, Datatilsynet (the Norwegian Data Protection Authority) recently issued a ban on processing after receiving a complaint on the extension from a Norwegian individual marked as trans-hostile by Shinigami Eyes (Datatilsynet, 2022a).

After publishing their intent to impose a ban, Datatilsynet received no replies from the publishers of Shinigami Eyes, several more e-mails from the original complainant, and on June 16, 2022, they proceeded to state that

> Even without a response from Shinigami Eyes we have concluded that we have sufficient information in order to conclude on the legality of the browser extension, i.e. that the processing of personal data in question is in violation of Article 6(1), Article 12(2) and Article 14 GDPR.
>
> (Datatilsynet, 2022a)

The legal details are of less interest in this context than the overarching arguments used to justify the ban. First of all, they argue that marking trans-hostile and trans-friendly people online might constitute protecting people from harm, and consequently that Shinigami Eyes pursues a legitimate interest (Datatilsynet, 2022a). Furthermore, they agree that data processing might be necessary to pursue this legitimate interest. However, they argue that the processing done through the extension constitutes "profiling" according to Article 4(4) GDPR, and that being marked entails potentially significant negative consequences for the ones marked. They can "lose their job, or friendships, and the individual could be the target of hate and mistreatment", and the individuals in question are not notified of being marked or provided any insight into *why* they are labeled (Datatilsynet, 2022a).

This led Datatilsynet to issue a ban, as the potential harms are found weightier than the potential benefits and the strength of the legitimate interest pursued through Shinigami Eyes. They also mention the issue that such extensions can create "chilling effects" online, seemingly mainly noting the chilling effects on those who fear marking, and not on trans people who fear harassment and encountering anti-trans sentiment online if they express themselves and their identities openly (Datatilsynet, 2022a). The potential chilling effects associated with social media filtering as here discussed should not be underplayed, as there are good reasons to be wary of a development where the debate about LBGTQ+ issues, and trans-related issues in general, is fully closed. Our purpose in this chapter, however, is to more thoroughly explore the implications for those being discussed – and how open debate influences them.

Datatilsynet seems to state that individuals have the option not to engage with whomever they want on an individual basis, and indicate that trans people should do so instead of relying on a "collective decision-making and categorization" which "could strengthen echo chambers found online" (Datatilsynet, 2022a).

In closing, they state that "the legitimate interest pursued by Shinigami Eyes cannot be assessed as one of significant strength or importance", and they consequently impose a ban, the effect of which at the time of writing is unsure. It might be added that the users of the extension do not share traffic with the developers, as the username check is performed locally on each computer. Furthermore, the data stored locally are not accessible in readable format but are represented as a bloom filter (Bloom, 1970), which can be used to test whether a particular account is in the set or not, but not list the entries in the set in a human-readable format (Kiran, 2022).

5.4 DISCUSSION: ONLINE PUBLIC SPHERES AND THE NEED FOR SAFE SPACES

The previous section shows how technology can be used by individuals and groups to change and mediate mainstream media and platforms. The chosen example demonstrates how a particularly vulnerable group can use Shinigami Eyes to make mainstream platforms feel safer and more comfortable, enabling them to potentially continue using and taking part in mainstream social media platforms instead of withdrawing from such platforms completely or having platforms tailored exclusively for such groups, such as Trans Time or similar alternatives (Haimson et al., 2020). While such technologies could indeed be good, and could be called "real trans technology" (Haimson et al., 2021), the benefits of maintaining contact between the trans community and non-trans people seem significant. The positive potential of using technologies similar to Shinigami Eyes is consequently also significant, as it potentially enables marginalized individuals to feel safer and included, and to take part in mainstream arenas and platforms of communication considered important for learning, engaging, and constructing identities (Lucero, 2017). In order to evaluate whether their means of achieving such participation is acceptable, we must begin by recognizing that marginalized individuals can experience online participation as riskier than "more privileged counterparts" (Kruse et al., 2018).

Nevertheless, Shinigami Eyes is subject to significant criticism, and we now proceed to discuss the main potential pitfalls related to the use of such technologies. The main objections are connected to (a) ideals related to public spheres, free and open debate, and (b) the rights of those being labeled by such tools. By and large, the objections mainly relate to SDG 16, and we restrict our analysis to the potentially sustainability-related negative effects of Shinigami Eyes in the following.

5.4.1 Public Spheres and the Free Exchange of Ideas

The main objections to Shinigami Eyes under this heading could be summarized as follows:

1. Labeling individuals as friendly/hostile through a non-universal technology distorts and undermines the idea of social media as Habermasian public spheres.
2. The marketplace for free exchange of ideas is undermined by labeling of individuals instead of arguments and messages.

3. Community-exclusive technologies such as Shinigami Eyes promote the emergence of echo chambers online.

5.4.1.1 Habermasian Public Spheres

The first objection suggests that social media labeling undermines the ideal of an open and universally accessible *public sphere* (Habermas, 1991). A public sphere must be open to all and allows for society to engage in "critical public debate" (Habermas, 1991). The public sphere is a realm of society where individuals and groups can freely partake in a special form of public debate where arguments, and not status or tradition, govern the discourse. The public sphere relates to *political* communication (Bruns & Highfield, 2015; Fuchs, 2015; Kruse et al., 2018) and must be independent of economic and political power with a particular emphasis on the absence of state censorship (Fuchs, 2015). This relates to Hanna Arendt's notion of the public *realm* "as the common world that gathers us together", creates bonds, and makes people interested in common affairs and each other (Arendt, 1998).

An ideal public sphere would be conducive to foster common understanding and the foundations for a functioning deliberative democracy (Kruse et al., 2018). This could consequently contribute to promoting SDG 16 and inclusive political institutions, but by extension it could also arguably contribute to increased awareness of and a reduction of differences (SDG 10) and safer and more inclusive societies (SDG 11), for example. The question, then, is whether Shinigami Eyes undermines the public sphere as an arena for communicative action and deliberative democracy through what could be categorized as "avoidant social media behaviors" such as labeling and potentially blocking/hiding/unfollowing other participants in the sphere in question (Kruse et al., 2018).

First, the public sphere is arguably an ideal that has never existed (Bruns & Highfield, 2015). According to Fuchs (2015), Habermas engages in immanent critique when he describes the public sphere in ways that draws out the shortcomings of capitalist society. Assuming that the ideal of the public sphere currently exists, and that trans people labeling trans-hostile people destroys the public sphere, entails critiquing Shinigami Eyes based on a fantasy. There are multiple *publics*, and these different publics do – and arguably should be able to – follow their own norms and use the affordances provided through extensions and add-ons – and through the platforms themselves – according to their interests and needs (Bruns & Highfield, 2015).

Second, social media has arguably never constituted a public sphere (Kruse et al., 2018). People use social media for a variety of purposes, and many legitimately avoid political topics, seek to avoid harassment, and prefer to use these media as a "happy" place for connecting with like-minded people (Kruse et al., 2018). There is also an important difference between capitalist media and civil society media, and it is the latter that arguably carries the potential for engendering a public sphere free of political *and economic* interests (Fuchs, 2015). Facebook, Twitter, Google, YouTube, and the other media and platforms Shinigami Eyes deals with are all private media, and despite formal equality of access these are arenas where "elites remain elites" (Dagoula, 2019). In Habermas' account elites communicated politically with "ordinary people" as an audience (Bruns & Highfield, 2015), and while social media is different from earlier mass media in this respect, there are still disparities in power and reach.

Third, power and exclusion matter. The notion of the public sphere is arguably also blind to many forms of marginalization and exclusion (Fraser, 2021; Lunt & Livingstone, 2013). While liberal ideals suggest that universal access and individual's freedoms constitute the foundation for justice and at times equity, both material and non-material inequalities lead to important differences in access to and ownership of the arenas of discourse and organization (Fuchs, 2015; Habermas, 1991; Kruse et al., 2018). The public sphere as an ideal has been a space for "educated, rich men", whereas women, "gays and lesbians" have been relegated to the private sphere (Fuchs, 2015), and "digital gatekeepers" and others with power construct systems in which heteronormative content is prioritized and trans-hostile content is accepted (Jenzen, 2017). If Shinigami Eyes helps those most vulnerable to enter and use mainstream media, rather than relying on their own separate platforms (Haimson et al., 2020), this arguably takes us *towards*, rather than away from, the ideal of the public sphere.

While some are also wary of the chilling effect of Shinigami Eyes (Datatilsynet, 2022a), it seems highly unlikely that this potential chilling effect is of more consequence than the chilling effect that occurs from trans people abandoning mainstream platforms because of unbridled and unmoderated trans-hostility. That one of the most marginalized groups in society is supposedly able to undermine free speech and chill free discourse through an extension that helps them identify individuals that are hostile to them seemingly relies on the liberal notion of the parity and natural equality of individuals. We have shown, however, that this is little more than a fantasy, and that marginalization is both real and must be dealt with in order to achieve sustainable development.

5.4.1.2 Marketplace of Ideas

The second objection is closely related to the first and is based on the idea that the "marketplace of ideas" should be as free as possible to generate a broad set of ideas that allows us to consider all options freely in order to more easily arrive at something close to "truth". The idea is often attributed to John Stuart Mill (1985), who argued that even the "ramblings of madmen" are important inputs in such a process. As seen through the other objections, some fear that Shinigami Eyes can contribute to chilling effects, arguing that the marketplace of ideas becomes less free with such a tool, and that people will refrain from taking part in debates, or will moderate themselves, when fearing that they could be marked as trans-hostile.

We posit that few historical media or arenas can be characterized as simultaneously more accessible and unmoderated than, for example, Twitter. As a medium, then, it could serve as an example of what results from a relatively free marketplace of ideas. Social media platforms have been wary of heavily moderating content on their platforms, fearing both accusations of interfering with free speech, but also to be held liable as editorially responsible for content on the platforms (Bozdag, 2013; Southerton et al., 2021). Rather than making these platforms the incarnation of Habermas' public sphere, this has led to a situation in which a large number of users do not engage in politics on these platforms as they there experience incivility, harassment, lack of privacy, and a general disinterest in seeking communicative action and genuine discourse aimed at becoming better informed about others and others' perspectives (Kruse et al., 2018). Furthermore, non-moderation does not ensure some form of equity or justice, as non-interference tends to promote the interests of those already privileged (Dagoula, 2019).

5.4.1.3 Arguments, Not People

In a press release accompanying the ban on Shinigami Eyes' processing of data, Datatilsynet's acting director Janne Stang Dahl states that privacy is intended to safeguard a space for free speech, and that "the assessment of the quality of an utterance shall not be dependent on the unknown labels that follow a person across the internet" [our translation] (Datatilsynet, 2022b).

This relates to old ideals of separating an argument from those who make the argument (Sætra & Fosch-Villaronga, 2021). One example is Merton's norms of science, of which *universalism* is of particular relevance. The norm states that within the scientific discourse, characteristics of the individual putting forth an argument or making a knowledge claim should not be considered; theories, discoveries, and ideas should be assessed on what their contents are, and according to accepted principles of the scientific method (Merton, 1942). In his seminal 1942 paper on his norms of science, Merton stakes out the same principles for the exchange of ideas and arguments, in line with the quote from Datatilsynet earlier.

In a discussion of *careful consumption* in media, Edmond (2022) describes how markets are increasingly moralized while morality is marketized. We are currently seeing how individuals are increasingly doing just what Merton warned against, namely evaluating products or messages in light of their creators and the process of creation (Edmond, 2022). Rejecting the debates about wokeness and cancel culture as too simplistic, they describe how individuals in efforts of self-improvement and self-care exercise self-control and increasingly screen and filter content. Trans youth in particular need coping strategies and have found measures for protecting themselves (Jenzen, 2017), such as Shinigami Eyes.

Datatilsynet (2022a) referred to their fear that screening and filtering behavior will lead to echo chambers online – a phenomenon related to filter bubbles – in which people are not exposed to divergent ideas (Sætra, 2021). This connects to the preceding objections, and we here merely refer to how research shows that such phenomena, while theoretically feasible, seem to be exaggerated (Bruns, 2019). Bruns and Highfield (2015) show how the existence of multiple publics does not necessarily lead to echo chambers, and that such publics are partly overlapping and not isolated. Furthermore, in the case of Shinigami Eyes, the question is not only whether it creates echo chambers in a particular medium, such as Twitter, but whether it could actually contribute to a *wider diversity* of users finding the platform bearable, and consequently *countering* echo chamber tendencies.

5.4.2 The Clash of Individual Rights of Protection and Privacy Online

A key reason for Datatilsynet's ban on Shinigami Eyes' processing of data was the purported need to protect those being labeled. They state that being labeled could be stigmatizing and lead to loss of friends, work, etc., that there is no way to respond to being labeled, and that those who are labeled are not informed about why they are labeled or that they are labeled at all (Datatilsynet, 2022a).

These are all good reasons to question Shinigami Eyes, and they must be weighed against the potential benefit of the extension. Much of the criticism is, however, aimed at the technical details of how Shinigami Eyes achieves its objectives. This is not something we explore here, and we also argue that most of these concerns could in theory be solved with technological changes while preserving the main functionality of the extension. For example, if Shinigami Eyes was a registered EU entity which provided reasons and insight into the categorization process and a channel for challenging the categorizations made, many of Datatilsynet's concerns would in fact be moot. We believe that informing individuals of why they are listed and providing a way to appeal categorization would be beneficial, and we assume that the reason this is not done is that Shinigami Eyes is nonprofit and seemingly made and maintained by one or very few individuals.

The first concern earlier, however, would remain. It *is* potentially stigmatizing to be labeled as trans-hostile. Simultaneously, while freedom of speech is an important principle, there are very few who argue that one must be able to utter whatever one likes *and* be free from the social consequences of such actions. Taking Twitter as an example, making a trans-hostile post could easily lead to other users quote retweeting the post and calling the user trans-hostile. Furthermore, it could lead to the person being put on a manually assembled "ban list", for example, put together by individuals concerned with informing other trans people of which accounts to potentially avoid if they do not want to encounter trans-hostile content online. In addition, individuals could of course ban the user directly. Being put on such a list, and being quote retweeted and called trans-hostile, is also stigmatizing, but it is also something that most people would accept as necessary costs associated with other individuals' freedom to refrain from being exposed to other people and what they perceive as harmful content. Freedom of speech does not come with an associated duty for everyone else to listen and pay attention to the speaker. Neither does it come with a duty for others to agree with and not challenge the speaker.

SDG 16 refers to equal access to justice and appeals to the rule of law and transparency are clearly valid when discussing the potential negative aspects of Shinigami Eyes. As we have stated, however, it would be theoretically possible to face all legal requirements and still provide categorization of trans-hostile and friendly people in a manner similar to Shinigami Eyes. The chilling effects and effects on public debate have been discussed earlier, and the remaining concern – that it is stigmatizing to be characterized based on one's utterances – seems to us to be outweighed by the needs of others to be informed about and potentially take precautions against people and messages that are hateful and make them feel unsafe.

5.5 CONCLUSION

When weighing diverging interests as we have done in this chapter, it is imperative that we consider the disparities in power and vulnerability of those involved, and we argue that sustainable development often requires us to side with those least well-off. Agenda 2030 is ripe with such emphasis (Sætra, 2022), and if we must choose between the rights of the privileged and those most vulnerable, the SDGs suggest we *leave no one behind*. The internet is a "crucial lifeline" for many LGBTQ+ youth (Jenzen, 2017), and this should urge us to take their concerns and initiatives to increase resilience seriously.

Some of the key dilemmas we seemingly face when evaluating a tool like Shinigami Eyes are:

1. The need to feel safe versus the desire for unmoderated and unfiltered debates.
2. The desire of a community to share information regarding friends and foes versus an ideal in which there are no groups – only individuals.
3. The need to filter messages by sender due to the massive amount of information available versus the ideal of only evaluating arguments as detached from sender.
4. Untraditional forms of self-protection and resistance from the marginalized versus calls for adherence to rules and established institutions by the privileged.

The public sphere and communicative action are indeed important, but we argue that it seems misdirected to require of the trans community that they spearhead such a development and simultaneously support media in which they are marginalized and encounter massive hostility without taking measures of self-care and self-protection (Edmond, 2022). In any society, some will have more power than others, and the extant structures of media and politics will be better aligned with the needs, interests, and capabilities of some groups of people than others. That this leads to "counter-publics" and fosters resistance through various means and tactics is quite natural (Bruns & Highfield, 2015; Jenzen, 2017), and we believe this is not necessarily something to be combatted.

There are *multiple* public spheres (Bruns & Highfield, 2015), and we should do our best to encourage the participation of the most vulnerable and marginalized in the mainstream media and arenas where politics and culture are debated and shaped. Communication technology has provided new media and arenas, but these are privately owned, largely unmoderated, and not exempt from extant social structures and power. The use of technology to create the conditions for overcoming and navigating such landscapes, like Shinigami Eyes or other tools, demonstrates how individuals and groups can to some degree resist and protect themselves. While such technologies should ideally adhere to prevailing laws, we also argue that prevailing laws could perhaps go further in protecting and promoting the interests and needs of those least well-off, rather than relying on ideals of the free market of ideas and a public sphere – one that has arguably never existed – in ways that arguably protect the interests of those already privileged. As forcefully argued by Ongsupankul (2019), without taking the rights of sexual minorities seriously the prospects for achieving sustainable development on all three dimensions are undermined.

NOTES

1 The definitions related to SOGIESC used in this chapter follow the Yogyakarta principles and the Yogyakarta Principles plus 10 (YP plus 10), found at https://yogyakartaprinciples.org/principles-en/

2 https://twitter.com/CaseyHo/status/1452007901585809411?s=20&t=OgwPxEeQdf3uYbXzV1i3Ag

5.6 REFERENCES

Arendt, H. (1998). *The human condition*. The University of Chicago Press.

Bloom, B. H. (1970). Space/time trade-offs in hash coding with allowable errors. *Communications of the ACM, 13*(7), 422–426.

Bozdag, E. (2013). Bias in algorithmic filtering and personalization. *Ethics and Information Technology, 15*(3), 209–227. https://doi.org/10.1007/s10676-013-9321-6

Bruns, A. (2019). *Are filter bubbles real?* Polity Press.

Bruns, A., & Highfield, T. (2015). Is Habermas on Twitter? Social media and the public sphere. In *The Routledge companion to social media and politics* (pp. 56–73). Routledge.

Ciszek, E., Haven, P., & Logan, N. (2021). Amplification and the limits of visibility: Complicating strategies of trans voice and representations on social media. *New Media & Society*. https://doi.org/10.1177/14614448211031031

Crenshaw, K. (1990). Mapping the margins: Intersectionality, identity politics, and violence against women of color. *Stanford Law Review, 43*, 1241. https://doi.org/10.2307/1229039

Dagoula, C. (2019). Mapping political discussions on Twitter: Where the elites remain elites. *Media and Communication, 7*(1), 225–234.

Datatilsynet. (2022a). *Ban on processing – Shinigami Eyes* [Press release]. www.datatilsynet.no/contentassets/2d7e7acc2e4548fa8aca6f17dc3267b8/21_02455-11-ban-on-processing-shinigami-eyes-295856_7_1.pdf

Datatilsynet. (2022b). *Vedtak om forbud mot "Shinigami Eyes" i Norge* [Press release]. www.datatilsynet.no/regelverk-og-verktoy/lover-og-regler/avgjorelser-fra-datatilsynet/2022/vedtak-om-forbud-mot-shinigami-eyes-i-norge/

Divan, V., Cortez, C., Smelyanskaya, M., & Keatley, J. (2016). Transgender social inclusion and equality: A pivotal path to development. *Journal of the International AIDS Society, 19*, 20803.

Dorey, K., & O'Connor, J. (2016). *The sustainable development goals and LGBT inclusion*. Stonewall International.

Edmond, M. (2022). Careful consumption and aspirational ethics in the media and cultural industries: Cancelling, quitting, screening, optimising. *Media, Culture & Society*, 01634437221099615.

Fraser, N. (2021). Rethinking the public sphere: A contribution to the critique of actually existing democracy. In *Public space reader* (pp. 34–41). Routledge.

Fuchs, C. (2015). *Culture and economy in the age of social media*. Routledge.

Habermas, J. (1991). *The structural transformation of the public sphere: An inquiry into a category of bourgeois society*. MIT Press.

Haimson, O. L., Buss, J., Weinger, Z., Starks, D. L., Gorrell, D., & Baron, B. S. (2020). Trans time: Safety, privacy, and content warnings on a transgender-specific social media site. *Proceedings of the ACM on Human-Computer Interaction, 4*(CSCW2), 1–27.

Haimson, O. L., Dame-Griff, A., Capello, E., & Richter, Z. (2021). Tumblr was a trans technology: The meaning, importance, history, and future of trans technologies. *Feminist Media Studies, 21*(3), 345–361. https://doi.org/10.1080/14680777.2019.1678505

Hines, S. (2019). The feminist frontier: On trans and feminism. In *The Routledge handbook of contemporary feminism* (pp. 94–109). Routledge.

Inter-American Commission on Human Rights. (2018). *On the occasion of International Transgender Day of Visibility, the IACHR and a UN expert urge States to guarantee the full exercise of the human rights of transgender persons* [Press release]. www.oas.org/en/iachr/media_center/PReleases/2018/069.asp

Jenzen, O. (2017). Trans youth and social media: Moving between counterpublics and the wider web. *Gender, Place & Culture, 24*(11), 1626–1641. https://doi.org/10.1080/0966369X.2017.1396204

Kiran. (2022). *Shinigami eyes*. https://shinigami-eyes.github.io

Kruse, L. M., Norris, D. R., & Flinchum, J. R. (2018). Social media as a public sphere? Politics on social media. *The Sociological Quarterly, 59*(1), 62–84.

Logie, C. H. (2021). Sexual rights and sexual pleasure: Sustainable development goals and the omitted dimensions of the leave no one behind sexual health agenda. *Global Public Health*, 1–12. https://doi.org/10.1080/17441692.2021.1953559

Lucero, L. (2017). Safe spaces in online places: Social media and LGBTQ youth. *Multicultural Education Review*, *9*(2), 117–128.

Lunt, P., & Livingstone, S. (2013). Media studies' fascination with the concept of the public sphere: Critical reflections and emerging debates. *Media, Culture & Society*, *35*(1), 87–96. https://doi.org/10.1177/0163443712464562

Majeedullah, A., Wied, K., & Mills, E. (2016). *Gender, sexuality and the sustainable development goals: A meta-analysis of mechanisms of exclusion and avenues for inclusive development.* https://opendocs.ids.ac.uk/opendocs/bitstream/handle/20.500.12413/12636/ER206_GenderSexualityandtheSustainableDevelopmentGoals.pdf?isAllowed=y&sequence=1

Matthyse, L. (2020). Achieving gender equality by 2030: Transgender equality in relation to Sustainable Development Goal 5. *Agenda*, *34*(1), 124–132. https://doi.org/10.1080/10130950.2020.1744336

Merton, R. K. (1942). Science and technology in a democratic order. *Journal of Legal and Political Sociology*, *1*(1), 115–126.

Mill, J. S. (1985). *On liberty*. Penguin books.

Mills, E. (2015). *'Leave no one behind': Gender, sexuality and the sustainable development goals.* https://www.ids.ac.uk/publications/leave-no-one-behind-gender-sexuality-and-the-sustainable-development-goals/

Nakatani, T., Tamura, M., & Maruyama, M. (Writers). (2006–2007). *Death note* [TV]. Shueisha.

Ohba, T., & Obata, T. (2003–2008). *Death note*. Shueisha.

Ongsupankul, W. (2019). Finding sexual minorities in United Nations Sustainable Development Goals: Towards the deconstruction of gender binary in international development policies. *LSE Law Review, 5*.

Sætra, H. S. (2021). *Big data's threat to liberty*. Academic Press.

Sætra, H. S. (2022). *AI for the sustainable development goals*. CRC Press.

Sætra, H. S., & Fosch-Villaronga, E. (2021). Research in AI has implications for society: How do we respond? *Morals & Machines*, *1*(1), 60–73.

Southerton, C., Marshall, D., Aggleton, P., Rasmussen, M. L., & Cover, R. (2021). Restricted modes: Social media, content classification and LGBTQ sexual citizenship. *New Media & Society*, *23*(5), 920–938. https://doi.org/10.1177/1461444820904362

United Nations. (2015). *Transforming our world: The 2030 agenda for sustainable development.* Division for Sustainable Development Goals.

Vaidhyanathan, S. (2018). *Antisocial media: How Facebook disconnects us and undermines democracy*. Oxford University Press.

CHAPTER 6

Lessons to Be Learnt? Education, Techno-solutionism, and Sustainable Development

Neil Selwyn

CONTENTS

6.1 INTRODUCTION

At first glance, education might seem an obvious area where digital technology can be a major enabler of sustainable development. Indeed, many people presume *any* aspect of education to inherently support sustainable development – as Kioupi and Voulvoulis (2019, p. 13) reason, "the road to sustainability is paved by education, and unless we invest on it, we might never get there". This chapter therefore addresses the widely held belief that digital technologies are an integral part of establishing high-quality educational conditions for all. With digital technologies now entwined with the generation and sharing of knowledge throughout contemporary society, growing numbers of policymakers, industrialists, and education professionals continue to promote the imminent "digital transformation of education" (OECD, 2021, p. 5) as a key element of sustainable development in the 2020s and beyond. This is especially evident through the UN Sustainable Development Goals (SDGs) –

DOI: 10.1201/9781003325086-6

as UNESCO's framing Incheon Declaration put it: "information and communication technologies must be harnessed to strengthen education systems, knowledge dissemination, information access, quality and effective learning, and more effective service provision" (UNESCO, 2015, p. iv).

In contrast, this chapter argues that we need to radically reconsider the relationship between education, sustainable development, and emerging digital technologies. Belying the prevailing hype, there is little justification to expect the digitally driven reversal of long-standing inequalities in education. Indeed, the continued reform of education around excessive digital technology consumption might well exacerbate social inequalities, political divisions, *and* environmental degradation. In short, this chapter argues that discussions around sustainable development need to move away from default assumptions that the continued digitization of education is an inherently "good thing". This is not to say that digital technologies have no place *at all* in desirable future forms of education – rather this is a call to re-imagine education technology in scaled-back and slowed-down forms that might be deemed appropriate for an era of continuing social upheaval and climate collapse.

6.2 A BRIEF HISTORY OF THE DIGITAL (NON) TRANSFORMATION OF EDUCATION

Notwithstanding the ongoing hype around digital education, the past 40 years provide scant evidence that the implementation of digital technologies straight-forwardly "enhances" or "transforms" education systems in widespread and/or lasting ways (see Selwyn, 2021a). Of course, education systems in the Global North have invested heavily during this period in the adoption of computers, the internet, platforms, and most recently a range of Artificial Intelligence systems and tools. Yet, while these waves of "innovation" have all prompted notable surface-level changes to the nature and form of what takes place in classrooms and other education settings, any sense of progress is tempered with a general acknowledgment that long-standing structural inequalities and inefficiencies of education persist (Rafalow & Puckett, 2022). In this sense, it could be argued that we now have education systems that might well appear to be increasingly technologically dependent, but are still *not* technologically empowered.

Any discussions of education technology and sustainable development therefore need to pay close attention to the long-standing complexities and intricacies of digital technology and educational change. Take, on one hand, the ways in which current digitally driven forms of school and university education in high-income countries continue to be structured by entrenched educational logics. For example, it seems that digital technology has done little to disrupt resilient hierarchies of time, space and place, as well as curricularized assumptions of knowledge, the primacy of individual assessment, and other established institutional forms of schooling. On the other hand, the dominant forms of digitization that *have* taken hold in education seem primarily to reinforce and amplify "corporate reforms" of schools and universities – supporting the creeping standardization of practice, a reliance on metrics and data-driven accountability, and increased emphasis on market-led efficiencies. In all these ways, then, the impacts of digital "innovations" and "disruptions" in most

Global North education systems remain firmly in lockstep with vested institutional interests and the maintenance of administrative power.

At the same time is a long-standing – but largely unrealized – faith in the capacity of education technology initiatives to redress social, economic, and cultural disparities in low-income and middle-income regions. In policy terms, the UN 2030 Agenda for Sustainable Development is framed around the notion that "the spread of information and communications technology and global interconnectedness has great potential to accelerate human progress" (UN, 2015, para 10). In practical terms, the past 20 years or so of work in the field of "information and communications technologies for development" (ICT4D) is littered with high-profile instances of such tech-driven hubris. Toward the end of the 1990s, for example, the much-celebrated "Hole In The Wall" initiative saw the placement of unattended computer kiosks in some of the poorest communities around India – ultimately rewarded by a $1 million TED prize to further develop the idea of technology-driven "self-organized learning". As was subsequently claimed by the initiative's founder (then chief scientist at the global "IT learning solutions" corporation NIIT): "what children achieve routinely in hundreds of 'Holes-in-the-Walls' in some of the remotest places on earth is nothing short of miraculous – a celebration of learning and the power of self-motivation" (Mitra & Dangwal, 2010, p. 680). In contrast, the impact of these interventions was acknowledged widely by disinterested observers to have fallen well short of any such claims. As Mark Warschauer (2003, p. 45) reflected: "in short, communit[ies] came to realize that minimally invasive education was, in practice, minimally effective education".

Another salient illustration of these unfulfilled ambitions was MIT's much-celebrated "One Laptop Per Child" initiative, which worked throughout the 2000s to saturate educationally impoverished communities across sub-Saharan Africa and South America with rugged "$100 laptop" devices deemed capable of supporting self-directed learning. In contrast to its grand ambitions of "empowering the world's poorest children", OLPC ceased operations after a few years and was soon judged to have comprehensively failed (Keating, 2009), with independent observers noting "how little the project's vision [was] taken up in practice" (Ames, 2016, p. 95).

Other high-profile digitally driven social renewals of education have similarly failed to materialize. During the 2010s, for example, the emergence of Massive Open Online Courses (MOOCs) prompted much enthusiasm – touted as a means of providing free university-grade tuition for masses of online learners around the world regardless of their local circumstances or prior educational experience. Yet, belying promises of radically democratized tuition, most MOOCs proved to merely advantage those who were *already* well-resourced and educationally successful (Rohs & Ganz, 2015). As Tressie McMillan Cottom (2012, n.p) reasoned at the time:

> The people who would benefit most from online learning are not necessarily where these programs are moving. That means really ambitious autodidacts – the kind who have long benefited most from innovative education models – will take advantage of MOOCs to become, well, more autodidact-ish. Already privileged

> elite students with broadband, iPads, Macbooks, and time (the greatest luxury of all) will simply have more spaces in which to be privileged and elite.

Such high-profile failures illustrate the stubborn trend across the past 40 years for even the most well-resourced and innovatively designed digital education "solutions" to flounder in terms of lack of fit with local contexts and needs, a lack of interest in the long-term sustainability of such ventures, and what Kemp Edmunds (cited in Madda, 2022, n.p) describes as "an unadulterated hubris that [technology actors] can solve complicated layered problems in extremely complex industries with many challenging stakeholders, [and] financial, bureaucratic and time constraints". Above all, these unrealized attempts at technological transformation warn against the folly of assuming that structural, societal, and deep-rooted issues surrounding education are easily "fixable" via technology. As Megan Erickson (2015, n.p.) puts it, "education is not a design problem with a technical solution".

6.3 CURRENT FRAMINGS OF ED-TECH AND SUSTAINABILITY INTO THE 2020S

Set against this background – and drawing on the key concepts underpinning this book (see Chapter 2) – why should we presume that further applications of digital education during the 2020s under the banner of "sustainable development" will be any different? In practical terms, one of the major reasons why people continue to anticipate the digitally driven improvement of education is the UN "Sustainable Development Goals" (SDGs) – in particular, SDG 4: "Ensure inclusive and equitable quality education and promote lifelong learning opportunities for all". Digital technology is clearly aligned with at least six of the ten targets in SDG 4, which confidently foreground digital technology as a key area of skill development, as well as a means of equitable delivery, and system-wide planning and monitoring. As such, tech firm Huawei (2018) reckoned SDG 4 to have some of the highest levels of correspondence with digital technology across *all* 17 SDG goals, concluding that "even slight technological improvements could result in better performance" in terms of progress toward SDG 4 criteria. All told, SDG 4 continues to boost the idea of digital technology as a potentially powerful enabler of sustainable education during the 2020s. As Tyagi et al. (2019, p. 441) contend in a report titled "ICT Goals for SDG 4":

> rolling out ICT solutions will not only transform education and deliver social and economic benefits to people but also provide substantial benefits for the education sector globally.

In more specific terms, the ten SDG 4 targets and associated indicators boldly frame digital technology use in education as potentially supporting economic, social, *and* environmental sustainability. These include various expected economic benefits in terms of increasing economic growth and workforce capacity. For example, bolstering the long-standing justifications around supporting young people to develop work-related computer skills is the continued promotion of specific forms of "digital literacy", "learning to code", and similar forms of technology-centered "up-skilling" within most levels of education (see also

Chapter 8). More recently, schools have been positioned as key sites in the development of "on-demand" technological competencies required by young people entering the "gig economy" and other forms of precarious labor. We have also seen digital education initiatives in middle-income and low-income regions seeking to develop "digital entrepreneurism" among young people – that is, programs designed to encourage technology-driven entrepreneurial skills, "digital livelihoods" and provide digital skills development opportunities with a focus on using digital technologies to start businesses, secure employment, and access financial products and services (UNESCO, 2018).

Alongside these presumed economic benefits are strong claims over social sustainability outcomes. In particular is the idea of digital technologies enhancing the accessibility (and therefore inclusivity) of education provision. This is seen to be the case, for example, for learners with neurodevelopmental conditions, and other mental and physical disabilities (see Chapter 7). As Megan Crawford (2017) reasons in terms of the SDG 4 goals:

> With connectivity, students can access learning resources and opportunities even in remote or low-income areas. Teachers can prepare for classes anytime or anywhere. ICT opens up access to education to underserved populations for whom improved educational opportunities lead to improved economic opportunities.

Specific social benefits include diversifying pathways into computer science and "STEM" careers for groups traditionally underrepresented in these employment sectors, as well as establishing schools, libraries, and other educational institutions as centers for shared community access to computers, internet, and other digital resources. All told, enthusiasm abounds for the capacity of most new forms of digital technology to enhance equality of educational opportunities and educational outcomes. As Ward (2020, n.p.) puts it, "rich nations can contribute immensely to SDG 4 by using the next generation of educational technology to democratize access to advanced learning on a global scale".

Finally are burgeoning claims that the continued digitization of education can contribute to environmental sustainability. For example, it is suggested that digital technologies are a ready means of developing environmental education, and fostering global communities of students centered around climate change awareness and action (Gismondi & Osteen, 2017). Alongside these benefits is the alignment of digital education with "green tech" principles – where increased use of digital technologies in education contributes to the pursuit of forms of eco-growth. In particular is the presumption that online education can support the reduction of carbon emissions associated with campus-based travel and education – not least by lowering emissions of students and teachers otherwise commuting to-and-from classes (Versteijlen et al., 2017) alongside the reduction of on-campus power consumption (Caird et al., 2015). If education technology is considered at all in terms of its environmental impact, this has tended to be in wholly positive claims of "protect[ing] global environmental resources" (Caird & Roy, 2019, p. 107). As Becker and Otto (2019, p. 8) conclude:

> Digital learning... saves resources and CO_2 emissions, thus contributing to the protection of the climate and to the goal of responsible consumption and production....

> it helps to connect people from different cultures by allowing for intercultural exchange among students without additional travelling… it facilitates a self-regulated learner-centered style of learning that is well-suited to empower learners to become agents of a sustainable development.

6.4 EDUCATION AND SUSTAINABILITY AS A FOCUS OF ONGOING TECHNO-SOLUTIONISM

All told, the received wisdom is that the continued digitization of education across the 2020s remains will play a pivotal role in advancing communities, countries, and regions toward a range of sustainability outcomes. Indeed, in reviewing the likely impact of digital technology across the 17 SDGs, Neves (2020, p. 266) takes time to highlight a particularly "strong linkage between education and digital solutions". Looking beyond such speculative expectations, however, the push for continued digitized education during the 2020s is clearly extending (if not amplifying) the spirit of techno-solutionism that underpinned previous decades of education technology. Indeed, if anything, expectations of an educational "technical fix" appear to be further exaggerated with each successive wave of emerging technology. Take, for example, recent expert assertions that all ten SDG 4 indicators can be positively influenced by emerging innovations around mobile internet, Big Data, and cloud computing (Gupta et al., 2020). Similarly, so-called 4IR (Fourth Industrial Revolution) technologies such as AI, biotech, mixed reality, quantum computing, and robotics are now inspiring heightened rhetoric such as "wealthy nations developing transformational 4IR Ed-Tech have a moral imperative to use it to achieve SDG 4" (Ward, 2020, n.p.). At this point, the only concerns being raised seem to relate to the speed and scale of the technology rollout. For example, when considering the potential of e-books and digital libraries, Evans (2016) reflects,

> Whilst they have their merits and arguably are having a deep and transformative impact on particular communities, the reality is that these solutions cannot scale quickly enough to reach the 250 million children who are not learning basic literacy. Time is not on our side if we are to achieve the SDGs by 2030.

Of course, as acknowledged throughout this book, such hype and hubris are not unique to education. Indeed, technological solutionism is integral to mainstream thinking around most aspects of sustainability and pervades discussions around most of the SDGs (see Chapter 2). Indeed, current discussions around sustainable development have been described as imbued with an "ambient promeatheanism" which foregrounds the problem-solving capabilities of human innovation, and "advances technological solutions developed by capital and energy-intensive projects" (Dillet & Hatzisavvidou, 2022, p. 1). Nevertheless, the persistence of this solutionist mentality within sustainability discussions appears particularly pronounced in terms of education and SDG 4. In many ways, the appeal of this enduring magical thinking around technology and education is understandable. Morgan Ames's (2019) notion of "charismatic technology" describes the ways in which new technologies gain traction within education discussions through the promised association of impressive technical progress with equally far-reaching social progress. Seen in this light, techno-solutionism offers a high-profile and expedient means for governments, education authorities, and other stakeholders to appear responsive to complex sustainable development imperatives in a forthright and innovative manner. There are obvious self-interested reasons for actors concerned with matters of sustainable development and education to be acting in this manner. The main question that remains far *less* obvious, however, is whether they are justified in doing so.

6.5 PROBLEMATIZING ONGOING TECHNO-SOLUTIONISM AROUND EDUCATION AND SUSTAINABILITY

At this point, then, we need to consider the prospect of education technology simply continuing over the next few decades as a case of "business-as-usual", perhaps overlaid with a veneer of green growth. The ongoing efforts around SDG 4 certainly seem to presume the continued mass implementation of digital technologies in the service of sustainable development outcomes. If so, we need to also take seriously the likely adverse outcomes and limits of continuing to pursue this line of solutionist thinking in contrast to the likely lack of impactful change. As Katy Jordan (2020) notes, the presumed beneficial impacts of digital technology on SDG 4 remain woefully under-scrutinized, with academic researchers showing little appetite for investigating actual links between digital technologies and SDG 4. Yet, in light of previous iterations of digital (non)transformation of education, it is doubtful that the promise of future digital education solutions under the aegis of SDG 4 will ever be *fully* realized. As such, it is important to acknowledge that few of these inventions and initiatives are likely to address the complex societal and planetary factors underpinning sustainability problems in the ways that are being currently promised.

A few interrelated issues therefore need to be brought to bear on the current optimism that persists throughout mainstream discussions around digital technologies and SDG 4. First, it is worth stressing that this is *not* an argument that digital technologies are of no benefit *at all*. As with previous waves of education technology initiatives and interventions during the 2000s and 2010s, it is likely there will be plenty of specific localized instances of digital technology innovation that might be seen as cases of "best practice" in terms of supporting particular social, economic, and/or environmental sustainability. Digital technologies are likely to continue to "work" for some people – for example, advantaging already privileged students and teachers, and/or resulting in outcomes that "work" in support of institutional agendas. Yet, it is important to consider the scale and scope of any ostensibly positive applications of technology – especially in terms of the inevitably limited capacity to address inequalities and disparities between historically marginalized and disadvantaged groups and regions. As Sætra (2022) points out, we should not let micro-level impacts on individuals and small groups distract attention from (i) the differential impact of any digital technology use at a meso-level of classes, nations, and regions, as well as (ii) the long-term macro-level impacts of digital technologies "on our economies and societies in the broadest sense".

In addition, it is also important to consider the interlinked nature of any sustainable development outcomes – with any application of digital technology in education likely to have corresponding (and often conflicting) social, economic, political, and environmental impacts. Indeed, one of the limitations of the current techno-solutionist framings of digital education is how they over-simplify sustainability issues in ways that fit the technological solutions being proposed – therefore doing little more than give a surface appearance of addressing the problems that they purport to address (Nachtwey & Seidl, 2020). Moreover, as reasoned in Chapter 2, dominant narratives of technological change tend to frame perceptions of what is possible solely in terms of the technology to hand. This chimes with Christo Sims' (2017) account of how technology actors often frame (and underestimate) education "problems" through a technical mindset, resulting in what Sims describes as a "tunnel vision" where attention is paid only to aspects of education that fit with the tools that are being developed.

The implications of this "tunnel vision" in terms of the capacity of current forms of digital technology to meaningfully address sustainability goals need to be taken seriously. For example, Sims (2017) argues that this narrowness of perspective tends to marginalize concerns for wider social contexts that shape educational institutions and the communities that they serve, while often glossing over issues of wider structural inequalities altogether. This "tunnel vision" therefore leads to tautological conclusions where the implementation of a new digital technology is offered as a plausible response to problems that have been set in train by the use of a preceding digital technology. In addition is a worrying tendency for digital technology to fundamentally alter how issues of sustainability in education are perceived – not least, the "economisation and depoliticization of planetary environmental issues" (Dillet & Hatzisavvidou, 2022, n.p.; see also Chapter 17).

In this sense, it is well worth reconsidering the likely adverse outcomes of continuing to pursue reductionist forms of solutionist thinking in relation to sustainable development and SDG 4. In particular, this raises the need to consider the possibility that digital education developments are unlikely to address the complex social nature of sustainability problems in the ways that are promised. Instead, if anything, it might be expected that the current wave of techno-solutionism is most likely to perpetuate (and perhaps intensify) sustainability harms. Take, for instance, the ways in which virtual schooling was implemented across the COVID-19 pandemic shutdowns of school and university systems – demonstrating consistently how technology-based education remains predicated on elite consumption and uneven benefits (Sosa Díaz, 2021). As Susan Flynn (2021), writing from the perspective of an educator in a high-income country, reflects on the "online pivot" during the COVID-19 pandemic:

> The digital realm proffers ease of access, connectivity, easy dissemination of knowledge and information but it also produces and sustains inequalities and perpetrates assumptions. Our students may not be the digital natives we assume, nor is their access to technology necessarily equitable.

Alongside these disparities in educational outcomes are the ways in which digitally focused forms of skills education have also proven to exacerbate social inequalities in employment outcomes – at best facilitating access to low-status, low-paid, routine and precarious jobs within the new economy, or trapping young people into a subsistence self-employment along decidedly vulnerable and marginalized forms of digitally driven "entrepreneurialism". As Angela Dy (2019, pp. 11–12) contends:

> Despite the transformative potential of the Internet, the landscape of digital entrepreneurship is still shaped by social patterns of privilege and disadvantage . . . the current trend of encouraging digital entrepreneurship as a means to social mobility and economic independence for marginalized and disadvantaged people may in fact be encouraging greater economic vulnerability.

Perhaps most noteworthy (yet glossed over in most discussions of education and technology) are the ways in which a continued emphasis on digital education is likely to exacerbate growing ecological and environmental harms associated with the production, consumption, and disposal of digital technology. For example, picking up on Brevini's arguments in Chapter 3, any enthusiasms for the increased use of AI in education surely have to reckon with the materiality of this technology and its deleterious consequences for the planet. Striving to refashion education around AI, augmented reality, and other emerging digital technologies feeds directly into the problems that Brevini highlights regarding the depletion of scarce resources in manufacturing, usage and disposal processes, alongside the excessive amounts of energy used to support data processing and storage, *and* the exacerbation of waste and pollution issues. The continued excessive application of digital technology in any context – education included – makes little sense in terms of environmental sustainability.

All told, we need to take seriously the prospect that digital education is unlikely to be a key enabler of sustainable development and might well result in regressive and ultimately harmful outcomes for many people and regions. This corresponds with broader emerging concerns among environmental activists – what Jonathan Crary sees as the "obvious but unsayable reality" that digital technology is not likely to be an instrumental of radical change – instead, that "if there is to be a livable and shared future on our planet, it will be a future offline" (Crary, 2022, p. 1). Set against this bleak perspective, therefore, it seems increasingly important that we need to *think otherwise* when it comes to the forms of "digital education" that should be encouraged under the aegis of sustainable development.

6.6 THINKING OTHERWISE – TOWARD RE-IMAGINING MORE SUSTAINABLE FUTURE FORMS OF DIGITAL EDUCATION

In this spirit of thinking otherwise, we now need to reflect on the challenge of looking forward to what could (or should) be done differently when it comes to digital technology, education, and sustainable development. First is the need to reframe popular, professional, and policy discussions about education and digital technology along more socially, economically, and environmentally realistic lines. The arguments advanced in this chapter (and across most chapters in this book) follow what can be broadly understood as a "socio-technical" perspective – that is, considering the application of digital technologies "in context, and not as some isolated and neutral tool" (Sætra, 2022). As has been evident throughout all of this chapter's discussions, the application of digital technology in education needs to be seen as a coming together of technical *and* social issues – of devices, institutions, economic and political systems, scientific laws, *and* social relations.

Seen in this light it makes little sense to expect sustainable development outcomes to arise unproblematically from the application of digital technology in education. Instead, education technology needs to be reframed as a political issue – that is, guided by values that some actors wish to see realized, while clashing with the values of others. At present, the dominant values informing the implementation of digital education under the aegis of SDG 4 and "sustainable development" make sense only in terms of what has been described elsewhere in this book as the "shallow" ecological thinking that typifies the utilitarian pragmatism of Western governments, businesses, and industry (see Chapter 18). The subtext of many of this chapter's criticisms is that any such "business-as-usual" approach is likely not good enough. Instead, we need to re-imagine digital education along more radical lines. In short, we need to find more refined forms of sustainable digital education that do not perpetuate harms to marginalized people and are not yoked to presumptions of continued economic growth and/or the excessive harmful waste of planetary resources.

One such radical alternative might be reframing our ambitions for digital technology and education around an explicit eco-justice agenda (e.g., Ale, 2021). This might start with a collective "withdrawal" or "releasement" from unsustainable and harmful forms of education technology (Heikkurinen, 2018) – that is, not feeling compelled to be led by "gigantic" developments in corporate software development, but instead feeling confident to follow other lines of digital use. In this spirit, we might choose to re-imagine education technology along more "sustainable" lines, where the allocations of digital resources are apportioned fairly and do not excessively disadvantage the already disadvantaged. In its own small way, then, these are alternate values and principles that educators, policymakers, and other stakeholders would do well to take on board as they face up to the challenge of deciding what "sustainable" technology use in education might look like in times that are increasingly shaped by climate crisis, social upheaval, and political instability. From this perspective, rethinking sustainable forms of digital technology use in education requires us to focus on questions of ethics alongside questions of ecology.

In the immediate term, this reframing of education technology first compels us to consider current forms of digital technology that might be considered "sustainable" in nature, and therefore deserving of retention. These technologies and practices might fulfill a number of briefs. First is the encouragement of "low-impact" technology use sustained by re-use, repair, and recycling of digital resources within education settings and local communities (thereby bringing education technology into correspondence with the efforts around SDG 12 – "responsible consumption and production"). Second is a general emphasis on basic forms of digital technologies that can be easily produced, maintained, augmented, and repurposed on a local basis. This relates to the degrowth ethos of "voluntary simplicity" (Liegey & Nelson, 2020, p. 12) and the prioritization of technology that is "slower by design" (Kallis et al., 2020, p. ix). Echoes of these approaches are certainly evident throughout the recent history of educational computing – especially in terms of the renewed interest throughout the 2010s for mini-computers such as the Raspberry Pi and MicroBit, as well as the adoption of "Maker technology" and other forms of "tinkering".

Much of this alternate spirit of what might be termed "Ed-Tech within limits" (Selwyn, 2021b) is also evident within communities and contexts whose engagements with digital technologies are *already* substantially constrained by climate, hostile environments, and limited resources and infrastructure. In this sense, high-income countries might also learn from looking toward the forms of technology-based education innovation that have developed (and sometimes flourished) in low-income countries, regions, and contexts. Indeed, the history of locally driven education technology development in low-income contexts over the past 20 years or so can provide strong pointers to the forms of digital technology infrastructure and use that might be considered "appropriate" for increasingly resource-constrained and environmentally hostile circumstances elsewhere. These include "low-tech" technology such as radio, telecentres, SMS, and pre-loaded dumbphones. Indeed, the worldwide shutdowns of schooling during the successive waves of the COVID-19 pandemic saw local educators in many low-income and middle-income countries turn to analog technologies (such as radio, television, dumbphones, and paper-based resources) to redress the inequalities of online schooling (see Meinck et al., 2022). As such, all these examples point to a number of principles that might be taken forward into re-imagining education technology for an era of climate crisis.

Regardless of specific form and design, the different ideas just outlined all point to ways in which we might begin to rethink education technology in terms of distributive justice – fostering sustainable and socially appropriate forms of technology use for disadvantaged groups who stand to benefit most from the use of education technology (and, conversely, lose most from the enforced absence of education technology). This also suggests establishing norms of what constitutes "fair" forms of education technology access and use, as well as foregrounding principles of collective decision-making and communal ways of managing how technologies are developed and deployed in education (O'Sullivan et al., 2021). All told, these examples highlight that reframing forms of education technology that are appropriate for an age of climate crisis, increasing social instability, and other concerns that relate to "sustainable development" does not simply equate with making *less* use of technology in education. Instead, it requires engaging with the complex problem of how to make *less use* of technology for *more just* education outcomes. These are clear shifts and changes that will not arise from continuing to approach education and sustainable development along techno-solutionist lens.

6.7 REFERENCES

Ale, H. (2021, July 23). Why climate emergency and climate justice can't go hand in hand. *The Sociological Review*. https://doi.org/10.51428/tsr.ohpe8915

Ames, M. (2016). Learning consumption. *The Information Society*, *32*(2), 85–97.

Ames, M. (2019). *The charisma machine*. MIT Press.

Becker, S., & Otto, D. (2019). Digital learning and sustainable development. In W. Leal Filho (Ed.), *Encyclopedia of sustainability in higher education*. Springer.

Caird, S., Lane, A., Swithenby, E., Roy, R., & Potter, S. (2015). Design of higher education teaching models and carbon impacts. *International Journal of Sustainability in Higher Education*, *16*(1), 96–111.

Caird, S., & Roy, R. (2019). Blended learning and sustainable development. In W. Leal Filho (Ed.), *Encyclopedia of sustainability in higher education* (pp. 107–116). Springer.

Cottom, T. (2012). *MOOCs, Coursera, online education and performing innovation*. https://tressiemc.com/uncategorized/moocs-coursera-online-education-and-performing-innovation/
Crary, J. (2022). *Scorched earth*. Verso.
Crawford, M. (2017, May 30). *The role of technology in the UN SDGs*. www.adecesg.com/resources/blog/an-overview-of-the-role-of-information-and-communications-technology-ict-in-each-of-the-sustainable-development-goals-sdg-part-one/
Dillet, B., & Hatzisavvidou, S. (2022). Beyond technofix. *Contemporary Political Theory, 21*, 351–372.
Dy, A. (2019). Levelling the playing field? *Futures*, 102438.
Erickson, M. (2015). Edutopia. *Jacobin*, (17). www.jacobinmag.com/2015/03/education-technology-gates-erickson/
Evans, T. (2016, July 3). The role of technology in global goal 4: Quality education. *Global Citizen*. www.globalcitizen.org/en/content/technology-access-education-sdg-4/
Flynn, S. (2021). Education, digital natives, and inequality. *Irish Journal of Sociology, 29*(2), 248–253.
Gismondi, A., & Osteen, L. (2017). Student activism in the technology age. *New Directions for Student Leadership, 153*, 63–74.
Gupta, S., Motlagh, M., & Rhyner, J. (2020). The digitalization sustainability matrix. *Sustainability, 12*(21), 9283.
Heikkurinen, P. (2018). Degrowth by means of technology? *Journal of Cleaner Production, 197*, 1654–1665.
Huawei. (2018). *Accelerating SDGs through ICT*. Huawei.
Jordan, K. (2020). Beyond indicators. In *European conference on technology enhanced learning* (pp. 353–357). Springer.
Kallis, G., Paulson, S., & D'Alisa, D. F. (2020). *The case for degrowth*. Polity.
Keating, J. (2009, September 9). Why did one laptop per child fail? *Foreign Policy*. https://foreignpolicy.com/2009/09/09/why-did-one-laptop-per-child-fail/
Kioupi, V., & Voulvoulis, N. (2019). Education for sustainable development. *Sustainability, 11*(21), 6104.
Liegey, V., & Nelson, A. (2020). *Exploring edgrowth*. Pluto.
Madda, M. (2022, March 14). Who is the Theranos of education? *EdSurge*. www.edsurge.com/news/2022-03-14-who-is-the-theranos-of-education
Meinck, S., Fraillon, J., & Strietholt, R. (2022). *International evidence from the Responses to Educational Disruption Survey (REDS)*. UNESCO/IEA
Mitra, S., & Dangwal, R. (2010). Limits to self-organising systems of learning–the Kalikuppam experiment. *British Journal of Educational Technology, 41*(5), 672–688.
Nachtwey, O., & Seidl, T. (2020, October 29). *The solutionist ethic and the spirit of digital capitalism*. https://osf.io/preprints/socarxiv/sgjzq/
Neves, L. (2020). Digital solutions for SDG 4. *International Journal for Education Law and Policy, 16*, 265–270.
OECD. (2021). *Building the future of education*. OECD.
O'Sullivan, K., Clark, S., Marshall, K., & MacLachlan, M. (2021). A just Digital framework to ensure equitable achievement of the Sustainable Development Goals. *Nature Communications, 12*(1), 1–4.
Rafalow, M., & Puckett, C. (2022). Sorting machines: Digital technology and categorical inequality in education. *Educational Researcher, 51*(4), 274–278.
Rohs, M., & Ganz, M. (2015). MOOCs and the claim of education for all. *International Review of Research in Open and Distributed Learning, 16*(6), 1–19.
Sætra, H. (2022). *AI for the sustainable development goals*. CRC Press.
Selwyn, N. (2021a). *Education and technology* (3rd ed.). Bloomsbury.
Selwyn, N. (2021b). Ed-Tech within limits. *E-Learning and Digital Media, 18*(5), 496–510.
Sims, C. (2017). *Disruptive fixation*. Princeton University Press.

Sosa Díaz, M. (2021). Emergency remote education, family support and the digital divide in the context of the COVID-19 lockdown. *International Journal of Environmental Research and Public Health, 18*(15), 7956.

Tyagi, R., Vishwakarma, S., Alexandrovich, Z., & Mohammmed, S. (2019). ICT Skills for Sustainable Development Goal 4. In W. Leal Filho et al. (Eds.), *Quality education, encyclopedia of the UN sustainable development goals* (pp. 436–442). Springer.

UN. (2015). *Transforming our world: The 2030 agenda for sustainable development.* United Nations (A/RES/70/1)

UNESCO. (2015). *Education 2030: Incheon Declaration and Framework for Action: Towards inclusive and equitable quality education and lifelong learning for all.* World Education Forum, Incheon (ED-2016/WS/2)

UNESCO. (2018). *Digital skills for life and work.* UNESCO Broadband Commission for Sustainable Development.

Versteijlen, M., Salgado, F., Groesbeek, M., & Counotte, A. (2017). Pros and cons of online education as a measure to reduce carbon emissions in higher education in the Netherlands. *Current Opinion in Environmental Sustainability, 28,* 80–89.

Ward, J. (2020, August 8). 4IR for SDG 4: Societies must use 4th Industrial Revolution Ed-Tech to achieve Sustainable Development Goal 4. *Foreign Policy News.* https://foreignpolicynews.org/2020/08/08/4ir-for-SDG 4-societies-must-use-4th-industrial-revolution-ed-tech-to-achieve-sustainable-development-goal-4/

Warschauer, M. (2003). Demystifying the digital divide. *Scientific American, 289*(2), 42–47

CHAPTER 7

Virtual Reality and Autism

Anders Dechsling and Anders Nordahl-Hansen

CONTENTS

7.1 INTRODUCTION

Autism Spectrum Disorder is a neurodevelopmental condition affecting approximately 1% of the world's population directly (Lord et al., 2020; Zeidan et al., 2022), and many autistic individuals have a need for help in terms of special education and other support systems. There is not a broad consensus on how and how much help is needed, but it is indisputable that the amount of support is demanding, expensive, and time-consuming. Even though the diagnostic term and description of autism spectrum disorder (autism from hereon) somewhat indicate that this is a lifelong condition, there is no reason to believe that autistic individuals are not able to learn. Hence, they should be provided with the same possibilities as others – which is in line with the United Nations' Sustainable Development Goals (SDGs) (see Chapter 2). Early interventions can lead to huge differences later in life for autistic individuals, which in turn leads to the individual fulfilling their life in line with their own values.

There are already significant differences between service sites, schools locally, and countries (e.g., low- to middle-income countries [LMIC] vs. high-income countries). Even in high-income countries, autistic children do not necessarily receive the best possible service due to, for instance, a lack of trained personnel and resources. In addition, rural places might lack specialists making it necessary to have a more efficient system of service delivery. Interventions aiming to help autistic individuals in maneuvering society are not available for all. Recent technological advancements have opened new approaches to understanding

DOI: 10.1201/9781003325086-7

progress in developmental sciences, including intervention and support for groups that delineate from normative development (Marschik et al., 2022). Technology, such as virtual reality (VR), has the potential to be of help in reaching the SDGs related to equality in education and facilitation for all individuals in acquiring the help needed. Despite technology having potential in educational settings and interventions, there are important aspects to problematize. For instance, as Selwyn points out in Chapter 6, whether technological advancements will benefit LMIC countries and regions is not a straightforward matter and might even increase inequalities. Further, it is unlikely that VR can be a substitute for human resources in autism interventions, but rather a supplement.

After presenting important characteristics of autism and autism research, and VR autism research, this chapter will discuss advantages and disadvantages of applying VR in light of the SDGs. We will touch upon and elaborate on the availability of the tool, the support needed, the fact that it should be used as a supplement rather than a substitute, and that there are a variety of factors related to support needs, implementation, and development.

7.2 AUTISM

Autism Spectrum Disorder is a diagnosis characterized by differences in social communication, and repetitive patterns of thinking and behavior (American Psychiatric Association, 2013). A person can be diagnosed with autism at any point in life but usually the diagnosis is given in childhood and adolescence. The diagnosis is given on the grounds of presentation of the core characteristics mentioned earlier. It is important to note that autistic individuals are a highly heterogeneous group that includes people that need full care throughout their lives, but also those that can live independently with little need of support.

Autistic children and adolescents by definition of the diagnostic criteria experience social-communicative challenges and have fewer reciprocal relationships in kindergarten and school. It is common for autistic children to experience bullying which in turn lead to health issues like anxiety (Skafle et al., 2020), depression (Øien et al., 2019), loneliness (Mazurek, 2014), and school refusal behavior (Munkhaugen et al., 2017). Substantial efforts have been made to develop interventions that can help autistic individuals, in particular with a focus on social communication. During the last decade, there has been a considerable increase in the focus on enabling technologies and how technological support might be used within existing intervention frameworks, with further developments of these frameworks. In the following, we focus on the particulars of one technological support aid, namely VR, and how this might be used within the intervention framework viewed as one of the most effective evidence-informed approaches to helping autistic children in regard to the social-communicative domain of development, namely Naturalistic Developmental Behavioral Interventions (NDBIs). But first we give a general overview of the field of autism interventions from past to present.

7.3 AUTISM INTERVENTIONS

Interventions are delivered in a variety of formats and vary with regard to aims. The interventions may focus on particular skills or behaviors, while others have a broader scope and focus on multiple skills, behaviors, and domains. Depending on the choice of interventions, there is a significant range in intensity of hours. For instance, some short interventions are conducted for one hour per week in a short period, while more comprehensive interventions lie in between and up to 40 hours per week for several years. It also varies whether the interventions are led by trained personnel such as therapists or special education therapists working with the children in clinical or educational settings, or whether the interventions are conducted in natural settings led by therapists, or carried out by caregivers or other non-specialists under supervision from trained personnel (Kaale & Nordahl-Hansen, 2019).

Developmental intervention approaches focus on the fundamental developmental aspect that children explore and act on their environment, with support and guidance from significant persons in their surroundings. Building on theoretical and empirical accounts (Carpenter et al., 1998), reciprocity between child and significant others can be capitalized upon to spur further development and learning. Many influential autism interventions are built on such principles, and joint attention, that is, the sharing of a common focus toward an event or object, has been highlighted as important for development and thus an important target for early interventions (Lord et al., 2022).

Interventions based on Applied Behavior Analysis (ABA) approaches target a broad range of behaviors and skills considered important for the individual, such as communication and language skills, motor development, and adaptive behaviors. These interventions can be comprehensive both in time, frequency, and context. Many studies indicate an impact of ABA interventions in terms of changes in effect sizes, but it is difficult to conclude the effectiveness of ABA interventions due to shortcomings in design and sample size (Sandbank et al., 2020), mainly due to single-case designs without control groups.

Several issues must be considered when deciding on intervention approaches for autistic children. For instance, even though the characteristics of the diagnosis indicate a number of similarities, there is a significant heterogeneity among autistic individuals. The differences are shown not only in cognitive and other behavioral skills but also in how they might respond to various interventions. Further, people change due to development which also have implications for type of intervention or content for younger children as opposed to older adolescents. Still, the clinical effectiveness of interventions and evidence on best practice remain undecided. There are methodological challenges as well as measurement issues smudging the possibilities of clear-cut conclusions. However, NDBIs were considered the most promising intervention approach in a recent meta-analysis by Sandbank et al. (2020). NDBI is an overarching term describing various interventions using techniques from developmental psychology and behavioral theory in naturalistic settings (Schreibman et al., 2015). NDBIs are all using three core elements and have several common interventional strategies (Schreibman et al., 2015). These three core elements involve (a) training a wide range of skills instead of just training specific skills in discrete training environments, (b) an emphasis on naturalistic learning contexts, and (c) using combinations of various elements and strategies to boost learning.

The cost of many autism interventions is high in terms of both economic and practical feasibility. This is often the case because of the resources, requirements, and lack of training opportunities, in addition to challenges related to accessibility in rural areas where qualified personnel and resources might be scarce (Lang et al., 2010). Many of these challenges in autism interventions can be considered key strengths of what a VR intervention program can offer. In addition, caregivers and professionals have already made the call for suitable technological tools that support and educate autistic children (Putnam, 2019). However, the evidence on technology interventions is still scarce and research applying technology to the aforementioned and already established intervention approaches is warranted (Sandbank et al., 2020).

7.4 VIRTUAL REALITY

The use of technology to facilitate learning and skill development has undergone numerous iterations during its relatively short history, and its application has expanded to include VR. The term VR describes artificial environments displayed through visual and auditory stimuli by generating realistic images or other sensations that simulate a sense of presence. VR is considered suitable to simulate naturalistic situations and, since it is not necessarily restricted to geographical locations, it enables clinicians to create training environments otherwise not accessible (Dechsling, Shic, et al., 2021) and to deliver over distance. The virtual environment can be set up so that the user can do various tasks with a high level of flexibility and complexity, presenting a range of potential benefits to ecological validity. These environments and tasks (in VR) are presented with different kinds of hardware equipment with different levels of immersion. Some virtual environments are presented using desktop or tablet computers which are considered low-immersive, while Kinect and other interactive equipment are considered more immersive. The highest levels of immersion are presented in the now affordable consumer product head-mounted displays (HMD) or in Cave Automatic Virtual Environments (CAVE; Miller & Bugnariu, 2016). HMD is a wearable mobile headset equipped with projectors while also shutting out visual stimuli from the real environment in order to provide a sense of the virtual environment surrounding the user completely and enables them to directly interact and get sensory feedback. CAVE, on the other hand, presents a pseudo-three-dimensional environment without any wearables using several two-dimensional projected displays around the user or using motion sensor cameras.

7.5 VIRTUAL REALITY AND AUTISM

Acceptability, enjoyment, and high motivation in using HMDs and computer-technology platforms have been reported in many studies, including children with autism (Dechsling et al., 2020; Newbutt et al., 2016; Newbutt et al., 2020). In addition, several researchers have described VR as having advantages over other forms of training skills for autistic individuals (Yang et al., 2017). The possibility of simulating real-world environments with reduced stress or feeling of anxiety for negative real-world consequences when practicing various situations is seen as the major possible advantage.

The embedding of interventions in an HMD-VR environment might even decouple the need for an intervention from the need for highly trained and rare specialists providing 1:1 intervention with physical presence. This in turn could empower teachers and caregivers who are in daily contact with the children and makes use of centralized and digitally accessible expertise to a wider range of beneficiaries.

Sandbank et al. (2020) concluded that there was a lack of evidence for technological interventions for autism and one of the possible explanations was the absence of theoretical grounding in the assessed studies. They articulated the need for a better integration of technology and validated non-technological intervention techniques, and Schreibman et al. (2015) called for continued innovation in implementing NDBI in clinical practice in order for it to be more available and widely delivered in community settings. Despite agreement on the necessity of including NDBI elements, no highly immersive VR intervention on children with autism explicitly mentions NDBI and only ten studies using immersive VR incorporated some NDBI elements of the common intervention strategies (Dechsling, Shic, et al., 2021). Technology interventions building upon established effective interventions, such as NDBI, offer a potentially more effective, innovative, and efficient way of delivering evidence-based interventions (Dechsling, Shic, et al., 2021).

As one of the main characteristics of autism is associated with social communication, a common target for interventions focuses on skills and behaviors within this developmental domain. When investigating the scope of the literature on autism, VR, and social skills, Dechsling, Orm, et al. (2021) found 50 studies with a broad range of modalities, intervention strategies, and target behaviors. The main research gap within this line of research appeared to be low reporting of the acceptability, skewed male-to-female ratio, lack of theoretical grounding, few studies on autistic individuals with intellectual disabilities, and lack of rigorous and diversity in types of research designs (Table 2; Dechsling, Orm, et al., 2021; see also Nordahl-Hansen et al., 2020).

7.6 THE SUSTAINABLE DEVELOPMENT GOALS AND AUTISM

The SDG 4 seeks to “Ensure inclusive and equitable quality education and promote lifelong learning opportunities for all”. Education and possibilities for learning are key factors for psychological developmental processes, also for individuals with developmental disabilities of various kinds. Individuals with intellectual disabilities or developmental disorders seem to have a higher dropout rate compared to neurotypical peers (Adams et al., 2022). Hence, in order to reach the SDG 4, it is important to focus on the group of children and students that need special education resources and provide them with the possibilities of reaching relevant and effective learning outcomes, as well as increasing the chances for academic completion and success. Furthermore, the use of VR in educational settings can actually be considered an “education facility that is child, disability and gender sensitive and provides safe, non-violent, inclusive and effective learning environments for all” (SDG 4.a).

Targeting a wide range of skills and areas might be crucial in reaching the SDG for the autistic community. Later, we elaborate further on how VR has already addressed some of these issues, and how we suggest that VR can contribute further in reaching the goals of SDGs. We start off with the optimistic view of the technology’s possible solutions before pointing at the critical questions that need to be raised and elaborating on our concerns by thinking that technology alone can solve the issues raised.

The problems related to school refusal, dropouts, or lack of educational degrees (SDG 4, target 4.1.2) for autistic individuals are multifaceted. For instance, the school refusal behavior is often related to problems with social interaction and other demands, as well as motivational issues (Adams et al., 2022; Munkhaugen et al., 2019). Training social skills might have cascading effects on school attendance and academic skills (Kim et al., 2018). Furthermore, also in rural areas, this kind of technology could provide individual adjusted or alternative educational context that increases each individual's opportunity to complete their educational degree with *at least a minimum proficiency level in* academic skills (SDG 4, target 4.1.1; Parsons et al., 2019; Speyer et al., 2018). In addition, more experience using such tools might increase their competence within ICT (SDG 4, target 4.4.1).

Gender issues are also prevalent in autism research. First of all, there is a higher prevalence of diagnosed boys with autism than girls, with an estimated ratio of 3:1, respectively (Loomes et al., 2017). In VR, research samples appear to have a gender ratio of around 6:1 (Dechsling, Orm, et al., 2021). There might be several reasons why these gender ratios appear as they do (see, for instance, Øien et al., 2018), but the gender ratio in VR research on autistic populations is a gap that should be taken into consideration by researchers conducting such studies (Dechsling, Orm, et al., 2021) and should be given extra focus in order to reach the target of *eliminating gender disparities in education and ensure equal access to all levels of education and vocational training for the vulnerable, including persons with disabilities, indigenous peoples, and children in vulnerable situations* (SDG 4, target 4.5).

Target 4.5 also entails the importance of giving the same possibilities to persons with disabilities. Approximately, 70% of autistic individuals have co-occurrence of additional diagnoses such as intellectual disability (Øien et al., 2019), but autism and intellectual disabilities still appear as a research gap within most areas of autism research (Russell et al., 2019), including VR and autism research (Dechsling, Orm, et al., 2021). Therefore, it should be given more focus in order to reach the target.

Several studies utilize VR in training vocational skills (Kim et al., 2022; Speyer et al., 2021). These studies might be crucial in increasing the number of people that get relevant technical and vocational skills, thus reaching the SDG 4, target 4.4. These skills range from job interview skills to on-task training.

Intervention research for children with autism, including the NDBIs, has mainly focused on children of preschool age but VR studies for preschool children are few (Dechsling, Orm, et al., 2021), and thus making it challenging using VR to reach the SDG 4, target 4.2 on pre-primary education.

A concern is whether skills acquired through VR interventions can be generalized into skills in everyday life. Little evidence from interventions support the generalization of skills from different types of technological interventions. Further, there is a risk that the amount of interaction between teachers, caregivers, or peers is reduced, and thus reduces real-life social interactions (Dahl et al., 2021; Ramdoss et al., 2012). It is probably wise that VR should be used not as a substitute for traditional support and education but rather as a supplement.

7.7 ETHICAL CONSIDERATIONS

There are several ethical considerations that need to be made when implementing and applying technology. Virtual reality technology adds to the numbers of considerations when thinking of the implications of the use of hardware such as HMD (Sætra et al., 2022). For instance, General Data Protection Regulation plans need to be rigorously handled for the protection of the individual participating in both research and practical implications. Bearing in mind the reports on autistic individuals having an increased amount of sensory issues compared to the rest of the population (Glod et al., 2019; Thye et al., 2018), it is necessary to look into the reports from autistic individuals on whether they think VR is an acceptable and preferable tool compared to traditional intervention approaches. As autistic individuals represent a highly heterogeneous group, no study can account for every individual and it is important to always make the normative considerations individually. Tailoring can, of course, be done within traditional approaches, but this is also possible in VR. Dechsling et al. (2020) propose a systematic way of making these considerations based on the checklist by Løkke and Salthe (2012) and provide a systematic review of the acceptability reported by autistic individuals on using VR- and computer-based technology. They concluded that VR technology is considered acceptable, enjoyable, and even motivating (Dechsling et al., 2020). Additional research underlines these results by clinical investigations. For instance, Newbutt et al. (2016) found that VR was both enjoyable and motivating, and Newbutt et al. (2020) found that high-immersive VR tools are the most preferred among less immersive versions.

Additional ethical considerations are to be made in the developments of interventions or assessment programs. Participatory designs have been given a great deal of attention, and there is and should be a wide focus on involving the users and their representatives in developing and evaluating tools and content (Parsons et al., 2020). Newbutt and Bradley (2022) offered a case example and suggestions on how these issues can be solved within VR research and practice.

7.8 CONCLUSION

The challenges associated with the characteristics of the autism diagnosis should be targeted on a societal level (Gillespie-Lynch et al., 2017), but it is also important to ensure lifelong learning and vocational opportunities in line with the SDGs. Giving autistic children the support and help they need through interventions can be costly and require specific training, and accessibility varies due to geographical conditions and lack of qualified personnel or other resources (Lang et al., 2010). However, cost–benefit analyses indicate coming in early with help and support can lower costs drastically at later stages in life (Buescher et al., 2014). Thus, training and economically low-threshold qualification opportunities providing contact with persons with the skills necessary to conduct a beneficial and automatized intervention is often key. VR interventions in a virtual environment provide centralized and digitally accessible expertise to a wider range of beneficiaries. Studies are emerging that indicate feasibility of distanced interventions through, for instance, parent coaching, and also in low-resource regions. Methodologically sound investigations of intervention effects have been scarce, but there is an increase in available evidence and rapid implementation of potentially positive research findings can be put into practice (Sandbank et al., 2020). Implementation of research findings does potentially not require effortful systemic changes in the healthcare apparatus, but low economic investment, minimal, or no changes in regulatory or educational aspects, and a constant quality assurance by combining the dissemination of novel digital means with the centrally available expertise. However, it is of key importance that user involvement, meaning that autistic people themselves take part in the development and designs of VR intervention studies, is an inherent and active part of studies on VR for autistic individuals for quality assurance. Individual autistic users, user group representatives, and an autistic co-researcher should be a natural part of VR and autism studies for it to be inclusive and increase the applicability and acceptance.

Digital means, such as VR technology, can contribute to a new level of sustainability and reaching the SDGs in terms of interventions that are personally delivered (by, for instance, teachers), but centrally quality assured by scientists – providing novel and easy-to-use opportunities for more ease of access regardless of geographical position or socioeconomic level. VR interventions have the possibilities to acquire and exercise various skills in a safe environment even from an early age. Enhanced skills and the capacity to cope with daily challenges, for example, in school settings could prevent school refusal or drop out and thus provide the relevant skills also for autistic individuals. However, more emphasis is needed on providing females with the same possibilities as they are underrepresented in research regarding technology and autism. Providing autistic individuals with these opportunities has possible cascading effects toward reaching additional SDGs such as increased well-being in individuals, use of enabling technology to empower women and girls, work for all, and hopefully reduced inequalities. Still, although many studies are emerging testing efficacy and effectiveness of VR interventions for autistic individuals, the jury is still out as to what types of VR interventions are effective for what type of individuals. Also, an important point is whether VR interventions can be a supplement or delivered as total intervention packages. Further, although some results indicate acceptability within the autism community, that is the end users, the amount of research is still scarce.

7.9 REFERENCES

Adams, D., McLucas, R., Mitchelson, H., Simpson, K., & Dargue, N. (2022). Form, function and feedback on the school refusal assessment scale-revised in children on the autism spectrum. *Journal of Autism and Developmental Disorders, 52*(5), 2156–2167. https://doi.org/10.1007/s10803-021-05107-4

American Psychiatric Association. (2013). *Diagnostic and statistical manual of mental disorders (DSM-5).* American Psychiatric Publishing.

Buescher, A. V., Cidav, Z., Knapp, M., & Mandell, D. S. (2014). Costs of autism spectrum disorders in the United Kingdom and the United States. *JAMA Pediatrics, 168*(8), 721–728.

Carpenter, M., Nagell, K., Tomasello, M., Butterworth, G., & Moore, C. (1998). Social cognition, joint attention, and communicative competence from 9 to 15 months of age. *Monographs of the Society for Research in Child Development*, i-174.

Dahl, C., Sætra, H. S., & Nordahl-Hansen, A. (2021, July). Computer-aided games-based learning for children with autism. In *International conference on human-computer interaction* (pp. 145–158). Springer. https://doi.org/10.1007/978-3-030-77277-2_12

Dechsling, A., Orm, S., Kalandadze, T., Sütterlin, S., Øien, R. A., Shic, F., & Nordahl-Hansen, A. (2021). Virtual and augmented reality in social skills interventions for individuals with autism spectrum disorder: A scoping review. *Journal of Autism and Developmental Disorders*. https://doi.org/10.1007/s10803-021-05338-5

Dechsling, A., Shic, F., Zhang, D., Marschik, P. B., Esposito, G., Orm, S., Sütterlin, S., Kalandadze, T., Øien, R. A., & Nordahl-Hansen, A. (2021). Virtual reality and naturalistic developmental behavioral interventions for children with autism spectrum disorder. *Research in Developmental Disabilities, 111*, 103885. https://doi.org/10.1016/j.ridd.2021.103885

Dechsling, A., Sütterlin, S., & Nordahl-Hansen, A. (2020). Acceptability and normative considerations in research on autism spectrum disorders and virtual reality. In D. Schmorrow & C. Fidopiastis (Eds.), *Augmented cognition. Human cognition and behavior.* HCII 2020. Lecture Notes in Computer Science, 12197. Springer. https://doi.org/10.1007/978-3-030-50439-7_11

Gillespie-Lynch, K., Kapp, S. K., Brooks, P. J., Pickens, J., & Schwartzman, B. (2017). Whose expertise is it? Evidence for autistic adults as critical autism experts. *Frontiers in Psychology, 8*. https://doi.org/10.3389/fpsyg.2017.00438

Glod, M., Riby, D. M., & Rodgers, J. (2019). Short report: Relationships between sensory processing, repetitive behaviors, anxiety, and intolerance of uncertainty in autism spectrum disorder and Williams syndrome. *Autism Research: Official Journal of the International Society for Autism Research, 12*(5), 759–765. https://doi.org/10.1002/aur.2096

Kaale, A., & Nordahl-Hansen, A. (2019). Barn og unge med autismespekterforstyrrelse. In I. E. Befring, K. A. B. Næss, & R. Tangen (Red.), *Spesialpedagogikk* (Vol. 6, pp. 523–547). Cappelen Damm.

Kim, S. H., Bal, V. H., & Lord, C. (2018). Longitudinal follow-up of academic achievement in children with autism from age 2 to 18. *Journal of Child Psychology and Psychiatry, 59*(3), 258–267.

Kim, S. Y., Crowley, S., & Lee, Y. (2022). A scoping review of technology-based vocational interventions for individuals with autism. *Career Development and Transition for Exceptional Individuals, 45*(1), 44–56.

Lang, R., O'Reilly, M. F., Sigafoos, J., Machalicek, W., Rispoli, M., Shogren, K., Chan, J. M., Davis, T., Lancioni, G., & Hopkins, S. (2010). Review of teacher involvement in the applied intervention research for children with autism spectrum disorders. *Education and Training in Autism and Developmental Disabilities, 45*(2), 268–283. www.jstor.org/stable/23879811

Løkke, J. A., & Salthe, G. (2012). Sjekkliste for målrettet tiltaksarbeid. *Norsk tidsskrift for atferdsanalyse, 39*(1), 17–32.

Loomes, R., Hull, L., & Mandy, W. P. L. (2017). What is the male-to-female ration in autism spectrum disorder? A systematic review and meta-analysis. *Journal of the American Academy of Child & Adolescent Psychiatry, 56*(6), 466–474. https://doi.org/10.1016/j.jaac.2017.03.013

Lord, C., Brugha, T. S., Charman, T., Cusack, J., Dumas, G., Frazier, T., Jones, E. J. H., Jones, R. M., Pickles, A., State, M. W., Taylor, J. L., & Veenstra-VanderWeele, J. (2020). Autism spectrum disorder. *Nature Reviews Disease Primers, 6*(1), 1–23. https://doi.org/10.1038/s41572-019-0138-4

Lord, C., Charman, T., Havdahl, A., Carbone, P., Anagnostou, E., Boyd, B., Carr, T., de Vries, P. J., Dissanayake, C., Divan, G., Freitag, C. M., Gotelli, M. M., Kasari, C., Knapp, M., Mundy, P., Plank, A., Scahill, L., Servili, C., Shattuck, P., Simonoff, E., . . . McCauley, J. B. (2022). The Lancet Commission on the future of care and clinical research in autism. *Lancet (London, England), 399*(10321), 271–334. https://doi.org/10.1016/S0140-6736(21)01541-5

Marschik, P. B., Poustka, L., Bölte, S., Roeyers, H., & Nordahl-Hansen, A. (2022). Trajectories in developmental disabilities: Infancy–childhood–adolescence. *Frontiers in Psychiatry*. https://doi.org/10.3389/fpsyt.2022.893305

Mazurek, M. O. (2014). Loneliness, friendship, and well-being in adults with autism spectrum disorders. *Autism, 18*(3), 223–232. https://doi.org/10.1177/1362361312474121

Miller, H. L., & Bugnariu, N. L. (2016). Level of immersion in virtual environments impacts the ability to assess and teach social skills in autism spectrum disorder. *Cyberpsychology, Behavior, and Social Networking, 19*, 246–256. https://doi.org/10.1089/cyber.2014.0682

Munkhaugen, E. K., Gjevik, E., Pripp, A. H., Sponheim, E., & Diseth, T. H. (2017). School refusal behaviour: Are children and adolescents with autism spectrum disorder at a higher risk? *Research in Autism Spectrum Disorders*, *41*, 31–38. https://doi.org/10.1016/j.rasd.2017.07.001

Munkhaugen, E. K., Torske, T., Gjevik, E., Naerland, T., Pripp, A. H., & Diseth, T. H. (2019). Individual characteristics of students with autism spectrum disorders and school refusal behavior. *Autism*, *23*(2), 413–423. https://doi.org/10.1177/1362361317748619

Newbutt, N., & Bradley, R. (2022). Using immersive virtual reality with autistic pupils: Moving towards greater inclusion and co-participation through ethical practices. *Journal of Enabling Technologies*. https://doi.org/10.1108/JET-01-2022-0010

Newbutt, N., Bradley, R., & Conley, I. (2020). Using virtual reality head-mounted displays in schools with autistic children: Views, experiences, and future directions. *Cyberpsychology, Behavior and Social Networking*, *23*(1), 23–33. https://doi.org/10.1089/cyber.2019.0206

Newbutt, N., Sung, C., Kuo, H., & Leahy, M. J. (2016). The potential of virtual reality technologies to support people with an autism condition: A case study of acceptance, presence and negative effects. *Annual Review of Cybertherapy and Telemedicine*, *14*, 149–154.

Nordahl-Hansen, A., Dechsling, A., Sütterlin, S., Børtveit, L., Zhang, D., Øien, R. A., & Marschik, P. B. (2020). An overview of virtual reality interventions for two neurodevelopmental disorders: Intellectual disabilities and autism. In D. Schmorrow & C. Fidopiastis (Eds.), *Augmented cognition. Human cognition and behavior.* HCII 2020. Lecture Notes in Computer Science (Vol. 12197). Springer. https://doi.org/10.1007/978-3-030-50439-7_17

Øien, R. A., Nordahl-Hansen, A., & Schjølberg, S. (2019). Mental health and ASD. *Encyclopedia of Autism Spectrum Disorders*, 1–5. https://doi.org/10.1007/978-3-319-91280-6_102050

Øien, R. A., Vambheim, S. M., Hart, L., Nordahl-Hansen, A., Erickson, C., Wink, L., Eisemann, M. R., Shic, F., Volkmar, F. R., & Grodberg, D. (2018). Sex-differences in children referred for assessment: An exploratory analysis of the autism mental status exam (AMSE). *Journal of Autism and Developmental Disorders*, *48*(7), 2286–2292. https://doi.org/10.1007/s10803-018-3488-y

Parsons, D., Cordier, R., Lee, H., Falkmer, T., & Vaz, S. (2019). A randomised controlled trial of an information communication technology delivered intervention for children with autism spectrum disorder living in regional Australia. *Journal of Autism and Developmental Disorders*, *49*(2), 569–581. https://doi.org/10.1007/s10803-018-3734-3

Parsons, S., Yuill, N., Good, J., & Brosnan, M. (2020). Whose agenda? Who knows best? Whose voice? Co-creating a technology research roadmap with autism stakeholders. *Disability & Society*, *35*(2), 201–234. https://doi.org/10.1080/09687599.2019.1624152

Putnam, C., Hanschke, C., Todd, J., Gemmell, J., & Kollia, M. (2019). Interactive technologies designed for children with autism: Reports of use and desires from parents, teachers, and therapists. *ACM Transactions on Accessible Computing*, *12*(3), 1–37. https://doi.org/10.1145/3342285

Ramdoss, S., Machalicek, W., Rispoli, M., Mulloy, A., Lang, R., & O'Reilly, M. (2012). Computer-based interventions to improve social and emotional skills in individuals with autism spectrum disorders: A systematic review. *Developmental Neurorehabilitation*, *15*(2), 119–135. https://doi.org/10.3109/17518423.2011.651655

Russell, G., Mandy, W., Elliott, D., White, R., Pittwood, T., & Ford, T. (2019). Selection bias on intellectual ability in autism research: A cross-sectional review and meta-analysis. *Molecular Autism*, *10*, 9. https://doi.org/10.1186/s13229-019-0260-x

Sætra, H. S., Nordahl-Hansen, A., Fosch-Villaronga, E., & Dahl, C. (2022). Normativity assumptions in the design and application of social robots for autistic children. *SHI 2022*, 136. ISO 690

Sandbank, M., Bottema-Beutel, K., Crowley, S., Cassidy, M., Dunham, K., Feldman, J. I., Crank, J., Albarran, S. A., Raj, S., Mahbub, P., & Woynaroski, T. G. (2020). Project AIM: Autism intervention meta-analysis for studies of young children. *Psychological Bulletin*, *146*(1), 1–29. https://doi.org/10.1037/bul0000215

Schreibman, L., Dawson, G., Stahmer, A. C., Landa, R., Rogers, S. J., McGee, G. G., Kasari, C., Ingersoll, B., Kaiser, A. P., Bruinsma, Y., McNerney, E., Wetherby, A., & Halladay, A. (2015). Naturalistic developmental behavioral interventions: Empirically validated treatments for autism spectrum disorder. *Journal of Autism and Developmental Disorders*, *45*(8), 2411–2428. https://doi.org/10.1007/s10803-015-2407-8

Skafle, I., Nordahl-Hansen, A., & Øien, R. A. (2020). Short report: Social perception of high school students with ASD in Norway. *Journal of Autism and Developmental Disorders, 50*, 670–675. https://doi.org/10.1007/s10803-019-04281-w

Speyer, R., Chen, Y. W., Kim, J. H., Wilkes-Gillan, S., Nordahl-Hansen, A. J., Wu, H. C., & Cordier, R. (2021). Non-pharmacological interventions for adults with autism: A systematic review of randomised controlled trials. *Review Journal of Autism and Developmental Disorders*, 1–31. https://doi.org/10.1007/s40489-021-00250-1

Speyer, R., Denman, D., Wilkes-Gillan, S., Chen, Y. W., Bogaardt, H., Kim, J. H., Heckathorn, D. E., & Cordier, R. (2018). Effects of telehealth by allied health professionals and nurses in rural and remote areas: A systematic review and meta-analysis. *Journal of Rehabilitation Medicine*, *50*(3), 225–235. https://doi.org/10.2340/16501977-2297

Thye, M. D., Bednarz, H. M., Herringshaw, A. J., Sartin, E. B., & Kana, R. K. (2018). The impact of atypical sensory processing on social impairments in autism spectrum disorder. *Developmental Cognitive Neuroscience*, *29*, 151–167. https://doi.org/10.1016/j.dcn.2017.04.010

Yang, Y. J. D., Allen, T., Abdullahi, S. M., Pelphrey, K. A., Volkmar, F. R., & Chapman, S. B. (2017). Brain responses to biological motion predict treatment outcome in young adults with autism receiving Virtual Reality Social Cognition Training: Preliminary findings. *Behaviour Research and Therapy*, *93*, 55–66. http://dx.doi.org/10.1016/j.brat.2017.03.014

Zeidan, J., Fombonne, E., Scorah, J., Ibrahim, A., Durkin, M. S., Saxena, S., Yusuf, A., Shih, A., & Elsabbagh, M. (2022). Global prevalence of autism: A systematic review update. *Autism Research: Official Journal of the International Society for Autism Research*, *15*(5), 778–790. https://doi.org/10.1002/aur.2696

CHAPTER 8

The Technologically Sustained Digital Divide

Erlend Ingridsønn Nordrum

CONTENTS

8.1 INTRODUCTION

The SDGs emphasize technology-based growth to achieve social sustainability, but is there an inherent tension between the increased use of technology and social sustainability? This chapter argues that there is and that this is demonstrated in tensions between various goals in the United Nation's sustainable development goals (SDGs) (United Nations, 2015). A particular tension I focus on here is between SDG target 9.c which aims increasing access to information and communications technology (ICT), and SDG target 10.1 which aims raising the relative income of the lower 40% within nations. While these might appear to be relative eclectic minor parts of the SDGs, I will argue that this tension is one of the most central challenges in contemporary society: ICT is potentially going to change how we produce and distribute wealth. The tension also involves other goals like SDG 8 which aims at promoting "sustained, inclusive and sustainable economic growth, full and productive employment and decent work for all", and SDG 4 which promotes equitable quality education for all.

DOI: 10.1201/9781003325086-8

I begin this chapter by laying out the main SDGs and the UN's techno-optimistic view of ICTs. While I agree that ICTs can be developed to solve many problems and neutralize disadvantages, I point out one kind of inequality that does not seem to go down with technological advancement, namely the inequalities within countries. In the next section, I begin explaining why economic inequalities are maintained through three stages of digital divides. I then proceed to discuss the economy of digital commodities and how supply and demand will make the digital divide larger as the technology progress. I argue that this process can fundamentally change how we produce and distribute resources in society and draw a parallel to Karl Marx's concept of *mode of production* which he used to describe how mass production enabled the *capitalist* mode of production. Lastly, I discuss and conclude the consequences my analysis has for the pursuit of the UN's SDGs.

8.2 DIGITAL TECHNOLOGY HAS NOT FIXED INEQUALITY

In this section, I will place this chapter's research question within the frames of the UN's Sustainable Development Agendas and explicitly point out the relevant elements of the SDGs. Within these SDGs, I point out an overly optimistic expectation of ICTs to improve society which comes without sufficient justification. I then present some ways ICTs *can* reduce inequalities between people, and why some might believe that the technology will lead to more economically equal societies. I then show that this does not seem to align with the available evidence.

8.2.1 SDGs

SDG 10 takes aim at reducing inequalities, both within and among countries (United Nations, 2015). It is worth noting that this goal targets different levels of inequality; at the macro-level between countries and at the meso- and micro-levels within countries (see Chapter 2). These levels are also evident in the various targets. For example, target 10.1 concerns raising the relative income of the lower 40% of within nations. Thus, this target is aimed mainly at impacting the micro-level. Targets 10.2 and 10.3 are mainly targeted at the meso-level, as it seeks to empower and include various demographic groups. Target 10.6 is an example of a macro-oriented target, where it seeks to ensure enhanced representation of developing countries in international economic and financial institutions. In this chapter, I will primarily focus on the inequalities between individuals and socioeconomic groups, which then places this in a micro- and meso-perspective. And I will be primarily concerned with the UN's goal of reducing inequality within countries.

SDG 9 aims at building resilient infrastructure, fostering inclusive and sustainable innovation and industrialization. Promoting ICT and providing universal internet access to the least developed countries are explicitly mentioned as one of the targets (9.c) of this goal. Thus, the UN considers ICTs and particularly the internet as resilient, inclusive, and/or sustainable technology – at least *potentially*, as discussed further in Chapter 18. In Agenda 2030, it is claimed that

> [t]he spread of information and communications technology and global interconnectedness has great potential to accelerate human progress, to bridge the digital

> divide and to develop knowledge societies, as does scientific and technological innovation across areas as diverse as medicine and energy.
>
> (United Nations, 2015, p. 5)

However, very few details on the mechanisms for how this is achieved are provided. This optimistic view of the potential for ICTs to improve society ignores potential adverse effects and relies on an overly simplistic view of the *digital divide*, as I will show later in this chapter.

8.2.2 Techno-solutionism and the Notion That Access to Digital Technology Can Reduce Inequality

Information and communication technology certainly be used to empower the less privileged. Some of these solutions have been highlighted in other chapters of this book. For example, Deschling and Nordahl-Hansen (Chapter 7) use virtual reality to help people with autism disorder develop social skills, and Tschopp and Salam (Chapter 9) explore the potential impact of AI on gender and discrimination. Education is another prime area discussed by Selwyn in Chapter 6. He deals with the idea that it is often assumed that harnessing information and communication technologies is a requirement for strengthening "education systems, knowledge dissemination, information access, quality and effective learning, and more effective service provision" (OECD, 2021, p. 5).

Despite the potential improvements that digital technology brings to the life of people with various disadvantages, these are generally solutions in the form of products to be consummated by the disadvantaged. In this chapter, I focus instead on how ICTs influence the production process and employment of people. Some might think that providing access to digital tools will obviously bring the developing world up the economic speed of the developed. After all, the digital technology sector is a highly lucrative sector with high-paying jobs in many parts of the world right now. If the poorer workers of the world could participate in this economy, they could significantly improve their financial situation. At least that is how the story goes. Next, I show that this does not seem to be the case as most people in even highly digitalized societies don't possess advanced digital skills and that inequalities within countries seem to be increasing.

8.2.3 Lagging ICT Skills and Increasing Inequalities Within Countries

The SDG target 9.c related 2021 report from the International Telecommunication Union (ITU) shows the potential benefits of being online. We find that access to the internet is increasing rapidly across the globe (especially in the least developed countries) with a current estimate of 63% of the world population being online and 95% covered by a mobile broadband network (Bogdan-Martin, 2021). Further, the ITU claims that

> ICTs and the Internet have been vital in maintaining continuity in business activity, employment, education, provision of basic citizens' services, entertainment, and socializing. Digital platforms and services have enabled countless innovations

> that helped mitigate the health, social and economic costs of the tragedy, and build resilience against future crises.
>
> (Bogdan-Martin, 2021, p. iii)

However, the report also emphasizes that digital skills (or lack thereof) constitute an important bottleneck for getting people online when they have coverage, and a hazard in terms of vulnerabilities to cyberattacks, scams, fake news, or harmful content when online (Bogdan-Martin, 2021, p. iv).

The UN tracks ICT skills as a part of the SDG 4 (target 4.4) (United Nations, 2021a) and divides them into three different levels: Basic, Standard, and Advanced.[1] The ITU 2021 rapport (Bogdan-Martin, 2021, p. 20) shows that basic skills are quite well dispersed in countries with high internet access. The standard skills distribution is lower. Only three (out of 76) economies have more than 60% of their population performing computer tasks like installing new software, creating presentations, or using spreadsheets. No economy was reported with a standard skill proportion higher than 80%. And only 11 of the economies had advanced skills over 10%, and no economy with advanced skills over 50%.

So, what is the status of SDG 10 with regard to the inequality between and within countries? Developing countries have in recent decades been catching up somewhat, which lowers the inequality *between* countries. However, economic inequality *within* most countries is growing (Chancel et al., 2021). Despite the increasing availability of ICT, we do not see any evidence that it has reduced the inequality between individuals. On the contrary, as I will show in the next section has ICTs likely contributed to the increased inequality between people.

8.3 WHY DIGITAL TECHNOLOGY HAS NOT FIXED INEQUALITY

In this section, I will show how ICT inequalities extend beyond access to hardware and software and are working to sustain and increase socioeconomic inequalities. I first focus on the individual perspective where the accumulation ICT resources (including skills) tends to be stronger among the already advantaged, which creates a growing divide in terms of these resources. Then I focus on market mechanisms of supply and demand, and how ICTs are changing the demand for various labors. These two mechanisms are complementing each other and pose a substantial threat to the UN's goal of reducing inequalities within countries (SDG 10).

8.3.1 3 Levels of Digital Divide Theory

The UN states that the rationale for target 9.c is to "highlights the importance of mobile networks in providing basic as well as advanced communication services, and will help design targeted policies to overcome remaining infrastructure barriers, and address the digital divide" (United Nations, 2021b). This formulation witnesses an understanding of the digital divide as a distinction between those that have access to the internet from those who do not. The digital divide is consequently portrayed as an infrastructure problem that can be solved by laying cables and building towers.

In contrast to this, I will now present a broader alternative view from sociological research where the digital divide is a *layered problem* where new divides surface as more individuals and groups get physical access (van Deursen et al., 2017; van Deursen & van Dijk, 2014; van Dijk, 2020; Farooq, 2015; Norris, 2001; Selwyn, 2004).

Three levels of the digital divide have been identified (van Deursen et al., 2015). The first level is the classic division between the digital "haves" from the digital "have-nots", defined by who has access to hardware. This is where we find the UN formulation mentioned earlier (United Nations, 2021b). It was noted early on that this access followed existing socioeconomic divides. The educated and wealthier part of the population got access first (van Dijk, 2020, p. 1; NTIA, 1995). As hardware access becomes more universal, a second-level digital divide emerges (van Dijk, 2020, p. 9). This denotes disparities in digital skills and usage patterns. Again, already advantages people tend to have the most advanced digital competencies and display the most advantageous use patterns (Cottom, 2012; Rohs & Ganz, 2015; Scheerder et al., 2017). The third level of the digital divide closes the causal loop, as it concerns how the aforementioned disparities in ICT use influence socioeconomic outcomes (van Deursen et al., 2015; Scheerder et al., 2017).

To explain the relationship between the three levels of the digital divide, van Dijk (2020) proposes a *resource and appropriation theory*, where personal and positional inequalities lead to resource inequalities that in turn determine the process of technology appropriation. The theory is summarized in the five following points (van Dijk, 2020, p. 31):

1. Inequalities in society produce an unequal distribution of resources.
2. An unequal distribution of resources causes unequal access to digital technologies.
3. Unequal access to digital technologies also depends on the characteristics of these technologies.
4. Unequal access to digital technologies brings about unequal outcomes of participation in society.
5. Unequal participation in society reinforces inequalities and unequal distributions of resources.

Further, van Dijk (2020, p. 117) remarks that to understand the workings of the third-level digital divide we need to consider the characteristics of contemporary society. Production of information is becoming more important than material goods, and new digital media facilitates the network society and increase relative inequality. This stratifies society into the information elite, the participatory majority, and the unconnected and excluded.

There is reason to believe that these digital divides exacerbate economic inequality. The rationale is straightforward and simple; high-skilled workers develop and use new technology that replace repetitive tasks previously performed by low-skilled low-income earner. This makes the high-skilled workers more productive, earning them more income. And low-skilled workers are left unproductive, losing their basis for income (Acemoglu, 2002; Brynjolfsson & McAfee, 2014; Danaher, 2019).

Thus, the three levels of digital divide demonstrate how ICTs can *prevent* the achievement of some of the targets in SDG 10. More specifically, it means that hitting target 9.c can come with a cost to target 10.1 and the reduction of relative inequality within countries. That is, while providing universal internet access to the least developed countries will give everyone access to extra technological means, it is likely that it is the already relatively advantaged people within these countries that will benefit the most from this technology. This could then reduce the relative income of the bottom 40 %, which is directly opposed to target 10.1. In the next section, I will go a little deeper into the economic mechanisms that show how digital technology is likely to drive economic growth, which reduce the inequality between countries but simultaneously increase inequality between individuals.

8.4 WHY DIGITAL TECHNOLOGY WILL NOT FIX INEQUALITY

Achieving target 9.c and providing internet access to developing and least developed countries will undoubtedly open up new economic opportunities. One can view the emergence of the digital technology as the modern economic frontier. If the new opportunities are seized, we get economic growth, which is also one of the 17 SDGs (8). Providing digital infrastructure (reaching target 9.c) could then be an opportunity for developing economies to catch up to developed economies while simultaneously hitting target 8.1 of achieving 7% per capita economic growth and SDG 10 reducing inequality between countries. However, digitalization-driven economic growth could also have undesirable effects on inequality *within* countries.

The impact that digital technological progress has on inequality has been subject to much debate. For example, the influential book *The Second Machine Age: Work, Progress, and Prosperity in a Time of Brilliant Technologies* (Brynjolfsson & McAfee, 2014) advances the notion that digital technology is driving a new industrial revolution. This revolution has two main effects on society that the authors refer to as *bounty* and *spread*. The bounty refers to promising economic benefits in terms of increased quality, variety, and volume of production. The spread refers to the potential socioeconomic inequality that ensues "unless we intervene" (Brynjolfsson & McAfee, 2014, p. 15). Inspired by Brynjolfsson and McAfee (2014), I will in the following paragraph lay out my understanding of how digital technological progress influence the economic conditions.

Digital technologies are rapidly and suddenly capable of performing new tasks. Similar to the way that the steam engine replaced manual work in the first industrial revolution are computers replaced cognitive work in the second machine age. Today machines are capable of driving cars and simple clerical tasks. In general, tasks that rely on a predictable set of actions are well suited for automation. Because the technological change in the first industrial revolution automated manual labor, it not only required the physical construction of factories to implement but also limited the speed at which replacement could happen. The predictable set of actions in cognitive work can be written in software code and distributed over the internet. The implication is that there is almost no marginal cost of production. That is the extra cost of producing an extra unit. Or, in other words, the cost of producing a computer program on computer compared to the cost of producing the same computer program and installing it on a 1 million computers are practically indistinguishable. Suppliers of these programs can sell these programs for nearly nothing and still make profits. This is great for the suppliers of the program and the consumers who get high-quality products at very low price and often free. But the humans who used to perform these tasks by the power of their own cognition are now faced with a dramatic increase in competition, that is not possible to meet. This means that a small team of developers can outcompete whole sectors of workers overnight. The laid-off workers would need to find new jobs in other industries or remain unemployed. Workers that relocate to other industries introduce extra supply of labor in the other industries. If the other industry does not experience at least equivalent increase in demand for labor, the extra supply of workers will bring extra competition and drive down wages in these industries. Because it would require less training to relocate to low-skilled employment, the automation of one low-skilled employment will bring down the wages of all low-skilled employment, all else equal. On the other side are workers who pose the right skills that are commentary to digital technology, either by further developing the machines or seemingly unrelated professions like athletes and authors. The market value of these latter mentioned professions can be enhanced many times with new digital broadcasting technology. In any case is it the skills that are hard for computers to replicate that is the future.

Then, it is perhaps not that surprising that the recommendation, for both individuals and policymakers, is to develop these skills (Brynjolfsson & McAfee, 2014, p. 167) and to unlock "billions of innovators" (Brynjolfsson & McAfee, 2014, p. 84). SDG target 10.1 explicitly specifies that the goal is to raise the income of the lower 40% faster than the national average. This be challenging, given that the UN digital skills indicator data show that it is rare to find economies with an above 60% proportion displaying moderate digital skills (Bogdan-Martin, 2021, p. 20). And as the three levels of the digital divide suggest, it is usually the lesser economically advantaged that are also lacking the digital skills. The implication of this is most likely that the already relatively advantaged laborers will primarily benefit in terms of wages from the digitalization of the economy. Laborers that don't keep up to speed with their skills to match the pace of technological progress get disconnected from an accelerating skill-demanding labor market with accelerating demands for skills, and catching up becomes harder with time. The economist Markovits (2019) argues that this mechanism has already dismantled the middle class in contemporary United States and makes the professional elite work with crushing intensity to stay ahead.

Now, higher wages for the top earners could cause them to spend more on commodities that are produced by the lower earners, and thereby indirectly raise the wages of the lower earners. This is the standard assumption in certain market-based economic theories based on the "trickle-down" principle and that rising tides raises all boats (Arndt, 1983; Sowell, 2013). But such mechanisms would not raise their wages at a higher rate than the national average as it would raise the average at a faster rate.

8.4.1 Mode of Production

As noted earlier, Brynjolfsson and McAfee (2014) see the emergence of digital technologies as the dawn of a new (second) industrial revolution. In the former industrial revolution, mass production emerged and transformed the way people worked, lived, and related to one another.

Karl Marx observed this and coined it the *capitalist mode of production* (Marx, 1959, Chap. 51). Every society has a particular mode of production which captures the relations of production and distribution. One of the distinguishing features of the *capitalist* mode of production was the dominant role of production of commodities, with wage labor as the dominant role of labor.

The steam engine was a technology that made mass production in factories possible, which also relied on wage labor. Hence, new technologies have the potential to disrupt the mode of production by rendering social arrangements obsolete and spawn new social relations compatible with the given technology (Barley, 2020). If digital technology instigates an industrial revolution as Brynjolfsson and McAfee (2014) suggest, then it should also be asked; are we also facing a new *mode of production?* Answering this question is beyond the scope of this chapter, but reflecting on the question is nonetheless useful if we want to understand how ICTs shape inequality. With the production of information becoming more important than the production of material goods, the mode of production of information also becomes more important for understanding inequality.

To shed light on the mode of production of information, I will now give a simplified description of it. Hence, the reader should not consider this the full description of contemporary economy, but instead as tendencies that might only become more pronounced as the digitalization of the economy progress.

Contemporary information and communication technology allow for information to be gathered, transformed (computed), and distributed with very little human effort. Thus, an information product, for example, a software program or a text, can be copied and sold with practically no marginal cost of production. It does not make much difference in terms of labor hours spent if the computer program is downloaded by 1 person or 5 million persons. Hence, this technology has the potential to make individuals extremely productive.

Digital technology enables a few to produce and provide products and services that can be used by increasing numbers – sometimes almost unlimited – users. Simultaneously, the application of digital technology replaces lower-skilled labor. In combination, we see that the technology has the potential and the economic incentives to produce both economic growth and economic inequality between individuals with different skills.

When there is competition between producers, this will then quickly drive down the unit price to a near-irrelevant level for buyers. These buyers can then choose the product that provides the highest utility at very little cost, while the producer (or seller of the product) can earn massive profits fast. This is perhaps most evident in the mobile app market where everyone with some programming skills and a computer can launch their own software in a fast and streamlined way.

While this is potentially great for both consumers and successful producers, for others this can pose a threat to their income, access to capital, and relative value in the markets. Increasingly sophisticated robots and software, for example, can reduce the value of labor. Every task that has a predictable set of procedures can in principle be automated (Danaher, 2019). The displaced workers either have to learn new (often more advanced) skills or find lower-skilled jobs that have yet to be automated for some reason. While learning new advanced skills can be feasible for some, for others it can be near impossible. And if they migrate to other lower-skilled jobs, they will compete with others that are already in these jobs and thus bring down their wages with their extra labor supply. And even if we could retrain everyone to be product developers, it is hard to see how everyone could produce something that was both unique and useful enough to be in demand.

Marx (1959, Chap. 15) famously predicted the overthrowing of capitalism as a mode of production by the proletariat as a consequence of falling rate of profits. However, this prediction has thus far failed, and subsequent theories have argued that capitalism is far more adaptive than Marx gave it credit for. Josef Schumpeter (1942) argued that the process of *creative destruction* sustains long-term economic growth and keeps capitalism as the optimal mode of production. Schumpeter instead predicted instead that it would be the success of capitalism that would eventually lead to its collapse (Schumpeter, 1942).

Borgebund revisits Schumpeter's analysis of capitalism in Chapter 12 and argues that said mode of production has the potential to survive also into the digital future. The argument forwarded by Borgebund is that capitalism has proven highly adaptive and this adaptability is likely to carry capitalism into the future. But this begs the question; at what time has capitalism adapted itself into something other than capitalism?

In Marx's theory of social change, a society's mode of production often contains remnants of earlier modes of production together with seeds of new modes of production (Marx, 1959, Chap. 51). Today we still see relics of the feudal mode of production, like the nobility in the Great Britain. Simultaneously are we witnessing new occupations social media influencers (Edwards, 2022; Khamis et al., 2017) and sprouts of new social classes like the precariat (Standing, 2014). While the former appears like the utopian vision for many individuals, most people are more likely to find themself in the latter situation, characterized by temporary and part-time employment without job security in the gig economy (Woodcock & Graham, 2020).

Hence, it is not unreasonable to argue that we are perhaps witnessing the dawn of a new social order brought about by technological change into the digital mode of production. If the capitalist mode of production will be sustained in the second machine age remains to be seen. While the central tenet of both Marx's and Schumpeter's evolution of capitalism is that socialism will follow. But, perhaps socialism is not inevitable next mode of production and it is not likely that it will be the same across the globe. Marx emphasized the historical preconditions in the development of any mode of production (Marx, 1959, Chap. 51) and commented on his contemporary society which was Western Europe in the 19th century. This capitalist society emerged from the feudal society that preceded it (Marx, 1959, Chap. 36) and was marked by mass production. Today, some Marxist scholars are advocating that we must consider a myriad of interconnected causes to grasp the evolution of a social system, which complicates any general future for capitalism (Althusser, 1969; Burczak et al., 2018). For example, some have argued that democracies are more likely than autocracies to redistribute resources (Acemoglu & Robinson, 2006; Boix, 2003), but empirical studies have also found significant heterogeneity which complicates this relationship (Albertus & Menaldo, 2014; Dorsch & Maarek, 2019; Houle, 2009; Knutsen & Rasmussen, 2018). Following the argument for an analysis which considers interconnected causes will any evaluation of the change in mode of production need to be society specific and account for the specific historical precondition and the specific production process in that society. This implies that the changing mode of production could influence different parts of the world differently, and lessons learned from one part do not necessarily translate directly into another part. It is beyond the scope of this chapter to evaluate all possible trajectories that the mode of production can take in various circumstances. However, it seems clear that digital technology is likely to change how things are produced and consequently how resources are distributed, and how this relation of production and distribution will look in the future is uncertain and likely to depend on local contexts. With that in mind, I will proceed to explore how ICT can affect the unequal distribution of resources in developed and developing parts of the world and close to the implications I see that this has for the pursuit of the SDGs.

8.5 CONSEQUENCES OF THE SDGS AND SUSTAINABILITY

Full broadband coverage has not yet been achieved by much of the developing world, and the third-level digital divide is a phenomenon that is primarily studied in the Global North. It remains to be seen if lessons learned from European welfare states are transferable to the developing economies, and we can only speculate how this will pan out. However, as I have argued in this chapter, digital technology has particular properties that lend itself to a mode of production which primarily rely on a skilled elite. Sociological research on the digital divide shows that ICT resources beget ICT resources and tend to reinforce socioeconomic inequalities. Inequalities within countries are rising, and individual differences are becoming more important. The production of ICT favors highly skilled labor, and many are struggling to keep up. ICTs are also finding evermore task to automate, which generally reduce the demand for low-skilled labor and punish them that do not adapt. When developing economies connect to the more digitally matured economies of the Global North, it can crack open new digital divides both across and between countries. Digital elites in developed countries have a head start in developing their strategic and technical skills which can put them in advantageous positions to profit from the new business opportunities that emerge. However, it is also likely that there will develop digital elites within the developing economies, like they have in the Global North. These can create new business opportunities and further advance technology within and abroad. This can accelerate the economic growth of the developing countries to partly catch up with the developed economies. But, where there are elites, there are also likely to be non-elites that struggle to acquire digital skills in developing nations, just as in the developed nations. Low-skilled labor is likely going to suffer extra pressure from automation, and much low-skilled labor is currently performed in developing nations. These are also often nations that don't offer the same degree of welfare benefits as many of the developed nations, which could dampen the consequences for these individuals. Thus, the digital divide could have more severe consequences in the developing world than what we have observed in the developed nations.

Then as for the effects of target 9.c on economic inequality within and between countries, we can see that hitting target 9.c can help to sustain economic growth and by extension reduce economic inequality between countries. However, target 9.c can simultaneously produce higher income inequality within countries. This inequality can be hard to counteract, and relying on education (SDG 4) to bridge the gap ignores the accumulative tendency of digital skills and how the marginal cost of production of digital commodities affects labor demand. While catch-up economics might work for countries, it will likely not work the same way for individuals.

NOTE

1 The digital skill level is operationalized by the ability to conduct specific tasks. Basic: Copying or moving a file or folder. Using copy and paste tools to duplicate or move information within a document. Sending e-mails with attached files between a computer and other devices.
   - **Standard:** Using basic arithmetic formula in a spreadsheet. Connecting and installing new devices. Creating electronic presentations with presentation software. Downloading, installing, and configuring software.
   - **Advanced:** Writing a computer program using a specialized programming language.

8.6 REFERENCES

Acemoglu, D. (2002). Technical change, inequality, and the labor market. *Journal of Economic Literature*, *40*(1), 7–72.
Acemoglu, D., & Robinson, J. A. (2006). *Economic origins of dictatorship and democracy*. Cambridge University Press.
Albertus, M., & Menaldo, V. (2014). Gaming democracy: Elite dominance during transition and the prospects for redistribution. *British Journal of Political Science*, (3), 575–603. https://doi.org/10.1017/S0007123413000124
Althusser, L. (1969). Contradiction and overdetermination. *For Marx*, *114*.
Arndt, H. W. (1983). The "Trickle-down" myth. *Economic Development and Cultural Change*, *32*(1), 1–10.
Barley, S. R. (2020). *Work and technological change*. Oxford University Press.
Bogdan-Martin, D. (2021). *Measuring digital development – facts and figures 2021*. International Telecommunication Union.
Boix, C. (2003). *Democracy and redistribution*. Cambridge University Press.
Brynjolfsson, E., & McAfee, A. (2014). *The second machine age: Work, progress, and prosperity in a time of brilliant technologies*. Norton.
Burczak, T., Garnett, R., & McIntyre, R. (2018). *Knowledge, class, and economics*. Routledge.
Chancel, L., Piketty, T., Saez, E., & Zucman, G. (2021). *World inequality report 2022*. World Inequality Lab.
Cottom, T. (2012). MOOCs, Coursera, online education and performing innovation. *Tressiemc* [Web Log].
Danaher, J. (2019). Automation and utopia. In *Automation and utopia*. Harvard University Press.
Dorsch, M. T., & Maarek, P. (2019). Democratization and the conditional dynamics of income distribution. *The American Political Science Review*, *113*(2), 385–404. https://doi.org/10.1017/S0003055418000825
Edwards, S. (2022). Branded dreams, boss babes: Influencer retreats and the cultural logics of the influencer para-industry. *Social Media + Society*, *8*.
Farooq, M. (2015). Towards a renewed understanding of the complex nerves of the digital divide. *Journal of Social Inclusion*, *6*(1), 71–102.
Houle, C. (2009). Inequality and democracy: Why inequality harms consolidation but does not affect democratization. *World Politics*, (4), 589–622. https://doi.org/10.1017/S0043887109990074

Khamis, S., Ang, L., & Welling, R. (2017). Self-branding, 'micro-celebrity' and the rise of social media influencers. *Celebrity Studies*, *8*(2), 191–208.

Knutsen, C. H., & Rasmussen, M. (2018). The autocratic welfare state: Old-age pensions, credible commitments, and regime survival. *Comparative Political Studies*, (5), 659–695. https://doi.org/10.1177/0010414017710265

Markovits, D. (2019). *The meritocracy trap*. Penguin UK.

Marx, K. (1959). *Capital volume III: The process of capitalist production as a whole (1894)*. USSR, Institute of Marxism-Leninism.

Norris, P. (Ed.). (2001). Social inequalities. In *Digital divide: Civic engagement, information poverty, and the internet worldwide, communication, society and politics* (pp. 68–92). Cambridge University Press.

NTIA. (1995). *Falling through the net: A survey of the "have nots" in rural and urban America*. U.S. Department of Commerce.

OECD. (2021). *Building the future of education*. OECD.

Rohs, M., & Ganz, M. (2015). MOOCs and the claim of education for all: A disillusion by empirical data. *International Review of Research in Open and Distributed Learning*, *16*(6), 1–19.

Scheerder, A., van Deursen, A., & van Dijk, J. (2017). Determinants of internet skills, uses and outcomes. A systematic review of the second- and third-level digital divide. *Telematics and Informatics*, *34*(8), 1607–1624. https://doi.org/10.1016/j.tele.2017.07.007

Schumpeter, J. A. (1942). *Capitalism, socialism, and democracy* (2nd ed., pp. IX, 381). Harper.

Selwyn, N. (2004). Reconsidering political and popular understandings of the digital divide. *New Media & Society*, *6*(3), 341–362. https://doi.org/10.1177/1461444804042519

Sowell, T. (2013). *Trickle down theory" and" tax cuts for the rich*. Hoover Press.

Standing, G. (2014). The precariat. *Contexts*, *13*(4), 10–12.

United Nations. (2015). *Transforming our world: The 2030 agenda for sustainable development*. United Nations.

United Nations. (2021a). SDG metadata-04-04-01.

United Nations. (2021b). SDG metadata-09-0C-01.

van Deursen, A. J. A. M., Helsper, E., Eynon, R., & van Dijk, J. A. G. M. (2017). The compoundness and sequentiality of digital inequality. *International Journal of Communication*, *11*, 452–473.

van Deursen, A. J. A. M., Helsper, E. J., Robinson, L., Cotten, S. R., & Schulz, J. (2015). The third level digital divide: Who benefits most from being online? *Studies in Media and Communications*, 29–52.

van Deursen, A. J. A. M., & van Dijk, J. A. G. M. (2014). *Digital skills: Unlocking the information society*. Springer.

van Dijk, J. A. G. M. (2020). *The digital divide*. Cambridge University Press.

Woodcock, J., & Graham, M. (2020). *The gig economy: A critical introduction*. Polity.

CHAPTER 9

Spot on SDG 5

Addressing Gender (In-)equality Within and With AI

Marisa Tschopp and Hanan Salam

CONTENTS

9.1 INTRODUCTION

No matter where we humans are and what we do, solving problems is part of our core existence. How do we humans solve problems? Every generation has its own way in dealing with problems arising in society. It seems to us that our generation is particularly inclined towards choosing a technical way for solving problems. In the wake of the great hype surrounding the advances of AI, this major player is now moving into the spotlight, which is hoped to go down in human history as a panacea for many of our problems. We call this phenomenon AI solutionism, which in our opinion deserves some special attention.

DOI: 10.1201/9781003325086-9

9.1.1 The Dual Fate of AI

AI is increasingly being used as a technological solution in various business and management domains (e.g., marketing, finance, retail) (Sestino & De Mauro, 2022) due to its potential in automating repetitive tasks and increasing business revenues via personalization and adaptation (Latinovic & Chatterjee, 2022). The potential of AI for sustainable development and achieving the SDGs is not negligible as well. However, the potential success of AI solutions can be accompanied by risks and challenges that can have a negative impact on sustainable development. For instance, it is not entirely clear whether the actual impact of AI on labor or other fields, such as education, can be sufficiently understood or measured (Frank et al., 2019). Furthermore, despite the promises, there is a growing amount of evidence showing that machine learning data, algorithms, and design choices that shape AI systems, reflect and amplify existing cultural biases[1] and prejudice (Buolamwini & Gebru, 2018; Crawford et al., 2019; Zou & Schiebinger, 2018), thus hindering the achievement of SDGs targeting equality such as SDG 5, SDG 10, and SDG 16. In other SDG critical areas, heated discussions are evolving around the negative environmental impact of certain AI technologies (Labbe, 2021). For example, while AI can help combat climate change by introducing low-emission infrastructure, it is significantly emitting carbon on the other side (Dhar, 2020). AI seems fated to take on a dual role; it helps in some areas and harms in others. At present, we are all still in the dark as to how to deal with this ambiguity. When is an AI system a good solution to a problem without harming other areas of life?

9.1.2 A Special Kind of Ethics Washing: Fem Washing

The call of the United Nations has probably sparked many ideas and ethical rationales for funding opportunities, which also marks the first reason why we want to explore the topic of AI and gender equality: *Enhancing the use of enabling technology, in particular information and communications technology, to promote the empowerment of women*[2] is a target set by the United Nations under the overall goal of SDG 5 to eliminate gender inequality and empower all women and girls. Numerous non-governmental or non-profit as well as pro-profit organizations have also committed to make "AI a force for social impact" and aim to "unlock AI's potential towards serving humanity ".[3]

Without soft and hard regulations, all "Tech for Good" initiatives and products risk to fail, despite their altruistic intentions. However, comprehensively understanding soft and/or hard regulations is a discipline on its own. 634 AI soft laws were made public between 2001 and 2019 (Clarke, 2021), comprising a variety of ethical guidelines, principles, and standards for the regulation of AI use, with no binding obligation to commit and no binding consequences for failing to adhere to the voluntary set guidelines. The latter makes the overall situation fragile and vulnerable to ethics washing. It is not farfetched that companies "strategically 'shop' for the principles that limit one's action as little as possible while simultaneously presenting oneself as contributing towards the common good" (van Maanen, 2022, p. 194). With no binding obligations, providing these guidelines publicly could already be criticized as serving as a gift for those wishing to take an easy way out of hard problems.

Despite the honorable efforts, researchers have also identified and criticized a sub-branch of ethics washing called "fempower-washing", where companies or people claim to promote gender equality, but in reality, efforts and tangible results are strikingly low (Sterbenk et al., 2021). So, it may not be the case, as mentioned earlier, that many are just about "shopping" for a good ethical reputation by promising to adhere to standards or investing in a decent ethics label, or that it is just about increasing the chances of receiving funding. It may furthermore be the case that the solution simply does not solve the problem as planned, or just contributes marginally, or in worst case, creates more problems than it solves.

In short, it is plausible that AI for social good initiatives serve more as a diversionary tactic to soften the negative news (ethics washing) than serve as a real contribution to improving the socio-economic problem of gender inequality. The latter we deem the even bigger problem in practice as it is hard to measure and define what real contribution is. This is, furthermore, complicated by the fact that a myriad of such tech solutions are already on the market. So, it must be clear that we are not talking hypothetically about future products and problems. We are already in the middle of this situation, without comprehensive mandatory regulations and without continuous standards for measuring success. We already have many AI systems in our lives and society, promising to solve a problem, but in fact, we are in the dark of the success of the enthusiastic technological fixes. This marks the second motivation why we are doing this research: To critically reflect upon whether AI is a suitable solution for problems in the field of gender equality.

9.1.3 Our Motivation: AI for Gender Equality

Research in the area of AI and gender has primarily focused on challenges linked to gender-related bias discovery and prevention in product design and machine learning data and algorithms (Mehrabi et al., 2019), with a great focus on solving rather mathematical (e.g., functions) or statistical problems (e.g., datasets). The question of female under-representation in STEM and AI careers and its influence on the development of AI systems is also largely investigated in the state of the art (Samuel et al., 2020). However, the question of how AI is currently or could be used to solve gender equality problems is (a) under-explored empirically in order to develop solutions and (b) under-evaluated practically in order to measure and understand the actual effect of the current solutions on the market (Lau et al., 2021). There is a gap of systematic knowledge in research, development, and success measurement about how AI (and other related technologies) solve or contribute to solving problems in the field of gender equality, especially when it comes to normative analyses. For example, what are the underlying values or benchmarks of creating a solution for a specific problem are still ambiguous. There is also a considerable lack of empirical evidence for measuring the actual impact of tech solutions in practice (Cowls et al., 2021).

Several examples provide evidence of how AI maintains, strengthens, and reproduces bias (Feine et al., 2019; Howard & Borenstein, 2018; Lambrecht & Tucker, 2019; McDonnell & Baxter, 2019; Otterbacher et al., 2017; Parra et al., 2021; Raji & Buolamwini, 2019) which presents major ethical concerns and contributes to amplifying gender inequality in society. On the other hand, product development and design choices can contribute to transmitting gender-related stereotypes to AI products by reflecting the thinking patterns of developers behind the AI system (Lee et al., 2019). Specifically related to gender issues, there is a growing concern about the dangers posed by the potential of AI to reinforce societal biases, where the wrong promotion of AI technology may simultaneously accelerate its negative impact. Knowledge of the negative impact of AI on society is specifically relevant, as some may be under the impression that "tech for good" solutions are somewhat free from these dangers, which is presumably not the case. For instance, an AI-based application which potentially solves one problem does not make up for all the negative implications of AI. It is not like a tit-for-tat kind of relationship where "AI for good, makes up for the bad" (Bryson, 2021). So again, the dual destiny of AI makes this situation utterly complex, yet fascinating. Thus, we have one overarching mindset or question we want to answer in the long run with our work. Is AI for gender equality more than just fem washing? Can AI serve as a real (significant) problem-solver? Or is it even causing more harm than benefits?

9.1.4 The Present Study in a Nutshell

Our first step to answering these questions was to conduct the study presented in this chapter. We critically discuss the results of our study which derived qualitative data on AI for gender equality: Existing and potential products and most pressing issues. The study is discussing existing and potential AI solutions in the field of gender equality in five critical areas derived from the call for action from the UN framework: Gender-related violence, gender-related inequalities in healthcare, gender pay gap, unpaid work, and uneven funding.

9.2 ADDRESSING GENDER INEQUALITY WITHIN AI: THE EXAMPLE OF ALGORITHMIC BIAS

Gender inequality within the field of AI has been primarily manifested through biased and consequently discriminatory behavior against females, mainly due to unfair decisions made by biased AI systems and algorithms. Examples of bias in AI applications and products span different areas. These include AI-powered chatbots (Feine et al., 2019; Lee et al., 2019; McDonnell & Baxter, 2019), employment matching (Dastin, 2018), search engines (Kay et al., 2015; Otterbacher et al., 2017), advertising algorithms (Lambrecht & Tucker, 2019), face recognition applications (Raji & Buolamwini, 2019), recommender systems (Parra et al., 2021; Schnabel et al., 2016), and voice recognition systems (Howard & Borenstein, 2018; Osoba & Welser IV, 2017).

In the area of advertisement, research has shown that in some settings, Google online ads for higher-paid jobs were targeted more to male job seekers than female job seekers (Datta et al., 2015). Gender discrimination was also evident in a STEM job advertisement algorithm (Lambrecht & Tucker, 2019) which was initially designed to deliver gender-neutral advertisements. However, a gender imbalance in the training dataset has led to less women seeing the advertisement compared to men. Another example that show gender bias and stereotypes emergence through algorithms is "Google Translate" that translates gender-neutral phrases related to certain professions in certain languages (such as Hungarian or Turkish) to gender-specific professions (for example, she is a nurse, he is a doctor) (Prates et al., 2018). Biases were also found in image web search results for occupations (Kay et al., 2015; Otterbacher et al., 2017). Recent research examined the existence of biases, including gender bias in emotion recognition systems (Domnich & Anbarjafari, 2021), and evidence showing discrimination in AI was found for Facial Emotion Recognition (FER) approaches (Howard et al., 2017). A recent work on bias and fairness investigation in FER systems is that of (Xu et al., 2020), which investigated gender, race, and age bias in two benchmark FER datasets. The research showed that the model is biased toward females.

On the other hand, product development and design choices can contribute to transmitting bias to AI products. A famous example is designing personal assistants (e.g., Alexa, Google Assistant, Siri) to respond in a very loose manner to sexual harassment instead of a serious way. Another area of research, specifically in the field human–AI interaction and perception, is geared toward the evocation of stereotypes, depending on specific voice features (Tolmeijer et al., 2021). For instance, in one study, male AI agents were given higher competence scores and were trusted more in utilitarian context, while female assistants scored higher in warmth scores and were trusted more in hedonic contexts (Ahn et al., 2022). This seems to be reflected in practice, where virtual assistants with basic tasks were assigned female voices and impersonations compared to giving assistants with complicated tasks male impersonations (e.g., Apple's Siri and Amazon's Alexa vs. IBM's Watson and Microsoft's Einstein).

These prime examples are live evidence of the pitfalls of AI solutionism. Indeed, when data-driven AI products are developed and fed with data, we should be careful that we are neither simply confirming existing biases nor introducing new ones, thus contributing to the persistence of gender inequalities in the society. Moreover, a fundamental understanding of bias and fairness from a technical perspective is necessary, no matter what the designed products are intended to do, but especially, if we aim to develop and deploy AI systems to tackle gender inequality. In other words, isn't it paradox that we aim to fix a problem in a society or individual with a broken technology?

Researchers around the globe are working toward detecting and fixing or at least mitigating the bias problem. However, the underlying mechanisms of how bias happens from the computational and design perspectives are not yet thoroughly explored. The literature shows that discriminatory or biased behavior in AI algorithms occurs mainly due to (1) biased or unrepresentative training data or (2) a mathematical model that does not account for existing bias in data. Increasing the participation and representation of females in the design and development process of AI systems is debated as a non-tech solution to the AI bias problem as well. So, do we have to wait to use "AI for Good" until the technology is ready? Yes, of course! would be the answer in an ideal world. But reality paints a different picture. Many immature AI systems are on the market already.

9.3 AI SOLUTIONISM: PROMOTING GENDER EQUALITY WITH AI

Therefore, we find it quite urgent to look at these existing systems and at the same time investigate potential improvements or ideas for new AI systems that could promote gender equality. The core of our study is to explore the question of how AI is and can be used to promote gender equality, considering potential pitfalls of existing or potential systems. However, we still believe it is valuable to start with an optimistic mindset: In other words, how can this technology be used as a solution to a societal problem of crucial significance? Adopting a multidisciplinary view and assessment framework for projects tackling the SDGs (Cowls et al., 2021), we conducted a globally oriented qualitative study to explore the potential of AI tools to enhance gender equality. We focus on five specific areas within the area of gender equality (SDG 5), deemed most relevant based on our review of the literature and the UN SDG targets. The areas of focus include:

Target 5.2: Gender-related violence. "Eliminate all forms of violence against all women and girls in the public and private spheres, including trafficking and sexual and other types of exploitation. (Target 5.2)"

Violence against women comprises physical, psychological violence, threats, and femicide (the intentional killing of women or girls because of their gender) (Krantz & Garcia-Moreno, 2005). The World Health Organization (WHO) reports that worldwide, an estimated 1 in 3 women experience physical or sexual abuse over the course of their lives. Violence against women is considered a human rights violation of global significance with limited knowledge of successful interventions (Krantz & Garcia-Moreno, 2005).

Target 5.1: Gender-related inequality in healthcare. "End all forms of discrimination against women and girls everywhere (Target 5.1)"

Women, and especially women of color, do not have equal access to healthcare and standards of healthy living (Okoro et al., 2020). The clinical practice continues to rely on evidence collected mostly from men and assumed to represent the other half of humanity with negative outcomes for women (as well as for other genders or ethnic minorities). Efforts to improve the situation are gaining great momentum but much more transdisciplinary work needs to be done to get significant results (Buvinic et al., 2014; Cirillo et al., 2020).

Target 5.1/5.5: Gender pay gap. "Ensure women's full and effective participation and equal opportunities for leadership at all levels of decision-making in political, economic and public life (Target 5.5)"

Women face economic inequalities when it comes to wages, also known as the pay gap or gender pay gap. This gap has not closed despite many efforts and law enforcement efforts taking place in different locations of the globe, such as the United States or Europe. The topic remains to be utterly complex, for instance, due to the cultural differences across countries and the differences across economic sectors. It is very unlikely that the gap will close within the next decades, as too many open issues remain. Moreover, research on how the gender pay gap is best calculated and represented statistically remains unsolved (O'Reilly et al., 2015; Velija, 2022).

Target 5.4: Unpaid work. "Recognize and value unpaid care and domestic work through the provision of public services, infrastructure and social protection policies and the promotion of shared responsibility within the household and the family as nationally appropriate (Target 5.4)"

A rather modern issue is the unequal distribution of unpaid housework, which includes caring for children, elderly, and relatives. According to the United Nations report[4], women do 2.6 times the amount of unpaid care and domestic work compared to men. This has not only a negative impact on women's mental health but also substantial economic consequences, for instance, when it comes to missing retirement provisions (MacDonald et al., 2005). There is a lack of systematic empirical data in this nascent field, and structural problems seem to be evolving and outpacing the rapid change in the workforce due to the growing emancipation of women in business (Lundberg, 1996).

Target 5.5: Uneven funding. "Ensure women's full and effective participation and equal opportunities for leadership at all levels of decision-making in political, economic and public life (Target 5.5)"

Women are treated differently regarding loans, venture capital, or funding. Only 3% of venture capital goes to companies founded by women, where the problem occurs at every stage of the allocation of venture capital. Recent research has also shown that investors ask different questions to women and men when pitching their ideas (Kanze et al., 2018). With the rise of female entrepreneurs, the existing gender bias in funding is considered an emerging area of research worthwhile investing in (Färber & Klein, 2021).

Within each of these areas, the research questions (RQ) we aim to answer are as follows:

- RQ1. What AI-based solutions already exist in this area?
- RQ2. What potential AI-based solutions could contribute to this area?
- RQ3. What are the associated challenges in this area?
- RQ4. Who are the stakeholders in this area?

To this end, a global committee of Women in AI[5] members supported by the Swedish Innovation Agency[6] was founded to investigate these questions empirically.

9.4 METHODS

To answer the previous research questions from a global perspective, we conducted 12 semi-structured expert interviews and 3 local focus groups with different local experts, which were open to the public in three different countries. Finally, all participants and stakeholders were re-invited to take part in a global, final workshop with the aim of consolidating the findings and collectively brainstorming on the research questions. Due to privacy restrictions, only participants of the interviews are described. Originally, the workshops were planned as in-person full-day workshops. However, due to the COVID-19 pandemic, all workshops and interviews took place online with a shorter duration.

- **Procedure and sample of expert interviews**: In the first half of the year 2020, 12 semi-structured interviews were conducted with experts from heterogeneous backgrounds: AI, engineering, gender studies, ethics, data science, and industrial specialists with business background. Most of the interviewees hold a master's or doctoral degree (see Table 9.1). The interviews were 30–40 minutes long and followed the same procedure and set of questions. However, the interviewees could choose one

TABLE 9.1 Overview of interview participants (N = 13)

Gender	Field	Country
Female	Computer science assistant professor at university	Canada
Female	Scientific advisor (health and philosophy) and senior research fellow at university	United Kingdom
Female	Computer scientist and philosopher and founder of corporate research lab	United Kingdom
Female	Social scientist and research fellow at a university	United Kingdom
Female	Management scientist at university	USA
Female	Technology ethics professor at university	Germany
Female	Gender political activist	USA
Female	Corporate director for responsible AI	United Kingdom
Female	Nonprofit director	Sweden
Female	Bachelor student and fem-tech founder	United Kingdom
Male	Policy in AI researcher	Australia
Female	Computer science professor	USA
Female	Senior researcher at European Commission	Belgium

or more of the stated gender inequality areas they preferred to talk about. The interviews focused on the barriers, suggested concrete actions, and identified who should collaborate in such an effort to ensure responsible development and use of a potential application. The interviews were conducted by one interviewer online. They were recorded and later transcribed by the same interviewer. To derive the findings, a different researcher summarized and analyzed the transcripts along with the questions of interest compared to current knowledge available in the field from desk research.

- **Local focus groups**: In parallel, three online workshops, organized by members of the global Women in AI committee, took place in different locations. The events took place in Ireland, Australia, and Mexico. The key research questions were discussed at the local workshops. Written summaries structured around the four research questions were provided by the respective organizer and then analyzed by the authors.
- **Global workshop**: After the analysis of all workshops and interviews, a final global two-hour workshop took place. The participants included all the experts and local workshop organizers (25 participants). During this workshop, the findings from the desk research, the individual expert interviews as well as the local workshops were presented, with the aim of discussing and consolidating the results. The event took place on an online platform.[7] Participants were separated into different working groups per target area. Within the breakout sessions, ideas and limitations were discussed and prepared on a collaborative online platform.[8] Afterward, results of the working groups were presented and discussed with all the participants.

9.5 RESULTS AND DISCUSSION

In the following, the main findings of the study are discussed. First, we provide a selection of AI-based applications in the field of gender equality which were derived from desk research. These applications were provided to the participants as a basis for discussions at the global workshop. Then, the findings of the local and global workshops, and the individual interviews (identified opportunities, challenges, and relevant stakeholders) are summarized. Finally, recommendations for research and practice within the chosen areas of interest are presented.

9.5.1 Overview of AI-Based Systems in the Area of Gender Diversity and Inclusion

During the first part of the global workshop, the participants were presented with concrete examples for a better understanding of how an AI system can be used within the scope of gender equality. Due to the limited amount of time, we presented two areas, which we deemed not only most informative but also conceptually translatable to the other areas. Namely, AI in human resources management and AI in the area of violence against women, because in these areas, novel tools are disrupting current procedures at a very high pace (Dattner et al., 2019) and both seem to attract great interest in mainstream media as well. Especially, since the COVID-19 pandemic, the topic of domestic violence against women has gained even more crucial importance. Forcing many families to stay at home has led to an increasing vulnerability of victims of abusive relationships, where violent acts often remain unreported and without consequences for the perpetrators (Rodríguez et al., 2021; Sánchez et al., 2020).

An ample number of AI systems are available within the area of violence against women (including general human trafficking), from helping women in emergency situations of abuse to detecting sexist written text (Rodríguez et al., 2021). Spotlight[9], URSafe[10] (personal safety apps), or the wristwatch Nibye[11], for instance, are using a range of Natural Language Processing (NLP), Computer Vision (CV) capabilities, and analytics to identify victims of sex trafficking or domestic abuse.

In the hiring process, more and more tools find their way into management which are directed at mitigating human bias in the recruitment process. Textio[12], an intelligent text editor, or Pymetrics[13], a talent matching platform, for instance, are tools that are designed to identify gender bias in job applications or other stages of the recruiting process.

Other examples of AI-based applications targeted at gender equality issues:

- **Applications targeted toward *economic equality*:** Ailira[17] (analyzing gender bias in job postings), Talvista[14] (analyzing bias in job postings), Enteleo[15](anonymous interview solution), Gapsquare[16] (analyzing pay levels), and PIHR[17] (fair Pay solution).
- **Applications targeted toward *women's health* solutions:** Gracehealth[18] (among others period and ovulation tracker), Bonzun[19] (among others IVF support), NaturalCycles[20] (birth control app), WildAI[21] (sports performance according to menstrual cycles), Babylon Health[22] (engaging with practitioners), Wysa[23](anonymous mental health), and MyCoachConnect[24] (automated telephone-based reporting system).
- **Applications targeted toward *personal safety*:** Sisbot[25] (helping victims of violence), Safetipin[26] (public safety), Traffic Jam[27] (finding human trafficking victims), Spotlight[28] (human trafficking investigations), URSafe[29] (personal safety app), and Nibye[30] (wristwatch for personal safety).

9.5.2 Identified Opportunities

The results revealed that ideas on how AI can contribute to promote gender equality can be organized into two categories: (1) micro-level, concerning the individual and (2) meso-level, concerning organizational/structural level. On the individual level, initiatives, research, or applications focus on empowering individuals at risk of gender inequalities. On the organizational level, initiatives, research, or applications are targeted toward AI-based applications that can be deployed in an organization or institution in public or private sectors. Across areas of intervention, the key opportunities of using AI to promote gender equality lie in *tools for intervention* through blind-spot detection, education, self-reflection, and *autonomous action.*

- **Micro-level:** There was a consensus among the experts that many AI-based systems aim to empower individual women to better understand a situation, make better decisions, and/or let a device (e.g., mobile phone) decide what to do in an emergency situation.
    1. Empowering the individual to help herself: Such systems provide access to relevant information or (rather active) self-help through interactive communication and assessment tools. These systems enable the individual to reflect on a given situation and evaluate it critically.
    2. Empowering the individual through autonomous action: Such systems provide a function in which individuals can get help, for instance, from authorities, autonomously (e.g., via an application, unnoticeable from the outside). This was especially dominant in the *area of gender-related violence.* This means that the functions often remain undetected by perpetrators, as no obvious action is needed from the victim.
- **Meso-level:** From the organizational-level perspective, AI-based systems support or help on a structural level and are interwoven into an administrative (management) and/or decision-making system (leadership). In these cases, AI-based systems support or help the organization by:
    1. Detecting patterns invisible to the human eye: This kind of AI-based system detects problems in data, for instance, health data or written doctoral notes, or text for internal or external communication, where problems are not visible at all (e.g., noise in the dataset) or have not been detected by humans (e.g., masculine language in job advertisements).
    2. Improving decision-making mostly targeted toward reducing human bias: This kind of AI-based system aims to improve the decision-making process, where the system takes more autonomous action, for instance, deciding who gets funding or who is rejected or accepted for the next hiring steps.

9.5.3 Identified Challenges

The experts in our study evaluated existing or potential solutions not only in terms of technical feasibility but also from a legal and ethical perspective. The main questions we asked were "can we?" for technical as well as for legal feasibility, and "should we?" for ethical feasibility.

For instance, one group focused on how a particular AI system can help women who are victims of domestic violence and how AI systems can contribute to making public areas (e.g., streets) safer for women. Various solutions were discussed, such as an IoT device that calls the police or an application built into Google Maps, which warns a woman if she is about to enter a potentially dangerous area because of very low street lighting. The technical feasibility of the idea was followed by assessing context-dependent factors. Critical requirements of the system concerning hardware and software were discussed: Is it accessible at every location? What happens if it gets lost? How can privacy be preserved? How to ensure such a system doesn't run the risk of being misused by stalkers? How to ensure that the device remains undetectable by a potential perpetrator?

Within the other areas, the following challenges were discussed: Is it possible to track household work without having to install cameras or tracking apps, posing a challenge for a relationship based on trust? Is it feasible to implement care robots without neglecting the complexity and responsibility of what it means to take care of a human? Can a biased/non-perfect AI-based system be used to help evaluate who is eligible for funding? If people analytics tools are employed to monitor wages, career tracks, job ads, CVs, and potentially even written communication and meetings, how to integrate these into a trustful team culture?

In summary, across the studied areas of intervention, key challenges of using AI to promote gender equality lie in privacy concerns, risk for abuse, as well as negative effects on human-to-human relationships.

Furthermore, from a legal perspective, we identified great cultural differences, as well as differences concerning the respective legal systems. With, for instance, the proposed EU AI Act[31] on Europe's doorstep (which will impact trading with Europe), it will be inevitable to consider legal expertise in every step of the development process, next to technical and ethical challenges.

9.5.4 Identified Stakeholders

Anecdotal evidence was discussed, that remaining in one's own echo chamber often happens when designers or managers throw themselves into technical solutions full of enthusiasm and under pressure from the market, causing them not to include the classic "nay-sayers" in their development work. The usual suspects here are lawyers, privacy activists, gender advocates, and so on. However, transdisciplinary work comes at a high cost. An assessment of available resources should be performed, and then a clear strategy is needed in regard to who to involve at what point in time from ideation, over testing to implementing and monitoring. Who to engage, at what point of time, and to what extent is a decision not to be taken lightly. Great investments must be expected here, if done thoroughly. Experts working in the field have argued that managers mostly fear engaging multiple stakeholders as this comes with massively underestimated costs, from more time to communicate to financial payments of expert involvement, as well as the fear of failure. However, it is a requirement to ensure a high-quality product and to preserve a company's image.

The experts of the global workshop created a list of critical stakeholders across fields:

- Computer and data scientists: Academics and practitioners, with expertise in software engineering, AI/ML, IT security;
- Gender study experts by area (e.g., violence, health);
- Legal and policy experts;
- Humanities experts (ethicists, philosophers, psychologists, etc.);
- Business and economics experts;
- Civil society (NGOs, charities, activists, etc.);
- Individuals affected by the system: Doctors, patients, administrative staff, victims, police, HR managers, and potential users;
- Individuals with knowledge of both AI and the respective area of expertise will be of most value (and also most hard to find).

These findings are in line with researchers and practitioners promoting the integration of various stakeholders (Butcher & Beridze, 2019). Nevertheless, concrete networks (e.g., where does a company find those people?) and resources (e.g., how much does this cost?) are still a large problem.

9.6 RECOMMENDATIONS FOR RESEARCH AND PRACTICE

In the following, we briefly summarize our findings and highlight exciting avenues for further research and recommendations for practice per area of interest.

- **Gender-related violence:** Within this area, many applications such as personal safety apps already exist and can be built upon (Miranda et al., 2022). Most opportunities lie within detection, autonomous help, area safety warnings, and education. Experts agree that the greatest impact is on empowering individual women. This is a pressing need, and it may be possible to achieve results (in terms of realizing projects) relatively fast due to the vast experience available, if smart stakeholder management and privacy protection measures are taken care of. Multi-stakeholder collaborations can lead to fast, high-impact results, and are thus deemed a very exciting avenue for research and practice.
- **Gender-related inequalities in healthcare:** AI in healthcare is on the rise and of critical importance (Cirillo et al., 2020). The most discussed issue was around how to create more diverse datasets. Furthermore, the roles in the doctor–patient technology context seem to lack clarity. It is suggested that the actors of such human–machine trust framework must be clearly differentiated to get more insights into understanding what influences decision-making (Tschopp et al., 2021). A promising area of future research – especially since the COVID-19 pandemic – is how remote diagnosis could decrease bias and how AI systems could help doctors make better decisions in online or hybrid settings.
- **Gender-related pay gap:** Discussions in this area focused on people analytics tools and how they can be used to detect and warn employers if inequalities exist. This

can serve as a basis for blind-spot detection and a persuasion tactic to not only create awareness but also motivate for change. It is possible to use certain people analytics tools without transforming the workplace into a surveillance machine (Weibel et al., 2016). Close collaboration with work psychologists and HR managers who are knowledgeable in the AI/data science field is the most promising track (S. D. Schafheitle & Weibel, 2017). A research group has developed a promising framework which seems suitable to scrutinize how technology will influence management and leadership, and traditional configurations of control (S. Schafheitle et al., 2020).

- **Unpaid work:** Unpaid work was the most difficult area to, first of all, understand and then identify the ways by which AI could help or intervene. Examples of robots or tracing data were discussed with great skepticism, as it is questionable how this can be built while preserving privacy and healthy relationships built on mutual trust. So much groundwork is missing in this area, and collaborations with social sciences are called for. Much more exploration on a theoretical and conceptual level must be done having technology assisting already in mind. As there is a lot of research dedicated to the future of work in general, such as hybrid and work-from-home settings, unpaid household work is too often lacking within the majority of studies (Hertog et al., 2022) and thus considered an exciting avenue for future research.
- **Uneven funding:** The experts agreed upon the existence of inequality in this area and welcomed various ideas, especially the idea of having algorithms as co-evaluators, which were deemed most promising. Notably, there was a consensus that in this area, the technical feasibility poses the greatest challenge. In other words, can an algorithm, which is nowadays most likely biased, be used to de-bias human decision-making? Can we combat human bias with machine bias? More studies are needed, which detect gender discrimination in funding proposals (Romei & Ruggieri, 2014).
- **Across areas:** Despite the importance of choosing very specific use cases, the discussions also implied that some of the areas are interrelated and thus improvements in one area may also lead to improvements in other areas. For instance, decreasing inequalities in unpaid work or wage gap may lead to higher financial status for women/families and could reduce domestic violence or violent acts in public as women are better equipped or do not need to work, travel, or live in certain places.

In summary, we urge researchers and practitioners to consider the following key points. By taking a systemic approach, one should consider that technology is not independent of society that AI is not isolated from "other" IT systems and that multi-stakeholder management is key. Furthermore, by acknowledging that the areas as well are not isolated but related in many aspects, one can gain from possible spill-over effects that knowledge and interventions that are beneficial for and within SDG 5 are also beneficial for other areas within the other SDGs and vice versa.

9.7 LIMITATIONS

This exploratory work although done with best intentions, of course, also has limitations. First, the online format may have a negative impact on the depth of discussions and solution findings. We assume that the original full-day in-person design thinking workshop may have resulted in other outcomes.

The second set of limitations is based on the study participants. We deem this specifically relevant for researchers who aim to plan similar studies and urge them to take these issues into account. The gender imbalance among the study participants (12 females: 1 male) is a limitation of the methodology. The participants were all chosen based on their expertise relevant to the present study, and on convenience, that is, who is willing and available to contribute. Since we were specifically looking for people with expertise in at least the technical field and one other field related to gender equality, it is obvious that more women are inclined to study gender equality issues, and less men. The question of whether having a more gender-balanced panel would have altered the outcomes of the study is worthy of investigation.

Furthermore, we identified a lack of expertise in legal topics. This would have been particularly useful as the legal framework gives the required limitations in which ideas can move. We identified great cultural differences, as well as differences concerning the respective legal systems. Due to the lack of legal expertise and the changing regulations in the current state, we were unable to incorporate this perspective into our discussions.

Lastly, qualitative studies, including expert interviews, all suffer from the similar methodological issues around lack of generalization and objectivity. The small sample size does not allow for generalization, and outcomes are biased for various reasons, for example, cultural reasons, or biased researchers, evaluators moderators that only ask questions they think are relevant. The small sample does not allow us to see it as a reflection of reality, and the sample most likely favors a specific outcome. However, it shall be kept in mind that this method is well used, if new beliefs shall be captured, and it is much better suited for an exploratory field, where peoples' – in this case experts' – opinion shall serve as the guiding pillars to drive the issue of gender equality forward.

9.8 CONCLUSION

This chapter has addressed gender equality within AI and reported the results of our qualitative study investigating the barriers and opportunities of gender equality with AI within five chosen areas: Gender-related violence and health inequalities, pay gap, unpaid work, and uneven funding.

The findings of our AI for gender equality study indicate that key opportunities lie in awareness and intervention through blind-spot detection, in other words, using AI to mitigate biased outcomes, furthermore, using interactive AI systems to empower the individual through, for example, education and self-help. Finally, especially in safety-critical situations, for example, a woman being attacked by a perpetrator, AI-based systems can be used to help the victim by taking autonomous action.

We identified areas that need to be addressed in different ways: We motivate researchers and practitioners to further develop and engage in activities addressing very specific issues in the areas of gender-related violence, health inequality, and pay gap as a lot of AI systems are already deployed. On the other hand, more conceptual, groundwork is needed for the areas of unpaid work and uneven funding, such as (academic) research proposals. In these areas, subject matter knowledge is very narrow and lacks an empirical foundation.

In all stages of the research as well as the development and deployment phase of an AI system used to tackle a problem of gender equality, a transdisciplinary, multi-stakeholder approach is key to ensuring the trustworthiness of AI, which comprises ethical, legal as well as technical- and safety-related dimensions.[32]

However, it was also made clear that no technology is a magic pill to cure our society of inequality. Beyond all enthusiasm, no AI technology was found to serve as a panacea for gender-related issues. Presumably, no AI software or hardware can ban violence against women for good, but it has potential to help. Furthermore, there is a substantial risk of ethics washing, in this case "fem washing" and that the proclaimed impact is much lower than expected, which unfortunately is very difficult to measure and prove.

During the workshops, we have experienced a great deal of enthusiasm for the topic, especially from the developers' side, regarding all the opportunities certain technologies offer to (potentially) solve a certain problem. However, when discussed in an interdisciplinary group, where not only the technical but also the legal and ethical feasibility were discussed, we observed a decrease in enthusiasm. Especially, problems around regulation and privacy evolved into emotional and eventually unresolved discussions. This kind of friction is though what we deem necessary, to evaluate and develop a solution aimed at promoting gender equality (or probably any other SDG), especially in highly delicate areas such as personal safety, as it is important to avoid creating more problems than we solve. The overall consensus of experts supports our endeavors to continue our work on how AI can promote gender equality. We also want to motivate others to invest in technology for good, but cautiously, with skeptical optimism so to speak, and setting the right expectations with ambitious visions but realistic goals.

NOTES

1 AI bias can be defined as "*the inclination or prejudice of a decision made by an AI system which is for or against one person or group, especially in a way considered to be unfair*" (Ntoutsi et al., 2020).
2 www.un.org/sustainabledevelopment/gender-equality/
3 https://https://ai4good.org/
4 www.unwomen.org/en/news/in-focus/csw61/redistribute-unpaid-work
5 www.womeninai.co
6 www.vinnova.se
7 https://zoom.us/
8 www.mural.co/
9 www.thorn.org/spotlight/
10 www.ursafe.com/
11 www.nibye.com/
12 https://textio.com/
13 www.pymetrics.ai/ [17] www.ailira.com/
14 www.talvista.com/
15 www.entelo.com/
16 https://gapsquare.com/
17 https://pihr.com/
18 www.grace.health/
19 https://bonzun.com/
20 www.naturalcycles.com/

21 www.wild.ai/
22 www.babylonhealth.com/en-gb/
23 www.wysa.io/
24 www.clinicaltrials.gov/ct2/show/NCT01749124
25 https://changefusion.org/initiatives/11kdhvc0ebab7mgr9d85rviwj9axan
26 https://safetipin.com/
27 www.marinusanalytics.com/traffic-jam
28 www.thorn.org/spotlight/
29 www.ursafe.com/
30 www.nibye.com/
31 https://artificialintelligenceact.eu/
32 Referring to the EU Ethics guidelines for trustworthy AI www.aepd.es/sites/default/files/2019-12/ai-ethics-guidelines.pdf

9.9 REFERENCES

Ahn, J., Kim, J., & Sung, Y. (2022). The effect of gender stereotypes on artificial intelligence recommendations. *Journal of Business Research, 141*, 50–59.

Bryson, J. (2021). *Two ways AI technology is like nuclear technology*. Retrieved August 3, 2022, from https://joanna-bryson.blogspot.com/2021/04/two-ways-ai-technology-is-like-nuclear.html

Buolamwini, J., & Gebru, T. (2018). Gender shades: Intersectional accuracy disparities in commercial gender classification. *Conference on Fairness, Accountability and Transparency*, 77–91.

Butcher, J., & Beridze, I. (2019). What is the state of artificial intelligence governance globally? *The RUSI Journal, 164*(5–6), 88–96.

Buvinic, M., Furst-Nichols, R., & Koolwal, G. (2014). *Mapping gender data gaps*. Data2X Report, United Nations Foundation.

Cirillo, D., Catuara-Solarz, S., Morey, C., Guney, E., Subirats, L., Mellino, S., Gigante, A., Valencia, A., Rementeria, M. J., Chadha, A. S. et al. (2020). Sex and gender differences and biases in artificial intelligence for biomedicine and healthcare. *NPJ Digital Medicine, 3*(1), 1–11.

Clarke, L. (2021). AI is mostly governed by 'soft law'. But that is set to change. *Techmonitor*. Retrieved August 2022, from https://techmonitor.ai/policy/ai-mostly-governed-by-soft-law-but-set-to-change

Cowls, J., Tsamados, A., Taddeo, M., & Floridi, L. (2021). A definition, benchmark and database of AI for social good initiatives. *Nature Machine Intelligence, 3*(2), 111–115.

Crawford, K., Dobbe, R., Dryer, T., Fried, G., Green, B., Kaziunas, E., Kak, A., Mathur, V., McElroy, E., Sánchez, A. N. et al. (2019). *Ai now 2019 report*. AI Now Institute.

Dastin, J. (2018). Amazon scraps secret AI recruiting tool that showed bias against women. In *Ethics of data and analytics* (pp. 296–299). Auerbach Publications.

Datta, A., Tschantz, M. C., & Datta, A. (2015). Automated experiments on ad privacy settings. *Proceedings on Privacy Enhancing Technologies, 2015*(1), 92–112.

Dattner, B., Chamorro-Premuzic, T., Buchband, R., & Schettler, L. (2019). The legal and ethical implications of using AI in hiring. *Harvard Business Review, 25*.

Dhar, P. (2020). The carbon impact of artificial intelligence. *Nature Machine Intelligence, 2*(8), 423–425.

Domnich, A., & Anbarjafari, G. (2021). Responsible AI: Gender bias assessment in emotion recognition. *arXiv preprint*, arXiv:2103.11436.

Färber, M., & Klein, A. (2021). Are investors biased against women? analyzing how gender affects startup funding in Europe. *arXiv preprint*, arXiv:2112.00859.

Feine, J., Gnewuch, U., Morana, S., & Maedche, A. (2019). Gender bias in chatbot design. *International Workshop on Chatbot Research and Design*, 79–93.

Frank, M. R., Autor, D., Bessen, J. E., Brynjolfsson, E., Cebrian, M., Deming, D. J., Feldman, M., Groh, M., Lobo, J., Moro, E. et al. (2019). Toward understanding the impact of artificial intelligence on labor. *Proceedings of the National Academy of Sciences, 116*(14), 6531–6539.

Hertog, E., Fukuda, S., Matsukura, R., Nagase, N., & Lehdonvirta, V. (2022). The future of unpaid work: Estimating the effects of automation on time spent on housework and care work in Japan and the UK. Available at SSRN 4031667. https://papers.ssrn.com/sol3/papers.cfm?abstract_id=4031667

Howard, A., & Borenstein, J. (2018). The ugly truth about ourselves and our robot creations: The problem of bias and social inequity. *Science and Engineering Ethics*, *24*(5), 1521–1536.

Howard, A., Zhang, C., & Horvitz, E. (2017). Addressing bias in machine learning algorithms: A pilot study on emotion recognition for intelligent systems. *2017 IEEE Workshop on Advanced Robotics and Its Social Impacts (ARSO)*, 1–7.

Kanze, D., Huang, L., Conley, M. A., & Higgins, E. T. (2018). We ask men to win and women not to lose: Closing the gender gap in startup funding. *Academy of Management Journal*, *61*(2), 586–614.

Kay, M., Matuszek, C., & Munson, S. A. (2015). Unequal representation and gender stereotypes in image search results for occupations. *Proceedings of the 33rd Annual ACM Conference on Human Factors in Computing Systems*, 3819–3828.

Krantz, G., & Garcia-Moreno, C. (2005). Violence against women. *Journal of Epidemiology & Community Health*, *59*(10), 818–821.

Labbe, M. (2021). Energy consumption of AI poses environmental problems. *Techtarget*. Retrieved August 2021 from www.techtarget.com/searchenterpriseai/feature/Energy-consumption-of-AI-poses-environmental-problems

Lambrecht, A., & Tucker, C. (2019). Algorithmic bias? an empirical study of apparent gender-based discrimination in the display of stem career ads. *Management Science*, *65*(7), 2966–2981.

Latinovic, Z., & Chatterjee, S. C. (2022). Achieving the promise of AI and ML in delivering economic and relational customer value in b2b. *Journal of Business Research*, *144*, 966–974.

Lau, V. W.-y., Scott, V., Warren, M. A., & Bligh, M. (2021). Moving from problems to solutions: A review of gender equality interventions at work using an ecological systems approach. *Journal of Organizational Behavior*, 1–20. https://doi.org/10.1002/job.2654

Lee, N., Madotto, A., & Fung, P. (2019). Exploring social bias in chatbots using stereotype knowledge. *WNLP@ ACL*, 177–180.

Lundberg, U. (1996). Influence of paid and unpaid work on psychophysiological stress responses of men and women. *Journal of Occupational Health Psychology*, *1*(2), 117.

MacDonald, M., Phipps, S., & Lethbridge, L. (2005). Taking its toll: The influence of paid and unpaid work on women's well-being. *Feminist Economics*, *11*(1), 63–94.

McDonnell, M., & Baxter, D. (2019). Chatbots and gender stereotyping. *Interacting with Computers*, *31*(2), 116–121.

Mehrabi, N., Morstatter, F., Saxena, N., Lerman, K., & Galstyan, A. (2019). A survey on bias and fairness in machine learning. *arXiv preprint*, arXiv:1908.09635.

Miranda, J. A., Rituerto-González, E., Luis-Mingueza, C., Canabal, M. F., Bárcenas, A. R., Lanza-Gutiérrez, J. M., Peláez-Moreno, C., & López-Ongil, C. (2022). Bindi: Affective internet of things to combat gender-based violence. *IEEE Internet of Things Journal*, *9*(21), 21174–21193.

Ntoutsi, E., Fafalios, P., Gadiraju, U., Iosifidis, V., Nejdl, W., Vidal, M.-E., Ruggieri, S., Turini, F., Papadopoulos, S., Krasanakis, E. et al. (2020). Bias in data-driven artificial intelligence systems–an introductory survey. *Wiley Interdisciplinary Reviews: Data Mining and Knowledge Discovery*, *10*(3), e1356.

Okoro, O. N., Hillman, L. A., & Cernasev, A. (2020). "We get double slammed!": Healthcare experiences of perceived discrimination among low-income African-American women. *Women's Health*, *16*, 1745506520953348.

O'Reilly, J., Smith, M., Deakin, S., & Burchell, B. (2015). Equal pay as a moving target: International perspectives on forty-years of addressing the gender pay gap. *Cambridge Journal of Economics*, *39*(2), 299–317.

Osoba, O. A., & Welser IV, W. (2017). *An intelligence in our image: The risks of bias and errors in artificial intelligence*. Rand Corporation.

Otterbacher, J., Bates, J., & Clough, P. (2017). Competent men and warm women: Gender stereotypes and backlash in image search results. *Proceedings of the 2017 Chi Conference on Human Factors in Computing Systems*, 6620–6631.

Parra, C. M., Gupta, M., & Dennehy, D. (2021). Likelihood of questioning AI-based recommendations due to perceived racial/gender bias. *IEEE Transactions on Technology and Society*, *3*(1), 41–45.

Prates, M. O., Avelar, P. H., & Lamb, L. C. (2018). Assessing gender bias in machine translation: A case study with google translate. *Neural Computing and Applications*, 1–19.

Raji, I. D., & Buolamwini, J. (2019). Actionable auditing: Investigating the impact of publicly naming biased performance results of commercial AI products. *Proceedings of the 2019 AAAI/ACM Conference on AI, Ethics, and Society*, 429–435.

Rodríguez, D. A., Díaz-Ramírez, A., Miranda-Vega, J. E., Trujillo, L., & Mejía-Alvarez, P. (2021). A systematic review of computer science solutions for addressing violence against women and children. *IEEE Access*, *9*, 114622–114639.

Romei, A., & Ruggieri, S. (2014). A multidisciplinary survey on discrimination analysis. *The Knowledge Engineering Review*, *29*(5), 582–638.

Samuel, Y., George, J., & Samuel, J. (2020). Beyond stem, how can women engage big data, analytics, robotics and artificial intelligence? An exploratory analysis of confidence and educational factors in the emerging technology waves influencing the role of, and impact upon, women. *arXiv preprint*, arXiv:2003.11746.

Sánchez, O. R., Vale, D. B., Rodrigues, L., & Surita, F. G. (2020). Violence against women during the Covid-19 pandemic: An integrative review. *International Journal of Gynecology & Obstetrics*, *151*(2), 180–187.

Schafheitle, S., Weibel, A., Ebert, I., Kasper, G., Schank, C., & Leicht-Deobald, U. (2020). No stone left unturned? toward a framework for the impact of datafication technologies on organizational control. *Academy of Management Discoveries*, *6*(3), 455–487.

Schafheitle, S. D., & Weibel, A. (2017). *Big data or big brother-HR analytics and employee attitudes inside Swiss companies*. Retrieved August 3, 2022, from https://www.alexandria.unisg.ch/252116/1/__SV-MONK_UNISG-Rfolder$_sschafheitle_Desktop_HR_Analytics_OMTF.pdf

Schnabel, T., Swaminathan, A., Singh, A., Chandak, N., & Joachims, T. (2016). Recommendations as treatments: Debiasing learning and evaluation. *International Conference on Machine Learning*, 1670–1679.

Sestino, A., & De Mauro, A. (2022). Leveraging artificial intelligence in business: Implications, applications and methods. *Technology Analysis & Strategic Management*, *34*(1), 16–29.

Sterbenk, Y., Champlin, S., Windels, K., & Shelton, S. (2021). Is femvertising the new greenwashing? examining corporate commitment to gender equality. *Journal of Business Ethics*, 1–15.

Tolmeijer, S., Zierau, N., Janson, A., Wahdatehagh, J. S., Leimeister, J. M. M., & Bernstein, A. (2021). Female by default?–exploring the effect of voice assistant gender and pitch on trait and trust attribution. *Extended Abstracts of the 2021 CHI Conference on Human Factors in Computing Systems*, 1–7.

Tschopp, M., Scharowski, N., & Wintersberger, P. (2021, September 2–3). Do humans trust AI or its developers? Exploring benefits of differentiating trustees within trust in AI frameworks [Conference Presentation]. Venice, Italy. https://www.unive.it/pag/36810/

van Maanen, G. (2022). AI ethics, ethics washing, and the need to politicize data ethics. *Digital Society*, *1*(2), 9. https://doi.org/10.1007/s44206-022-00013-3

Velija, P. (2022). A sociological analysis of the gender pay gap data in UK sport organisations. In *Gender equity in UK sport leadership and governance*. Emerald Publishing Limited.

Weibel, A., Wildhaber, I., Leicht-Deobald, U., Schank, C., & Busch, T. (2016). *Big data or big brother? Big data HR control practices and employee trust*. Swiss National Science Foundation.

Xu, T., White, J., Kalkan, S., & Gunes, H. (2020). Investigating bias and fairness in facial expression recognition. *European Conference on Computer Vision*, 506–523.

Zou, J., & Schiebinger, L. (2018). Design AI so that it's fair. *Nature*, *559*(7714), 324–326.

CHAPTER 10

A Legal Sustainability Approach to Align the Order of Rules and Actions in the Context of Digital Innovation

Eduard Fosch-Villaronga,
Hadassah Drukarch, and Marco Giraudo

CONTENTS

10.1 INTRODUCTION

The increased productivity and efficiency in the industrial and retail sectors, thanks to robotics and Artificial Intelligence (AI), have caused an emerging interest in realizing a comparable transformation in other sectors, including healthcare, farming, and pharmaceutical (Simshaw et al., 2015). While technology has fueled significant innovations geared toward making life and work more accessible, not all such advancements benefit society. Some of this progress has led us to experience social, economic, and environmental challenges (Brundtland, 1987; Johnston, 2018; Zuboff, 2019; Crawford, 2021, see also Chapter 3). For instance, screen-based technology leads humans to be less creative

DOI: 10.1201/9781003325086-10

(Zomorodi, 2017), tired (Genner & Süss, 2017), and more distracted, thus increasing traffic accidents (NHTSA, 2021). On another note, the increasing use of technology in patient care makes one wonder whether and to what extent all parts of society can be automated (Fosch-Villaronga, 2019; Sharkey & Sharkey, 2020). Breakthroughs at the expense of nature also have disastrous consequences for planet Earth's survival (Khakurel, 2018).

Since new technologies are novel, push the boundaries of the current understanding of how the world works and may have a broad range of consequences on people; the development of new technologies comes with the need for increased regulatory attention. Nevertheless, there is an increasing gap between the policy cycle's speed and technological and social change (Downes, 2009; Sucha & Sienkiewicz, 2020). If market participants are to coordinate their investments and economic plans in light of the emerging legal rules being affirmed by the judicial bodies, there might be a problem of unexpected conflict between them. Spontaneous coordination of activities within the market may clash with the normative order of legal rules established and applied by the court system and rule makers. This gap is becoming broader and more prominent in the field of robotics and AI, as existing policies were unprepared to deal with machine learning and autonomous agents adequately and, consequently, often lag and do not adequately frame or address the implications of such technologies (Liu et al., 2020; Custers & Fosch-Villaronga, 2022).[1]

The enlarging divergence between policy adaptation and public authorities' responses can result from information and knowledge gaps between developers and policymakers. However, sometimes it also results from stalling strategies deployed by technological firms to seize dominant market shares to become "too big to be banned" (Mazur & Serafin, 2022, see Pollman & Barry, 2017). For instance, scholars have highlighted for years the legal, ethical, and societal consequences arising from services provided by Google or Facebook, which go from privacy violations to democracy alteration (Choolhun, 2009; Isaak & Hanna, 2018; Milmo, 2022), or sharing economy platforms such as Uber or Airbnb that challenge workers' rights (Posen, 2015; Lutz & Newlands, 2018). As our dependence on these firms has reached unprecedented heights, these companies seem to have become too big to adequately regulate by public authorities (Beard, 2022). In such an uncertain and strategic scenario, a recurrent question is then how society, and more precisely how the law, responds to these events (Sætra & Fosch-Villaronga, 2021a), or, in other words, what direction of adaptation should be between the "order of actions" and the "order of rules" (see, e.g., Dold & Lewis, 2022; Rizzo, 1999).[2]

Although technological change is believed to be at the core of all major historical disruptions, revolutions, wars, and general development (Bailey & Barley, 2020), it may nevertheless offer a means to mitigate precisely those problems it has caused (Millar et al., 2018). At the same time, it is also true that institutional change and legal reforms have been conducive to and have proved essential preconditions for technological investments and economic growth (Hodgson, 2015). We must ensure a synergistic relationship between these two constitutive economic and social progress dimensions.

Given such a co-evolving and constitutive relation between legal, institutional, and technological innovations, it is all the more urgent to modernize the policy cycle on many levels and to ensure as smooth co-evolution as possible between the "order of actions" and the "order of rules". It is in this state of affairs that data-driven policy interventions appeared, evidence-based mechanisms promising to offer a more comprehensive understanding of the issues at stake concerning a particular technology or field (such as chemicals or pharmaceuticals), frame it accordingly, and provide more detailed guidance to developers (Höchtl et al., 2016; Athey, 2017; Calleja et al., 2022). Departing from the notion that information is power, we hold that data may offer a valuable means to policymakers to ensure adequate and sustainable policies for societal and technological developments (Sandersen & Kvalvik, 2014).

Legal sustainability is an interdisciplinary notion we derive from economic literature (Davies, 2013), aiming at singling out different postures toward the substitutability of legal capital with technological capital, bearing on the propensity toward gambling core legal interests and values to favor technological innovation patterns whose society-wide implications are unknown. For instance, the substitution of strong consent requirements and other legal guarantees for individual autonomy in favor of security or "efficiency" promises attached to new technologies. In this sense, and in the pursuit of the UN Sustainable Development Goals (SDG), we aim to contribute to the realization of legal sustainability through the encouragement of more robust and more capable institutions (SDG 16 – peace justice and strong institutions) in an age characterized by rapid developments, diminished accountability, and increased uncertainty.

By framing democratic public order, fundamental rights, and constitutional rules as core *legal* capital – at least in Western Legal Tradition (Gambaro, 2002) – the notion at hand aims at emphasizing *weak* and *strong* policy and legal approaches toward the danger of corroding the foundational constitutional structure of our society, as a result of unintended and unexpected consequences of technological adoption and diffusion. Advocates of weak approaches tend to see technological solutions as *substitutes* for constitutional rules and values. In contrast, proponents of *strong positions* see the corrosion of *legal* capital as conducive to the erosion of the ability of the legal order to reproduce itself and preserve the prevailing constitutional order in the future. In both cases, access to better information is a condition to make sense of how the ongoing social, economic, and political dynamics around technological adoption and diffusion square or imperil the prevailing constitutional order and the rule of law.[3]

This chapter is structured as follows. After this introduction, we lay the grounds for this chapter in section 2 by explaining notions of sustainability in an uncertain legal world. Since the current narrative focuses on how the order of actions (i.e., technology) shapes constraints and world problematics and how the order of rules (i.e., the law) is doomed to lag behind the order of actions, sections 3 and 4 provide examples on how technology disrupts the legal ecosystem and how an uncontrolled legal environment may provide *carte blanche* to techno-solutionism to cause further disruptions. We explain a three-step process to align the order of ideas with that of actions in section 5. Such a process aims at bridging information asymmetries by generating policy-relevant data, sharing knowledge among stakeholders to understand and make sense of such information, and creating opportunities for those ideas to turn into an "action" in the world of actions. This chapter concludes with some final remarks in section 6.

10.2 INTERDEPENDENT TRANSITIONS TOWARD HOLISTIC SUSTAINABILITY

We are living in a time of transition, especially the twin transition concerning the technological migration toward a digital environment and the ecological transformation, which are regarded as complementary, co-evolving dynamics toward the sustainability of human development (European Commission, 2021). Technological breakthroughs and dedicated policies are essential to counter various crises, including climate one, toward realizing a sustainable, fairer economic–social system (Fischer & Newell, 2008; De Cian et al., 2012). Equally, technological innovation is increasingly necessary to solve long-standing social and economic problems and inequalities (Johnston, 2018).

Any transformative and innovative activity involves the exploitation of some resources, thus generating unpredictable and undesirable effects on third parties and the environment (Addressed in SDGs 12–15) (Calabresi, 1985). The most notable and well-known example is the environmental externalities of industrial development based on fossil fuels and plastic materials. From the 1950s onward, it had already become evident that the remarkable improvements in terms of productivity, economic growth, and standards of living enabled by massive motorization of society and new cheap materials were based on the consumption of natural capital, which were polluting the environment (see Missemer et al., 2022). However, the extent to which these undesirable and unintended implications of technological innovation would have affected the environmental sustainability of the prevailing business models of the time was unknown (Davies, 2013), with authoritative – yet unheard – voices calling for early course correction (Meadows et al., 1972).

On the one hand, advocates of a *weak sustainability* approach claimed full substitutability of natural capital being destroyed by the externalities of industrial and technological development. They held that technological improvements would have made it possible to replace natural capital with functionally equivalent technological artifacts to maintain ecosystemic equilibrium and avoid climate change and environmental disasters (Neumayer, 2003). On the other hand, advocates of a *strong sustainability* approach claimed that full substitutability was impossible (Dobson, 1998). Thus, they warned about the danger of depleting core natural resources, for their destruction would have hindered the replicability of the ecosystem, thus eliciting climate changes with unknown and potentially destructive implications. Time has shown who was right, and the current climate crisis is a monumental warning toward techno-solutions. Indeed, the more time advances, the more we can see a parallel with the ongoing legal debates around digital transformation based on datafication of human experience (Zuboff, 2019).

In this respect, diverging positions are emerging as to the *legal sustainability* of the ongoing digital transition in the face of a compelling and growing body of information and knowledge about prevailing business models' legal implications in terms of fundamental rights and democratic order (see, e.g., De Gregorio, 2022). The more visible these implications become, the more pressing the question becomes how those undesirable and unexpected harms to the foundational "legal capital" of the prevailing technological trends shall be treated (Flórez Rojas, 2016). These can be characterized as "constitutional externalities" caused by the prevailing data-driven business model, which must be duly accounted for when assessing the efficiency and costs of the current digital transition (see Giraudo, 2022a). Moreover, increasing information makes it apparent that AI development also has an enormous environmental impact, for instance (Zuboff, 2019; Crawford, 2021; Wynsberghe, 2021). As a result, instead of Earth resources as in the previous industrial revolution, we see a broader sustainability question that embraces our legal–institutional "capital" being corroded by surveillance capitalism.

The issue of externalities is neither a surprise nor new (Pigou, 1920; Coase, 1960). What we see today is the need to understand the peril of pushing the prevailing constitutional and legal order out of equilibrium with unpredictable political, economic, and environmental consequences. However vital the twin transition is, it is not self-standing. For it to be lasting, there is also a need to consider a third overlooked transition taking place whose implications are as important as the previously mentioned two: The legal–institutional transition that spans the economy, society, and environment altogether (Sætra, 2022a). We have to ensure that by pursuing the twin transition, we do not do away with preserving the prevailing constitutional order or the rule of law. In fact, given the foundational matters involved, there is little chance that the twin transformation can be stably attained if the legal foundations of these activities are not shared and properly accommodated with the prevailing fundamental rights and constitutional order. However, the rapidly evolving order of actions in the digital environment is shattering the order of legal rules and the constitutional order, the implications of which in terms of the prevailing legal order are substantial and largely still unknown. We provide some illustrative examples of this disconnect in the following section.

10.3 TECHNOLOGICAL INNOVATION DISRUPTS REGULATION

In their 2019 report on regulatory effectiveness in the era of digitalization, the OECD noted that governments and regulators, in particular, play a significant role in encouraging digital innovation and incentivizing the development of digital technologies for the benefit of society (OECD, 2019). It is within their ability to foster public and consumer interest in the deployment of these technologies and to limit, where possible, any unintended negative consequences of their introduction and use by providing general rules that reflect societal values and preferences (OECD, 2019). However, this does not come without drawbacks (Sætra & Fosch-Villaronga, 2021a).

This approach is reflected in our current legal system, a *horror vacui* system, which aims to prevent legal lacunae from presenting themselves and ensuring legal certainty at all times (Bryson et al., 2017). Departing from this objective, our legal system has produced laws covering many phenomena and developments, including newly developed technologies such as robot and AI technologies. However, regulatory frameworks often lack the agility to accommodate the increasing pace of technological developments and deeply challenge how governments regulate (Downes, 2009; OECD, 2019). While the benefits abound, new technologies inevitably disrupt how we conceive reality, causing growth and innovation across the board (SDGs 8 and 9) and leading us to question and challenge existing norms and push us toward an increasingly louder call for legal and regulatory change.

An illustrative example of this newly introduced complexity can be found in healthcare robots and AI, which, despite its unprecedented potential, still generally fails to incorporate safety sufficiency comprehensively to ensure satisfactory performance of the resulting technologies in the wild (Gruber, 2019). Healthcare robots and AI challenge the timeliness of laws and regulatory standards that were unprepared for technologies that, for instance, would assist wheelchair users in becoming mobile (Tucker et al., 2015), perform surgeries autonomously (Shademan et al., 2016), or offer assistance and support to children under Autism Spectrum Disease in learning emotions (Scassellati et al., 2012; Fosch-Villaronga, 2019; Sætra et al., 2022).

Myriad examples can be found which indicate how such technologies have led us to seriously question and challenge existing norms, such as safety and security, autonomy and responsibility, and non-discrimination and equality (as addressed by SDG 10) (Fosch-Villaronga & Drukarch, 2021). For instance, current legal frameworks tend to overfocus on physical safety when addressing safety concerns. However, they fail to account for other essential aspects like security, privacy, discrimination, psychological aspects, and diversity, which nevertheless play a crucial role in ensuring the safety of such devices to the fullest extent possible and for a wide diversity of potential users. Moreover, these robotic and AI technologies are becoming increasingly autonomous and complex in their interaction with humans, blurring the existing roles and responsibilities and ulteriorly affecting society (Carr, 2011; Yang et al., 2017; Boucher et al., 2020; Fosch-Villaronga et al., 2021).

While technology's dramatically accelerating pace increasingly causes these disparities between the norms we had once established when our societies took on an entirely different architecture and the practical meaning and applicability of these norms in our current technology-driven age, however, (adequate) legal responsiveness does not always follow as a consequent step (Collingridge, 1980; Downes, 2009; Marchant, 2011; Newlands et al., 2020). Moreover, while the pace of digitalization and its impacts on society and markets have become an independent topic of research and debate, far less is clear on how the traditional regulatory functions of governments should evolve with these transformative changes, as will further be elaborated on in the following section.

10.4 A CALL FOR A STRONG LEGAL SUSTAINABILITY APPROACH TO AVOID LEGAL BUBBLES OR CONSTITUTIONAL CRISIS IN THE CONTEXT OF DIGITAL INNOVATION

The legal ecosystem faces similar issues and dangers when deciding how to deal with externalities eroding constitutional capital at the core of constitutional democratic market economies: Fundamental rights and democratic order (Büchi et al., 2020). In democratic countries, fundamental rights and democratic order preserve the ability of the system to reproduce itself, change ruling élites, or spontaneously adapt to unexpected events. The uncertainty about the actual capacity of economic agents and political actors to deploy technological solutions to substitute the constitutional values eroded by some technologies gives rise to opposite policy stances, thus paralleling those observed regarding the issue of climate or technological change.

Also, opposite positions are facing the newly emerging constitutional externalities: *Weak legal sustainability* advocates versus *strong legal sustainability* ones. The first position has long been dominant in academia and industry by supporting the prioritization of technological innovation over legal compliance in case of conflicts or undesirable constitutional externalities. By contrast, the second position is strengthening over time, thanks to the consolidation of EU case law prioritizing privacy, data protection, and fundamental rights over economic interests attached to the deployment of technological innovations (e.g., Custers & Malgieri, 2022). As the current climate crisis suggests, "over-trust on the ability of the technology sector to correct and mitigate by itself the externalities it generates may sound fragile" (Aroyo et al., 2021), if not naive. However, unlike the case of environmental pollution, the possible crisis looming ahead if *weak sustainability approaches prevail is* of legal, economical nature – if not constitutional.

As we will discuss in this section, the brewing crisis may come about either in the form of a legal bubble or a constitutional crisis affecting the effectiveness and preservation of the rule of law with unforeseeable political implications (Giraudo, 2022a). Systemic disregard for constitutional externalities may create political friction and spell unprecedented conflict. From that perspective, we argue that a strong legal sustainability approach might be needed to avoid unexpected crises of constitutional nature with all the disruptive consequences which may follow.

Due to the *favor libertatis* (i.e., the preference for freedom/liberty) that formally underpins democratic market economies, everyone is free to let others infringe upon one's fundamental rights unless that jeopardizes public security or other public interests in a democracy. Thus, in the face of newly emerging business models based on direct or indirect forms of commodification of technological innovation, the long-term legal sustainability question relates to the notion of individual autonomy and liberty as well as to the expected implications of these transactions (e.g., privacy implications, dignity violations) in terms of public democratic order. Hence, the level of legal transaction costs attending any transaction involving technology is highly influenced by the account of liberty, and democratic public order is going to be dominant, as well as by the expected implications on society at large of the emerging privately concluded transactions. The disregard for the role of liberty and individual autonomy may, in fact, eventually hamper the sustainability and attainment of the SDG in liberal–democratic countries.

The core question is then multifold and refers to different levels that can be individual (micro), intermediate or organizational (meso), and collective, social, and economic (macro) (Sætra & Fosch-Villaronga, 2021a). On the individual plane, and if we think about a technology that processes personal data, the question can boil down to broadly understood data protection issues. Thus, it deals with the issue of the extent to which contractual agreements, working as a legal basis for data processing, can be qualified as spontaneous and consensual rather than the result of coercion and private power relations and lack of alternatives (Sætra, 2022a). On the collective and macro-levels, the question relates to the expected benefits of surveillance capitalism with AI at its core and its compatibility with the constitutional democratic order. From the different balancing of these expectations follow other stands toward the ability of surveillance capitalism and the downstream technology to correct the constitutional externalities by itself (Zuboff, 2019).

Of course, there is a spectrum of possible ways of balancing fundamental rights, privacy, and freedom of enterprise bearing on various views on the compatibility of the legal foundations of the data-driven economy with the prevailing notion of liberty and autonomy at the core of EU constitutional order that also happen in the meso-level (Spano, 2021). Depending on which facet of liberty one chooses to prioritize and which of the multiple possible expectations about the future implications of the digital transition, the answer to the legal sustainability issue can vary remarkably (Wu, 2010; Sætra, 2022b). We may frame the ongoing debate regarding *weak* and *strong sustainability approaches* competing at the policy, judicial, and academic levels to shape the legal–institutional foundations of the continuous twin transitions.

Building on the experience with environmental pollution, whereby over-trust in the ability of technological innovation to mitigate – even solve – consequences relating to depletion of natural capital, we may call for a more conservative, strong legal sustainability approach to avoid – or reduce the number of – unexpected future extreme phenomena within the legal, economic ecosystem. At the moment of writing, however, there is no way to envision which approach will prevail, or which future each of them embeds will come to light. Only by taking a precautionary approach and through discovery processes inside top judicial courts in the EU, elsewhere, and other legal actors will we discover whether the conflict exposed by constitutional externalities can be accommodated within the prevailing constitutional order or, instead, substantial constitutional incompatibility is going to bring to a stop many of the personal data commodification practices.

What is already visible today is how economic agents and some departments of public agencies, driven by techno-optimism, are migrating and investing enormous resources into newly emerging technological ecosystems. They do so *as if* AI-powered tech solutions could replace lost shares of individual autonomy and democratic debate or persistent legal uncertainty did not exist. Economic agents' investment strategies joining the digital transition seem unfettered by the ongoing legal discovery processes and shifts, up to the point that legal bubbles may eventually come to light (Giraudo, 2022a). To some extent, such a systemic overlook of institutional dynamics results from legal over-optimism that has been dominant for decades within the industry.

In other words, it is hard to reasonably explain economic agents' tacit legal bets on the stability of the legal foundations of the rampant digital economy unless we acknowledge the fact that there are systemic and silent assumptions within the industry that, sooner or later, *weak legal sustainability* approaches will unavoidably prevail. These possibly misplaced expectations of securing stable legal entitlements over personal data lie at the core of innovative business models dominant in technological innovation, generating tension between the order of actions and the order of rules at the heart of various industries. Sooner or later, they shall adapt as we cannot expect a market economy to last without firm legal foundations (Deakin et al., 2017). Thus, they will either "generate an industry-wide collapse, once the loss of courts' protection substantially decreases the value of investments as the keepers of the legal system" (Giraudo, 2022a) or the adaptation will favor the order of actions, thus exposing a constitutional conflict between courts, Data Protection Authorities (where present), and the legislative bodies coming to the rescue of the digital industry. Time will tell who guessed correctly, however. As common wisdom has it, "it ain't what you don't know that gets you into trouble. It's what you know for sure that just ain't so" (McKay, 2015).

10.5 ALIGNING TECHNOLOGICAL INNOVATION AND REGULATION

The High-Level Expert Group on Artificial Intelligence (HLEG AI) highlighted that AI technologies demand "new legal measures and governance mechanisms . . . to be put in place to ensure adequate protection from adverse impact as well as enabling proper enforcement and oversight, without stifling beneficial innovation" (HLEG AI, 2019). Nonetheless, although something ought to be done, there is no understanding yet of what exactly can or should be done or what effective regulation might look like (Wischmeyer & Rademacher, 2020), an uncertainty that, unfortunately, is at the expense of user rights (Fosch-Villaronga & Heldeweg, 2018).

The fallacy of composition is a recurrent problem, that is, those circumstances in which the central planner (private or public) is unable to anticipate the consequences of specific individual choices, and the other way round, it is unable to predict the effects of choices done at the macro-level (Finocchiaro, 2015). Regulators often operate in a regulatory environment where it is difficult to enter such a conversation, let alone intervene adequately. Moreover, academic research is often not ready to provide usable and valuable knowledge either because it takes time to adequately accommodate the ongoing evolution of the order of actions into a sufficiently adapted order of ideas or because its conclusions prove affected by a form of bias in favor of entrepreneurial endeavors due to the prevailing economic and political incentives structure (Nightingale & Coad, 2014; Whittaker, 2021). Regulatory capture strategies have been adapted to the new institutional multi-level and epistemic order, whereby the role of academic narratives and think-thanks cognitive support is an essential part of the attempts of the industry to seize control of the regulatory framework applicable to newly emerging technologies (Wold, 2022). As such, regulators find themselves at a strategic disadvantage due to information asymmetries, a lack of knowledge to properly understand the implications of technologically enabled social relations as well for lack of resources and institutional mechanisms to intervene timely before technology has been developed and widely adopted (Calleja et al., 2022). These are challenges that UN SDGs 16 and 17 precisely aim to address.

Should these asymmetries and strategic disadvantages continue, technology companies "[will] have a lock on how their products work while underfunded and understaffed regulators will continue to struggle not only to understand the technology but to articulate their concerns" (Guihot & Bennett Moses, 2020; Calleja et al., 2022), thereby further destabilizing the already weak constitutional accountability structures in place. Moreover, developers will struggle to implement legal provisions into their designs, resulting in constant disconnects between policy goals and safe technology (Kapeller et al., 2021). Equipping regulators with technical knowledge of design and practices could help them understand the regulatory needs of a specific and novel technology. However, solving such information asymmetries among developers and regulators raises questions concerning (1) what kind of information is needed to bridge this gap, (2) what knowledge-sharing mechanisms exist for these different stakeholders to make that information intelligible, and (3) what opportunities do those groups have to act upon such knowledge, and foster a change in the order of actions.

10.5.1 The Information Needed to Combat Information Asymmetries

Departing from the notion that assessing risks through experimentation is essential to ensure the safety of new technologies and compliance with existing norms, recent research has proved how experimentation facilities can serve as a source for overcoming information asymmetries between developers and regulators (Fosch-Villaronga & Heldeweg, 2018). Anticipating hazards and reflections on appropriate safeguards for new technologies often happens in testing beds, where prototypes' characteristics are improved to meet safety standards. Some initiatives in Japan (via the Tokku zones, see Weng et al., 2015) and Europe (see Calleja et al., 2022) depart from the premise that these settings could also provide knowledge to improve regulations. The idea behind their method is that testing zones primarily used by developers (for instance, the ones created by the H2020 EUROBENCH Project) can be places where policy-relevant data can be generated and that policymakers could ulteriorly use to establish new safety requirements for uncovered challenges or reformulate existing criteria inconsistent with how technology works. The project PROPELLING showed how scholars could harness robot testing zones as a source for evidence-based knowledge interventions concerning diversity and inclusion for lower-limb exoskeletons, that is, because experiments showed that exoskeletons do not account for sex differences and that men and women experience exoskeleton use differently, something that further influences the device safety. As the saying goes, scientia potentia est: Generating policy-relevant, accurate, and representative data can help decrease existing information imbalances among policymakers concerning different communities in line with the SDG 4 objective of improving the quality and accessibility of knowledge. Consequently, legal action can, if necessary, be more efficient.

10.5.2 Inclusive Policymaking

Although data generation is essential, its value ultimately depends on which sources these data come from and how it is distributed among other levels and orders. The H2020 COVR project, which stands for "Being safe around collaborative and versatile robots in shared spaces",[4] aimed to present detailed safety assessment instructions to robot developers and make the safety assessment process clearer and more straightforward, which, in turn, may allow robots to be used in a more trustworthy and responsible way. In this sense, this EU-funded project sought to develop a tool to better equip robot developers with knowledge (in line with SDG 4) about various aspects, among which legal and regulatory, that are relevant for them throughout the development of their creations. To this end, they created the COVR Toolkit ("toolkit"), an online software application that, among other things, aimed at aiding developers in identifying legislation and standards relevant to them in framing their robot development process and eventual product outcome. More specifically, the toolkit compiles safety regulations for collaborative robots or cobots, that is, robots developed to work closely with humans (Surdilovic et al., 2011) in various domains, such as manufacturing, agriculture, and healthcare.

Although compliance tools, such as the COVR toolkit, represent a practical step toward bridging legal knowledge gaps among developers, new robot applications may fail to fit into existing (robot) categories. A "feeding robot", for instance, may be composed of a robotic wheelchair, an industrial arm, and a feeding function (Herlant, 2018) and may be difficult to classify in existing laws and regulations that cover wheelchairs and industrial arms, but not such a complex cyber-physical system. Moreover, current standards (e.g., ISO 13482:2014 Personal Care Robots), laws (e.g., Medical Device Regulation, 2017), and proposed regulations (e.g., AI Act, 2021) are often technology-neutral and were enacted when practices were at the early stages of implementation and impacts were still unknown, often resulting in dissonances about their protected scope (Fosch-Villaronga, 2019; Salvini et al., 2021). Providing developers with legal information that may be outdated or unclear may do little to help them integrate these considerations into their R&D processes and may have ulteriorly adverse effects once their technologies are put into practice.

In this state of affairs, and contributing to the SDG 9 objective of investing in scientific research and innovation to facilitate sustainable development, the LIAISON Project was set in motion by researchers at Leiden University. LIAISON stands for "Liaising robot development and policymaking to reduce the complexity in robot legal compliance" and was a Financial Support to Third Parties (FSTP) from the H2020 COVR Project. LIAISON departed from the idea that developers may identify legal inconsistencies among regulations or call new categories of devices that struggle to fit any legal categories established in the legal compliance process. At the same time, patient organizations and other actors may identify other safety requirements (physical and psychological alike) that remain uncovered in existing legislation but are nevertheless essential to cover to protect user safety. LIAISON realized that this currently uncaptured knowledge could be formalized and serves as data to improve regulation. To do so, LIAISON attempted to formalize a communication process between robot developers and public and private regulators from which different actors (and eventually policies) could learn, thereby channeling robot policy development from a bottom-up perspective fueled by partnerships between relevant stakeholders across the technology-policy ecosystem (Fosch-Villaronga & Heldeweg, 2019). This approach would align with SDG 17, which aims to contribute to realizing a more sustainable constitutional interaction with digital innovation.

10.5.3 Creating Opportunities to Raise Unheard Voices

The existence of an opportunity to timely and effectively intervene in the regulatory process for noncommercial entities and individuals – for example, associations and think tanks – is essential for implementing a (strong) legally sustainable transition. That is true concerning the two most relevant institutions directly or indirectly influencing policymaking: Law and science (Greif & Mokyr, 2017). If information is available and technical knowledge has been gathered, then there is the need to make sense of all the evidence regarding societal implications and constitutional consequences to translate it into a policy agenda. This happens in the legislative process, court litigation, and academic research programs, whereby scientific authority is attached to sponsor alternative policy agents. If one-sided views dominate legal and scientific policy assessments, then the strategic disequilibrium society experiences will be exacerbated and bear on regulatory delays and the enlarging pacing problem (Downes, 2009; Sætra & Fosch-Villaronga, 2021b).

Such a situation of strategic disequilibrium between those advocating for legal innovation and those defending the prevailing order of rights shall be leveled off through positive actions funded by public institutions. There is the need to create equal opportunities for corporate and non-corporate entities to timely participate in rulemaking and knowledge production through academic and research institutions in line with the SDG 10 goal, which aims, among other things, to empower and promote the social, economic, and political inclusion of all, ensure equal opportunity and reduce inequalities of outcome (see Chapters 5 and 8). If there is no equal opportunity in both dimensions, society risks perpetuating the distortive loop whereby policymakers use knowledge produced by corporate-sponsored research programs to critically appraise corporate-designed rules laid down with intensive lobbying investments and strategically litigated case law (see, e.g., Mazur & Serafin, 2022).

As to the legal dimension, AI-enabled ecosystems enjoy a strategic hedge in their ability to directly shape the regulatory landscape because they have the economic means to reach out to rule makers and win long-lasting lawsuits. There is the need to create institutional venues to match the strategic disequilibrium between those willing to deploy legal innovation instrumental to technological solutions and those willing to protect the prevailing order of rights and interests. On the procedural level, there might be the case to think of dedicated specialized judicial bodies within the judiciary enjoying independence, autonomy, and impartiality to address prompt requests and adjudication on newly emerging issues. One might think of prototypical rules being announced to be applied to a specific case, with no ambition to have general scope during a period of grace. That may combine the need to immediately protect users' rights and prevent society-wide harm while limiting legal decisions' impact on reliance interests at the same time. In fact, due to the prompt legal adjudication from the specialized body, there is less time for relying on interests to form.

As to the knowledge production level, there is the need to counter technological firms' power to fund complacent academic research to shape exclusively positive narratives about technology's future implications and social costs without objectively considering the risks and threats accompanying these developments. Such a knowledge advantage lets them have the ability to rapidly and effectively shape the order of ideas, which will be used to appraise the unfolding technological dynamics critically. To counter that, it might be possible to establish publicly funded research schemes for scientific knowledge production specifically aimed at creating knowledge about the negative implications and incompatible effects with the prevailing order of rules. If public research institutions over-fund pro-innovation research programs, there is a risk of having a distorted incentive structure for researchers to underestimate possible threats coming along with specific trends of technological change (Sætra & Fosch-Villaronga, 2021b). By ensuring a knowledge generation on the implications of technological innovation does not suffer the consequences of (hidden) bias while simultaneously leaving room for all relevant interests and incentives – both economic and social – to be considered, a step closer to sustainable economic growth and innovation, in line with SDGs 9 and 10, may be achieved.

10.6 CONCLUSION

Information imbalance is one of the main drivers behind policy goals and technological innovation dissonances. As science moves faster than moral understanding, people even struggle to articulate their unease with the perils novel technologies introduce (Sandel, 2007), and it is not uncommon to see inventors and users sidelining ethical considerations while focusing on the practical considerations of efficiency and usability (Carr, 2011). Regulation is not immune to those problems. On the contrary, information asymmetries between corporations and regulatory agencies are increasing, impeding the enactment of frameworks closer to reality and more attuned to the real problems that technology poses. In this context, among the 17 UN SDGs, the UN SDGs 16 and 17 are instrumental in promoting a more participatory, inclusive, and accountable institution for technology regulation.

We live in a time of regulatory comeback and technological turmoil after years of retrenchment of public rule makers to the advantage of self-regulation. In this respect, society must be sure not to let the State and the public institutions be trapped within a *knowledge bubble* whereby alternative voices to current deployments and concerns at different levels are not let in for lack of opportunity. If the body of knowledge used to make sense of available information is biased in favor of corporate interests, the effort of the State to mitigate constitutional externalities is doomed to fail. For this reason, this article reasoned from a precautionary perspective about the importance of having a strong legal sustainability approach and generating information from different sources and communities, establishing communication processes to share such information, and creating opportunities to raise unheard voices and learn from diverse communities to avoid legal bubbles or constitutional crisis in the context of digital innovation.

NOTES

1 In this contribution, we use the terms policy, regulation, and law as synonyms.
2 We refer to the order of actions as the practices carried out by developers and the industry and the order of rules as the set of norms, policies, and laws governing the behavior of the actors involved in the order of actions (including industries but also consumers and users).
3 According to the UN, the rule of law is "a principle of governance in which all persons, institutions and entities, public and private, including the State itself, are accountable to laws that are publicly promulgated, equally enforced and independently adjudicated, and which are consistent with international human rights norms and standards". See www.un.org/ruleoflaw/what-is-the-rule-of-law/. The UN has highlighted that rule of law and development are strongly interlinked, and that strengthened rule of law-based society should be considered as an outcome of the 2030 Agenda and SDGs. Especially in relation to SDG 16, the development of inclusive and accountable justice systems and rule of law reforms will build trust in the legitimacy of governments. In this context, SDG 17 also contributes to this goal by forging partnerships to share ideas and foster innovation.
4 See www.safearoundrobots.com/home.

10.7 REFERENCES

Aroyo, A., de Bruyne, J., Dheu, O., Fosch-Villaronga, E., Gudkov, A., Hoch, H., Jones, S., Lutz, C., Sætra, H., Solberg, M., & Tamò-Larrieux, A. (2021). Overtrusting robots: Setting a research agenda to mitigate overtrust in automation. *Paladyn, Journal of Behavioral Robotics, 12*(1), 423–436. https://doi.org/10.1515/pjbr-2021-0029

Athey, S. (2017). Beyond prediction: Using big data for policy problems. *Science, 355*(6324), 483–485.

Bailey, D. E., & Barley, S. R. (2020). Beyond design and use: How scholars should study intelligent technologies. *Information and Organization, 30*(2), 100286.

Beard, A. (2022). Can big tech be disrupted? A conversation with Columbia Business School professor Jonathan Knee. *Harvard Business Review.* https://hbr.org/2022/01/can-big-tech-be-disrupted

Boucher, P., Bentzen, N., Laţici, T., Madiega, T., Schmertzing, L., & Szczepański, M. (2020). *Disruption by technology. Impacts on politics, economics and society.* European Parliamentary Research Service. www.europarl.europa.eu/RegData/etudes/IDAN/2020/652079/EPRS_IDA(2020)652079_EN.pdf.

Brundtland, G. H. (1987). Our common future–Call for action. *Environmental Conservation, 14*(4), 291–294.

Bryson, J. J., Diamantis, M. E., & Grant, T. D. (2017). Of, for, and by the people: The legal lacuna of synthetic persons. *Artificial Intelligence and Law, 25*(3), 273–291.

Büchi, M., Fosch-Villaronga, E., Lutz, C., Tamò-Larrieux, A., Velidi, S., & Viljoen, S. (2020). The chilling effects of algorithmic profiling: Mapping the issues. *Computer Law & Security Review, 36*(10), 53.

Calabresi, G. (1985). *Ideals, beliefs, attitudes, and the law.* Syracuse University Press.

Calleja, C., Drukarch, H., & Fosch-Villaronga, E. (2022). Harnessing robot experimentation to optimize the regulatory framing of emerging robot technologies. *Data & Policy*, Cambridge University Press, 1–15, https://t.co/arDLOYqNvL

Carr, N. (2011). *The shallows: What the internet is doing to our brains.* W.W. Norton & Company.

Choolhun, N. (2009). Google: To use, or not to use. What is the question? *Legal Information Management, 9*(3), 168–172.

Coase, R. H. (1960). The problem of social cost. *The Journal of Law & Economics, 3*, 1–44.

Collingridge, D. (1980). *The social control of technology.* St. Martin's Press.

Crawford, K. (2021). *The atlas of AI: Power, politics, and the planetary costs of artificial intelligence.* Yale University Press.

Custers, B., & Fosch-Villaronga, E. (Eds.). (2022). *Law and artificial intelligence. Regulating AI and applying AI in legal practice.* Information Technology and Law Series (Vol. 35). T.M.C. Asser Press.

Custers, B., & Malgieri, G. (2022). Priceless data: Why the EU fundamental right to data protection is at odds with trade in personal data. *Computer Law & Security Review, 45*, 105683.

Davies, G. R. (2013). Appraising weak and strong sustainability: Searching for a middle ground consilience. *The Journal of Sustainable Development, 10*(1), 111–124.

De Cian, E., Bosetti, V., & Tavoni, M. (2012). Technology innovation and diffusion in "less than ideal" climate policies: An assessment with the WITCH model. *Climatic Change, 114*(1), 121–143.

De Gregorio, G. (2022). *Digital constitutionalism in Europe: Reframing rights and powers in the algorithmic society* (Cambridge Studies in European Law and Policy). Cambridge University Press.

Deakin, S. F., Gindis, D., Hodgson, G. M., Huang, K., & Pistor, K. (2017). Legal institutionalism: Capitalism and the constitutive role of law. *Journal of Comparative Economics, 45*(1), 188–200.

Dobson, A. (1998). *Justice and the environment: Conceptions of environmental sustainability and dimensions of social justice.* Oxford University Press.

Dold, M., & Lewis, P. (2022). F.A. Hayek on the political economy of endogenous preferences: An historical overview and contemporary assessment. *Journal of Economic Behavior & Organization, 196*, 104–119.

Downes, L. (2009). *The laws of disruption: Harnessing the new forces that govern life and business in the digital age*. Basic Books (AZ).
European Commission. (2021). *Green and digital 'twin' transitions also spurs inclusive 'eco-recovery' mindset in waste management*. Retrieved July 7, 2022, from https://ec.europa.eu/environment/ecoap/about-eco-innovation/policies-matters/green-and-digital-twin-transition-also-spurs-inclusive-eco_en
Finocchiaro, M. A. (2015). The fallacy of composition: Guiding concepts, historical cases, and research problems. *Journal of Applied Logic*, *13*(2), 24–43.
Fischer, C., & Newell, R. G. (2008). Environmental and technology policies for climate mitigation. *Journal of Environmental Economics and Management*, *55*(2), 142–162.
Flórez Rojas, M. L. (2016). Legal implications after Schrems case: Are we trading fundamental rights? *Information & Communications Technology Law*, *25*(3), 292–309.
Fosch-Villaronga, E. (2019). *Robots, healthcare and the law: Regulating automation in personal care*. Routledge.
Fosch-Villaronga, E., & Drukarch, H. (2021). On healthcare robots: Concepts, definitions, and considerations for healthcare robot governance. *arXiv pre-print*, 1–87, https://arxiv.org/abs/2106.03468
Fosch-Villaronga, E., & Heldeweg, M. A. (2018). "Regulation, I presume?" said the robot–Towards an iterative regulatory process for robot governance. *Computer Law & Security Review*, *34*(6), 1258–1277.
Fosch-Villaronga, E., & Heldeweg, M. A. (2019). "Meet me halfway," said the robot to the regulation. In *The international conference on inclusive robotics for a better society* (pp. 113–119). Springer.
Fosch-Villaronga, E., Khanna, P., Drukarch, H., & Custers, B. H. (2021). A human in the loop in surgery automation. *Nature Machine Intelligence*, *3*(5), 368–369.
Gambaro, A. V. (2002). Western legal tradition. In P. Newman (Ed.), *The new Palgrave dictionary of economics and the law*. Palgrave Macmillan. https://doi.org/10.1007/978-1-349-74173-1_397
Genner, S., & Süss, D. (2017). Socialization as media effect. *The International Encyclopedia of Media Effects*, *1*, 15.
Giraudo, M. (2022a). On legal bubbles: Some thoughts on legal shockwaves at the core of the digital economy. *Journal of Institutional Economics*, *18*(4), 587–604.
Giraudo, M. (2022b). Legal bubbles. In A. Marciano & G. B. Ramello (Eds.), *Encyclopedia of law and economics*. Springer. https://doi.org/10.1007/978-1-4614-7883-6_786-1
Greif, A., & Mokyr, J. (2017). Cognitive rules, institutions, and economic growth: Douglass North and beyond. *Journal of Institutional Economics*, *13*(1), 25–52.
Gruber, K. (2019). Is the future of medical diagnosis in computer algorithms? *The Lancet Digital Health*, *1*(1), e15–e16.
Guihot, M., & Bennett Moses, L. (2020). *Artificial intelligence, robots and the law*. LexisNexis Butterworths.
High-Level Expert Group on Artificial Intelligence (HLEG AI). (2019). *Policy and investment recommendations*. European Commission. Retrieved January 12, 2023, from https://www.europarl.europa.eu/italy/resource/static/files/import/intelligenza_artificiale_30_aprile/ai-hleg_policy-and-investment-recommendations.pdf
Höchtl, J., Parycek, P., & Schöllhammer, R. (2016). Big data in the policy cycle: Policy decision making in the digital era. *Journal of Organizational Computing and Electronic Commerce*, *26*(1–2), 147–169.
Hodgson, G. (2015). *Conceptualizing capitalism: Institutions, evolution, future*. University of Chicago Press.
Isaak, J., & Hanna, M. J. (2018). User data privacy: Facebook, Cambridge Analytica, and privacy protection. *Computer*, *51*(8), 56–59.

Johnston, S. F. (2018). The technological fix as social cure-all: Origins and implications. *IEEE Technology and Society Magazine*, *37*(1), 47–54.

Kapeller, A., Felzmann, H., Fosch-Villaronga, E., Nizamis, K., & Hughes, A. M. (2021). Implementing ethical, legal, and societal considerations in wearable robot design. *Applied Sciences*, *11*(15), 6705.

Khakurel, J., Penzenstadler, B., Porras, J., Knutas, A., & Zhang, W. (2018). The rise of artificial intelligence under the lens of sustainability. *Technologies*, *6*(4), 1–18, 100.

Liu, H. Y., Maas, M., Danaher, J., Scarcella, L., Lexer, M., & Van Rompaey, L. (2020). Artificial intelligence and legal disruption: A new model for analysis. *Law, Innovation and Technology*, *12*(2), 205–258.

Lutz, C., & Newlands, G. (2018). Consumer segmentation within the sharing economy: The case of Airbnb. *Journal of Business Research*, *88*, 187–196.

Marchant, G. E. (2011). The growing gap between emerging technologies and the law. In G. E. Marchant, B. R. Allenby, & J. R. Herkert (Eds.), *The growing gap between emerging technologies and legal-ethical oversight: The pacing problem* (Vol. 7, 19–33). Springer Science & Business Media.

Mazur, J., & Serafin, M. (2022). Stalling the state: How digital platforms contribute to and profit from delays in the enforcement and adoption of regulations. *Comparative Political Studies*, 00104140221089651.

McKay, A. (2015). *The big short*. Paramount Pictures.

Meadows, D. H., Meadows, D., Randers, J., & Behrens, W. (1972). *The limits to growth: A report for the club of Rome's project on the predicament of mankind*. Universe.

Millar, C., Lockett, M., & Ladd, T. (2018). Disruption: Technology, innovation and society. *Technological Forecasting and Social Change*, *129*, 254–260.

Milmo, D. (2022, September 2). Nobel Peace Prize winners call for action on online disinformation. *The Guardian*. www.theguardian.com/media/2022/sep/02/nobel-peace-prize-winners-call-for-action-on-online-disinformation

Missemer, A., Gaspard, M., & Ferreira da Cunha, R. P. (2022). From depreciation to exhaustible resources: On Harold Hotelling's first steps in economics. *History of Political Economy*, *54*(1), 109–135.

Neumayer, E. (2003). *Weak versus strong sustainability: Exploring the limits of two opposing paradigms*. Elgar.

Newlands, G., Lutz, C., Tamò-Larrieux, A., Fosch-Villaronga, E., Scheitlin, G., & Harasgama, R. (2020). Innovation under pressure: Implications for data privacy during the Covid-19 pandemic. *Big Data & Society*, *7*(2), 1–14.

NHTSA. (2021). *Newly released estimates show traffic fatalities reached a 16-year high in 2021*. www.nhtsa.gov/press-releases/early-estimate-2021-traffic-fatalities.

Nightingale, P., & Coad, A. (2014). Muppets and gazelles: Political and methodological biases in entrepreneurship research. *Industrial and Corporate Change*, *23*(1), 113–143.

OECD. (2019). *Regulatory effectiveness in the era of digitalisation*. Retrieved July 7, 2022, from www.oecd.org/gov/regulatory-policy/Regulatory-effectiveness-in-the-era-of-digitalisation.pdf.

Pigou, A. C. (1920). *The economics of welfare*. Macmillan.

Pollman, E., & Barry, J. M. (2017). Regulatory entrepreneurship. *Southern California Law Review*, *90*(3), 383–448.

Posen, H. A. (2015). Ridesharing in the sharing economy: Should regulators impose Uber regulations on Uber. *Iowa Law Review*, *101*, 405.

Rizzo, M. J. (1999). Which kind of legal order? Logical coherence and praxeological coherence. *Journal des Économistes et des Études Humaines*, *9*(4), 497–510.

Sætra, H. S. (2022a). *AI for the sustainable development goals*. CRC Press.

Sætra, H. S. (2022b). The ethics of trading privacy for security: The multifaceted effects of privacy on liberty and security. *Technology in Society*, *68*, 101854.

Sætra, H. S., & Fosch-Villaronga, E. (2021a). Research in AI has implications for society: How do we respond? The balance of power between science, ethics, and politics. *Morals & Machines*, *1*(1), 62–75. https://doi.org/10.5771/2747-5182-2021-1-62

Sætra, H. S., & Fosch-Villaronga, E. (2021b). Healthcare digitalisation and the changing nature of work and society. *Healthcare, MDPI, 9*(8), 1–15, 1007.

Sætra, H. S., Nordahl-Hansen, A., Fosch-Villaronga, E., & Dahl, C. (2022, August 22–24). *Normativity assumptions in the design and application of social robots for autistic children.* 18th Scandinavian Conference on Health Informatics. www.researchgate.net/publication/362849566_Normativity_assumptions_in_the_design_and_application_of_social_robots_for_autistic_children.

Salvini, P., Paez-Granados, D., & Billard, A. (2021). On the safety of mobile robots serving in public spaces: Identifying gaps in EN ISO 13482: 2014 and calling for a new standard. *ACM Transactions on Human-Robot Interaction (THRI), 10*(3), 1–27.

Sandel, M. (2007). *The case against perfection: Ethics in the age of genetic engineering.* Belknap Press of Harvard University Press.

Sandersen, H. T., & Kvalvik, I. (2014). Sustainable governance of Norwegian aquaculture and the administrative reform: Dilemmas and challenges. *Coastal Management, 42*(5), 447–463.

Scassellati, B., Admoni, H., & Matarić, M. (2012). Robots for use in autism research. *Annual Review of Biomedical Engineering, 14*, 275–294.

Shademan, A., Decker, R. S., Opfermann, J. D., Leonard, S., Krieger, A., & Kim, P. C. (2016). Supervised autonomous robotic soft tissue surgery. *Science Translational Medicine, 8*(337), 337ra64–337ra64.

Sharkey, N., & Sharkey, A. (2020). The crying shame of robot nannies: An ethical appraisal. In W. Wallach & P. Asaro (Eds.), *Machine ethics and robot ethics* (pp. 155–184). Routledge.

Simshaw, D., Terry, N., Hauser, K., & Cummings, M. L. (2015). Regulating healthcare robots: Maximizing opportunities while minimizing risks. *Richmond Journal of Law & Technology, 22*, 1.

Spano, R. (2021). The rule of law as the lodestar of the European Convention on Human Rights: The Strasbourg Court and the independence of the judiciary. *European Law Journal, 2021*, 1–17.

Sucha, V., & Sienkiewicz, M. (Eds.). (2020). *Science for policy handbook.* Elsevier.

Surdilovic, D., Nguyen, T. M., & Radojicic, J. (2011, July). Human-like variable-impedance control for life-cycle testing. In *2011 RO-MAN* (pp. 150–155). IEEE.

Tucker, M. R., Olivier, J., Pagel, A., Bleuler, H., Bouri, M., Lambercy, O., . . . Gassert, R. (2015). Control strategies for active lower extremity prosthetics and orthotics: A review. *Journal of Neuroengineering and Rehabilitation, 12*(1), 1–30.

Weng, Y. H., Sugahara, Y., Hashimoto, K., & Takanishi, A. (2015). Intersection of "Tokku" special zone, robots, and the law: A case study on legal impacts to humanoid robots. *International Journal of Social Robotics, 7*(5), 841–857.

Whittaker, M. (2021). The steep cost of capture. *Interactions, 28*, 50–55.

Wischmeyer, T., & Rademacher, T. (Eds.). (2020). *Regulating artificial intelligence.* Springer.

Wold, T. L. (2022). Not a foregone conclusion: The early history of Facebook's political economy of social media. *Artifact Analysis, 1*(1), 1–17.

Wu, T. (2010). Is internet exceptionalism dead? In B. Szoka & A. Marcus (Eds.), *The next digital decade: Essays on the future of the internet.* TechFreedom.

Wynsberghe, A. van (2021). Sustainable AI: AI for sustainability and the sustainability of AI. *AI and Ethics, 1*(3), 213–218.

Yang, G. Z., Cambias, J., Cleary, K., Daimler, E., Drake, J., Dupont, P. E., Hata, N., Kazanzides, P., Martel, S., Patel, R. V., Santos, V. J., & Taylor, R. H. (2017). Medical robotics-regulatory, ethical, and legal considerations for increasing levels of autonomy. *Science Robotics, 2*(4), eaam8638.

Zomorodi, M. (2017). *Bored and brilliant: How time spent doing nothing changes everything.* Pan Macmillan.

Zuboff, S. (2019). *The age of surveillance capitalism: The fight for a human future at the new frontier of power.* Profile Books.

CHAPTER 11

Governing Toward Sustainable Development

From a Path-Dependent Transition to a Disruptive One

Lilja Mósesdóttir and Ivar Jonsson

CONTENTS

11.1 INTRODUCTION

In 2019, an independent expert group appointed by the United Nations (UN) found that the world is not on track for achieving most of the 169 targets that comprise the SDGs due to "rising inequalities, climate change, biodiversity loss and increasing amounts of waste" (UN, 2019, p. xx). All member states have made a commitment to fulfill Agenda 2030 with its 17 SDGs designed to achieve UN's vision of sustainable development involving economic growth, social progress, and promotion of environmental sustainability (see Chapter 1). The SDGs are not legally binding but provide the member states with global sustainability

DOI: 10.1201/9781003325086-11

goals whose fulfillment requires a transition involving "a profound and intentional departure from business as usual" (UN, 2019, p. 5). The UN experts responded to the lack of progress on the SDGs by urging member countries to devise transition pathways containing fundamental changes in current socio-environmental–economic systems. As the layers of change needed to attain the SDGs vary across countries and regions, different transition pathways will be created (UN, 2019, p. xx).

The UN experts' recommendations shed the light on the importance of political efforts in accelerating sustainability transition. At the same time, climate emergency has started to cast doubt on the ability of techno-solutionism (see Chapter 1) and the "reformed" capitalist system to secure sustainable development (see Chapter 17). Political efforts accelerating sustainability transition do not operate in vacuum but interact with technology, the national context, and pressures from the regional/global level. Political actors seek, for example, to promote technological changes to solve what they see as a problem, and they react when technology creates new opportunities or threats (see Carstensen & Schmidt, 2016; Acemoglu & Robinson, 2010). Moreover, sustainability transition can be facilitated by changes in a broad range of interrelated technologies that create, for example, technical solutions to the negative environmental effects of economic growth involving unsustainable resource use and carbon emissions. Both the health crisis created by COVID-19 and the War in Ukraine with its negative effects on food and energy prices have caused performance problems on the SDGs in many countries (see Zhao et al., 2022).

Different transition models have been suggested as a means to achieve sustainable development. These models involve green growth, circular economy, and greening of the economy in a fair and inclusive manner (just transition). According to the green growth theory, technological change, substitution, and structural change toward fewer resource-intensive services will allow us to decouple fast enough GDP growth (SDG 8) from resource use (SDG 12) and carbon emissions (SDG 13) to meet environmental targets. However, Hickel and Kallis (2020) were unable to find empirical support for this assertion. Surprisingly, their findings also apply to countries following the OECD's and the World Bank's recommendation to deploy technological innovation and government policy to achieve this decoupling. Hence, Hickel and Kallis claim that "green growth is likely to be a misguided objective, and that policymakers need to look toward alternative strategies" (2020, p. 469). Similarly, Geels (2014, p. 25) and Kivimaa and Kern (2016) argue that policymakers need to put greater emphasis on the destruction part in what Schumpeter (1942) calls "creative destruction" involving not only creation of green industries but also phasing out of fossil fuel industries.

A circular economy model is a model of production and consumption that ensures economic growth (SDG 8), while materials and products maintain their values, waste is avoided, and resources are kept within the economy (SDG 12). This is achieved through sharing, leasing, repairing, and recycling existing materials and products as long as possible (see Corvellec et al., 2021; Geisendorf & Pietrulla, 2018, p. 779). Critics maintain that the model is just another technical fix to the challenges of climate change that depoliticizes sustainable growth by emphasizing the role of consumers, markets, and corporations in the development of recycling economy (see Corvellec et al., 2021; see also Chapters 12,

15, and 17). According to Korhonen et al. (2018, p. 41), the limitation of the model is that cyclical production systems still consume resources and create wastes and emissions given the current technological state of the economy. Moreover, the extent to which the circular economy model is able to ensure social sustainability needs thorough investigation (see Mies & Gold, 2021).

Studies of environmental and climate change demonstrate how countries with high level of GDP per capita who are governed through social goals and relative egalitarian redistribution such as the Nordic countries are more likely to be successful in achieving sustainable development (see Lockwood, 2015; Kuzemko et al., 2016, p. 100; Koch, 2020, p. 122). This is also manifested by Sachs et al. (2022) who compared the overall performance of 163 countries on the 17 SDGs, giving equal weight to each goal. The comparison for 2022 places Finland, Denmark, Sweden, and Norway at the top of the list with scores ranging from around 86.5 to 82.3 followed closely by Austria and Germany. Sachs et al. (2022) found that Finland, Denmark, Sweden, and Norway have achieved the goals of no poverty (SDG 1) and affordable and green energy (SDG 7), while they have major problems attaining reduction in greenhouse emissions (SDG 13) and sustainable production and consumption (SDG 12). The good performance of the four Nordic countries on the SDGs has given support to what the global labor movement calls "just transition".

According to the International Labor Organization (ILO), "just transition means greening the economy in a way that is as fair and inclusive as possible to everyone concerned, creating decent work opportunities and leaving no one behind" (ILO, n.d.). In other words, economic growth (SDG 8) is compatible with high level of environmental protection (SDG 13) and social progress (SDG 5 and SDG 10) as a shift to a low-carbon energy system can create jobs that are not only environmentally friendly but also decent (see Sabato & Fronteddu, 2020, p. 13). The labor movement's understanding of just transition has been criticized for embracing economic growth and techno-solutionism (see Xinxin Wang, 2021; see also Chapters 17 and 18). Moreover, the just transition framework fails to acknowledge the need to transform power relations and structures underlying unsustainable growth regimes (Xinxin Wang, 2021). Hence, just transition needs to involve a radical critique of capitalism and green growth (see Normann & Tellmann, 2021, p. 424).

As the three transition models of green growth, circular economy, and just transition demonstrate, the SDGs are putting the developmental trajectory of many countries under strain. The SDGs involve partially contradictory goals that green technologies and political efforts have, so far, not managed to solve (see also Chapter 1). In addition, transitional changes facilitated by, for example, disruptive technologies and policies create not only winners when new structures are created but also powerful losers when old unsustainable ones are destroyed (Turnheim & Geels, 2012; see also Chapter 15). Hence, various actors are seeking to inform and/or control the terms of the debate about the need for continuity, adaptation, and transitional change to meet them. At the same time, the climate urgency has increased awareness that much more needs to be done to accelerate the sustainability transition. However, the four Nordic countries, Finland, Denmark, Sweden, and Norway, scoring highest on the SDGs are not characterized by transitional change but stability and path-dependent development involving just transition or inclusive economic growth (see

NOU, 2021a; Khan et al., 2021). This puzzle creates a need to gain a better understanding of the impulses behind stability and disruption in sustainability transition.

In the following, the literature on the political dimensions of sustainability transition is reviewed to unwrap the meaning of sustainability transition, identify the role political actors play in sustainability transition, and how their efforts are shaped by the national context. Based on our literature review, we develop a framework for analyzing the role of politics and path dependency in shaping sustainability outcomes. Our contribution to scholarship on the political dimensions of sustainability transition is to explain why top performance on the SDGs can be achieved without transitional change. Moreover, our framework goes beyond policy outputs by considering sustainability outcomes and their implications for transition pathways, that is, whether they are path-dependent, lock-in, or disruptive. Thereafter, we illustrate our framework by a case study. We have chosen Norway as our case as it has attained a high overall score on the SDGs and played an important role in the global partnership for sustainable development (achieved SDG 17). At the same time, Norway has a petroleum-based economy and is characterized by strong institutions (achieved this part in SDG 16) that are sign of stability rather than disruption. Finally, we analyze the national context prevailing in Norway in view of our framework to improve our understanding of why a certain transitional pathway receives support, and why it is associated with stability rather than disruption.

The main assumption of our framework is that ideas underlying the global SDGs need to be compatible with underlying political values and supported by material conditions (technology and wealth) for a country to attain a high score on the goals without radical change. The ideas on which the SDGs are grounded are "destructive creation", inclusive development, and fair distribution of opportunities and income. The last two ideas are in line with Social Democratic values of solidarity and universalism that have to vary degrees become entrenched in institutional structures of countries with top score on the SDGs. Our case study highlights how political actors in Norway have been assisted by the Nordic/Norwegian model of tripartite collaboration and the welfare state (the Nordic/Norwegian model) and, especially its entrenched values in their efforts to secure a path-dependent development in a petroleum economy. These values are *solidarity* with oil workers and future generations, *universalism* involving the need to maximize employment and welfare (state income) and *decommodification* of future generations by saving for their pension in the sovereign wealth fund (oil fund). However, there exists a real risk of lock-in situation at the current high level of the SDGs due to the enormous wealth creation of the petroleum industry and lack of technological advances allowing "decoupling" of environmental harm from economic activity.

11.2 STATE OF THE ART ON SUSTAINABILITY TRANSITION AND ON POLITICAL DIMENSIONS

11.2.1 Sustainability Transition

An important assertion of studies in the field of sustainability transitions is that incremental improvements and technological fixes are not enough to achieve sustainability transition (Fuenfschilling & Binz, 2018; Köhler et al., 2019). Instead, transition requires considerable changes along institutional, technological, organizational, political, and socio-cultural dimensions that make up the so-called socio-technical system (Markard et al., 2012; Duygan et al., 2019). The drivers of such radical regime change can, for example, be a large and unexpected system shock (e.g., drought, flood, financial market crash) and changes incubated by transition actors that accumulate into a transition of the current socio-technical system (see Werbeloff et al., 2016, p. 120). The main role of political actors and public policy in sustainability transition is to influence the transition's direction and speed (see Edmondson et al., 2019). However, political actors will disagree about the directions of transitions, appropriate ways to govern such processes, and the extent to which transitions create potential winners and losers (Köhler et al., 2019, p. 6).

Few studies focus on the implications of the SDGs on sustainability transitions, although they are a set of stretch goals defining UN's vision of sustainable development in terms of outcomes. Hence, the commitment of the UN member countries made in 2015 to achieve the 196 targets underlying the 17 SDGs by 2030 settled disputes over how sustainability transitions can be accomplished (mission), while the choice of transition pathways to sustainable development (strategy) was left to the national level to decide on (cf. strategic management). We know from studies of sustainability transition that political efforts to govern or facilitate transitional change along certain pathways are contingent on the national context and influenced by socio-technical change (see Fuenfschilling & Binz, 2018; Edmondson et al., 2018). As the SDGs are global goals, the extent to which political and institutional settings at the national level support their attainment varies across countries. Hence, political pressures arising from performance problems on the SDGs and potential distributional effects of the transition differ across countries.

11.2.2 Politics

As pointed out by Köhler et al. (2019), studies of the political dimensions of sustainability transition have sought to identify the (potential) effects of sustainability transition on different actors (winners and losers) and their strategies to protect their interests (see, e.g., Kalt, 2022; Geels, 2014; Turnheim & Geels, 2012). The focus of these studies has also been on how different groups seek to use their power to shape discourses and build alliances (based on shared beliefs, discourse, or common interest) to influence policy processes, and thereby policy outputs (see, e.g., Edmondson et al., 2019; Kuzemko et al., 2016; Kern, 2011). As transition involves fundamental changes in the distribution of power, there will be multiple political actors with multiple interests and strategies seeking to influence policy outputs (see Köhler et al., 2019; see also Chapter 15).

Geels (2014) identified various ways in which incumbent actors such as businesses in coal, gas, and nuclear production in the United Kingdom used power and politics to actively resist fundamental transitions to new low-carbon systems. According to Geels (2014), incumbent actors resist fundamental system change, using instrumental, discursive, material, and institutional forms of power. Instrumental forms of power are exercised when regime actors use resources in interactions with other actors as a means to achieve their goals and interests. Regime actors can also apply discursive forms of power to resist change by shaping not only what issues are being discussed but also how they are discussed. Material strategies are adapted by the regime actors when technical capabilities and financial resources are used to improve the technical dimension of socio-technical regimes. Examples are technological solutions like Carbon Capture and Storage (CCS) to reduce emission problems created by fossil fuel industry. Institutional forms of power are embedded in political cultures, ideologies, and governance structures. Institutional power is wielded when regime actors are able to use the institutional context or norms and decision-making procedures to shape policymaking in their favor or away from alternative transition (Geels, 2014).

Kalt (2022) has, for example, identified four possible transition strategies that different trade unions can adopt. The choice of strategy is shaped by the extent to which the unions regard the transition as an opportunity to renew their power or fear a loss of power. First, oppositional strategies are practiced by actors when seeking to protect jobs in fossil fuel industries through subsidies and relying on technological solutions. Second, the reactive strategies are adopted when actors accept slow phase-out of fossil fuel. Third, actors apply affirmative strategies when supporting changes within the existing institutional framework. Fourth, the transformative transition strategies are pursued when actors reject green capitalism by demanding radical changes in power relations and institutional and economic structures. His study of union's strategies in Germany and South Africa showed that social justice-oriented unions are more likely to adopt transformative transition strategies. Moreover, unions are more likely to be on the offensive when neoliberal approach to green transitions is adopted as it is believed to have negative employment effects (Kalt, 2022).

11.2.3 Path Dependency

The interaction between politics and the institutional context is also the focus of research on sustainability transition policies (see Kuzemko et al., 2016; Kivimaa & Kern, 2016; Kern, 2011). Kern (2011) studied, for example, the relationship between ideas, institutions, and interests and its influence on policy initiatives to promote system innovations in the United Kingdom and the Netherlands. In line with Schmidt (2008), Kern (2011, p. 1120) maintains that we need to pay attention to the processes through which actors create new policy ideas, and to the extent to which these ideas are shaped by existing institutional factors such as formal rules (e.g., laws, regulations, and standards) and informal rules (e.g., norms, habits, and customs). Ideas are the substantive contents of discourse that is to some extent constrained by existing institutions through their influence on what can be said meaningfully (Schmidt, 2008; Kern, 2011). According to Kern (2011, p. 1129), "radical policy change is expected to occur only when a new discourse transforms existing interests and successfully challenges existing institutional commitments". Moreover, policies are path-dependent when a new discourse transforms existing interests, while a mutually supportive relationship exists between a new discourse and existing institutions. Finally, little or no policy change takes place when the new discourse, existing interests, and institutions are mutually supportive.

Kuzemko et al. (2016) give a detailed account of why it is difficult to change existing institutions that they attribute to their path-dependent qualities. These qualities are, for example, sets of ideas that have become embedded in institutions and are used to mediate between interest groups that in turn influence which voices are "heard" in political debates and which are not. Hence, embedded framework of ideas shapes choices of policy objectives and instruments used to achieve them (Kuzemko et al., 2016, p. 99). Moreover, the path-dependent qualities of institutions diverge across countries and regions resulting in different transition pathways. In countries with strong market-liberal ideas as, for example, the United Kingdom, the range of acceptable policy options in response to climate change has been narrowed down to market-based instruments, while countries characterized by core values of Social democracy and greater collective and/or coordinative capacities allow for a more active role for government actors in setting and meeting sustainability goals (see Kuzemko et al., 2016, pp. 102–103). Chapter 13 also demonstrates how the neoliberal state in the United Kingdom has implemented policies focusing on "nudges" that seek through framing rather than economic incentives to steer people's behavior in green direction as a means to maintain its policy of austerity.

Kivimaa and Kern (2016, p. 205) argue that sustainability transition requires innovation policy mixes aiming at "creative destruction" or facilitating both creation of the new and destabilization of the old. However, studies have found that most policy mixes have developed through *policy layering* including new policy goals and instruments added on top of existing ones or *policy drift* including changed policy goals without changing the instruments (Howlett & Rayner, 2013). Designing innovation policies aiming at phasing out/destroying wasteful or polluting industries has been challenging as they contradict the ideology of traditional innovation policies seeking to contribute to economic growth (Kivimaa & Kern, 2016, p. 214). Disruptive policies are also actively opposed by those who believe they will lose out or are uncertain about its economic advantages (see Turnheim & Geels, 2012; Geels, 2014).

11.2.4 Governing Sustainability Transition – A Framework

Based on the central issues in the literature reviewed earlier, we develop a framework for analyzing the role of politics and path dependency in shaping sustainability outcomes. The aim is to explain why sustainability outcomes associated with transitions coexist with stability and incremental policy changes, and what the implications are for transition pathways. Its basic assumption is that the SDGs are based on ideas offering political blueprints for how to achieve sustainable development. These ideas need to be compatible with dominant political values embedded in the institutional context and supported by material conditions (technology and wealth) for a country to attain a high score on the goals (see Kern, 2011; Geels, 2014). In other words, the more compatible these factors are, the more successful a country is in attaining the SDGs with only incremental changes.

The SDGs not only are global goals defining sustainability outcomes but also contain ideas that favor certain policy objectives and instruments. These embedded ideas are inclusive development (e.g., SDGs 1, 2, 7, 8, 11, 16, 17), fair distribution of opportunities and income (e.g., SDGs 3, 4, 5, 10, 16), and "creative destruction" involving destruction of wasteful and polluting industries and consumption patterns at the same time as new green ones are created (e.g., SDGs 6, 9, 12, 13, 14, 15). Interestingly, inclusive development and fair distribution are in line with the Social Democratic idea that all who are able and willing to work should have access to employment and should receive a fair share of the economic benefits produced (von Platz, 2020, p. 22). Hence, countries where Social Democratic values have become embedded in the political cultures and governance structures have a comparative advantage and will start off with a better performance on the SDGs than countries dominated by other political ideologies (see Mósesdóttir & Jonsson, 2020).

When the SDGs and the national context are based on compatible ideas, the political elite is able to deploy their instrumental, discursive, material, and institutional forms of power to implement their oppositional strategy to transitional change involving emphasis on technological solutions to environmental challenges (see Geels, 2014; Kalt, 2021). The sustainability pathway is then characterized by a changing discourse arising from the SDGs' emphasis on not only the economic and social dimensions of the SDGs but also its environmental dimension. The changing discourse puts a pressure on the domestic interest constellations, while the institutional context remains intact due to its path-dependent qualities (see Kern, 2011; Kuzemko et al., 2016). This external pressure on the domestic interests may push the dominant political actors to adopt reactive strategy or to accept slow phase-out of not only the consumption but also the production of fossil fuel (see Kalt, 2022). Policy outputs are then layered (new policy objectives and instruments) or drifted (changed policy objectives but old instruments) and added on top of existing ones facilitating path-dependent policy outcomes (see Kivimaa & Kern, 2016).

We assume that most transition pathways to sustainable development as defined by the UN are at first path-dependent or shaped by stable institutions such that transition actors only make small adjustments in strategies and instruments in response to changing global discourse on sustainability (see Werbeloff et al., 2016; Hanger-Kopp et al., 2022; Kern, 2011). These path-dependent transition pathways differ across countries as they are contingent on the national context. Moreover, they are less contentious and contradictory in countries with high scores on the SDGs than in countries with lower scores. However, political tensions will also grow in high-scoring countries when it becomes apparent that the SDGs cannot be fulfilled without radical changes.

Path-dependent transition pathways will eventually create a lock-in situation manifested in a persistent lack of progress on the SDGs and no prospect of an improvement through, for example, technological advances in the country's performance (see Hanger-Kopp et al., 2022). If the balance of power and structures remain inflexible at the same time as no new technological solutions appear, a country moves from its path-dependent pathway to a lock-in pathway characterized by minor changes that do not meet external and internal pressures for change (see Hanger-Kopp et al., 2022). The lock-in period is marked by a real critical juncture in which actors struggle for different alternatives in societal development after the long period of path dependency with relatively stable institutions and power relations. During the critical juncture, the actors attempt, despite their conflicting interests and visions, to build coalitions around a shared vision of how future society is to be constituted (Mósesdóttir & Jonsson, 2020, p. 113). External pressures arising from climate emergency and the domestic lock-in crisis may tilt the political balance of power toward more radical alternatives, paving the way for disruptive changes along institutional, technological, organizational, political, and socio-cultural dimensions (see Markard et al., 2012; Duygan et al., 2019). The move toward disruptive transition pathway will be evolutionary, if radical changes do not impact simultaneously all three dimensions of sustainable development: The economic, social, and environmental.

Later, we illustrate how our framework can be applied to a case example. The aim of our brief narrative description, which is based on secondary sources, is to highlight how political efforts, material conditions (e.g., oil revenues, technology), and institutional conditions prevailing in Norway underpin its high score on the SDGs. We then analyze and discuss the case using key assumptions of our framework to identify the sustainability pathway of Norway.

11.3 THE CASE OF NORWAY

11.3.1 The Global Ambition

Norway is an interesting case among the Nordic countries as it has been trying for the last three decades to find a solution to its paradox of being dependent on oil revenues at the same time as it has climate leadership ambitions (Lahn, 2019; Tellmann, 2012; Tjernshaugen, 2011). According to Asdal (2014), political actors tried to solve this paradox during the 1990s and 2000s by organizing climate and petroleum policymaking as two separate policy fields. Hence, political efforts focused on pursuing ambitions of climate leadership by being the first country to adopt a target for CO_2 emissions reductions, which was subsequently expected to be achieved through carbon taxes and then emissions trading (Lahn, 2019; Tellmann, 2012; Tjernshaugen, 2011).

The broad consensus in Norway around its leading role in framing climate change as a demand problem rather than a supply problem was fundamentally shaped by concerns about how climate policy might impact oil and gas production. This decoupling of climate and petroleum in policy discussions was strengthened by international commitment to climate change (Kyoto Protocol in 1997) that focused on carbon trading (Lahn, 2019). According to Bang and Lahn (2020, p. 1001), this emphasis on the demand side of fossil fuels allowed Norway to continue pursuing its contradictory interests as a major oil and gas exporter and its ambition of acting as an international climate leader.

In Norway, the effects of market solutions such as carbon taxes and carbon trading on the emission of CO_2 turned out to be limited. Hence, the focus of political efforts to reduce carbon emissions gradually shifted after 2000 to measures supporting the development of technological solutions such as Carbon Capture and Storage or CCS (Lahn, 2019; Tellmann, 2012; Tjernshaugen, 2011). While public investment in CCS projects has enjoyed widespread support, a full-scale project for capture, transport, and storage of CO_2 in Norway has not materialized due to the high costs involved (Lahn, 2019, p. 14; Tvinnereim & Ivarsflaten, 2016, p. 366).

11.3.2 A Mismatch Between the Global and National Discourse

When ambitious global climate goals were agreed on (cf. Copenhagen 2009 and Paris 2016), it became increasingly clear that measures to reduce consumption of fossil fuels were no longer sufficient. As a result, the political controversy around the future of the Norwegian oil and gas industry intensified. According to Lahn (2019, p. 6), this political controversy shifted the political discourse toward framing that connected the two policy areas together, that is, petroleum policy and the climate policy. Oil was framed as an object of carbon risk since petroleum extraction contributes to global greenhouse gas emissions. However, smaller political parties go further in linking the two policy areas together than the two largest political parties, the Labor Party (LP) and the Conservative Party (CP), as they have advocated restrictions on new oil and gas extraction and in due course managed decline of petroleum production (see Bang & Lahn, 2020; Lahn, 2019).

During the last two decades, governments led by either the LP or the CP have had to form coalitions with the smaller parties to gain parliamentary majority. When in government, the largest party has secured the basic features of Norwegian petroleum policy, while the smaller parties have managed to impose restrictions on oil and gas development in some specific cases (Lahn, 2019, pp. 20–21). According to Bang and Lahn (2020, p. 998), the political majority has justified its support for a continuation of the main aspects of Norwegian petroleum policy with reference to "oil as welfare" and the relatively low level of production-related emission of Norwegian oil. Oil creates welfare in Norway as its production provides employment opportunities for current generations, and income to sustain the welfare state for future generations and during crisis (cf. COVID-19). Since the mid-1990s, all of the revenue that comes to the state from oil and gas production has been paid into the sovereign wealth fund (oil fund). The fund is meant to secure the welfare of future generations through its financing of their pension. Only returns from the fund's investments are paid out annually and used to finance infrastructure investments, and to maintain the welfare state during crisis (see Arvin, 2021).

11.3.3 Pressures From Below

The Nordic tradition of extensive tripartite cooperation on employment and social issues has enabled broad consensus-building in different policy areas (see NOU, 2021a; Dølvik & Steen, 2019; Lidskog & Elander, 2012). The Norwegian Confederation of Trade Unions (LO) is the largest umbrella organization for trade unions in Norway, and it has been the most powerful in terms of political influence due to its close links to the LP (Normann & Tellmann, 2021, p. 425). Conventionally, LO has been a supporter of petroleum industry as it organizes many petroleum workers who fear job losses (Houeland et al., 2020). Normann and Tellmann (2021, p. 429) claim that LO has during the last 10 years increasingly embraced the concept of just transition in an attempt to reconcile the interests of both petroleum workers and its affiliated unions with members in the service and public sectors who have become increasingly concerned with climate change. LO achieved this reconciliation by defining the concept of just transition in terms of reduction of emissions from production within the oil and gas industry through green technologies and the creation of new green jobs (Houeland et al., 2020; Normann & Tellmann, 2021).

In 2017, LO showed a sign of willingness to accept transformation away from fossil fuel industry after intensified pressure from affiliated unions in the service and public sectors. This happened when LO adjusted its policy of a clear support toward impact assessment (seen as a first step toward exploration) to a compromise that included permanent protection of some environmentally sensitive areas (cf. Lofoten, Vesterålen og Senja) from new oil and gas extraction. In 2018, the minority government led by the CP announced that it also went for permanent protection of these areas and the LP followed suit in 2019 (Normann & Tellmann, 2021; Lahn, 2019). In the Spring of 2022, the controversy among the affiliated unions of LO over oil and gas production surfaced again at the LO's Congress. The policy compromise agreed on by the Congress stated that extraction of new gas and oil fields should only take place if economically feasible and climate-sustainable (Svenning et al., 2022).

11.3.4 The Piecemeal Approach to the SDGs

In 2020, the Office of the Auditor General assessed the Norwegian government's management and review of the national follow-up of the sustainable development goals from 2016. During this period, the governments were headed by the CP in coalition with smaller right-center parties. The Office concluded that political efforts to attain SDGs have been piecemeal and lacked coordination across various governmental departments. This has led to, on the one hand, low awareness of the national challenges posed by the various SDGs and, on the other hand, insufficient efforts to improve progress and attainment of SDGs (see Fosser, 2020, p. 6).

Norway is not alone among the Nordic Social Democratic countries in adopting a piecemeal approach to greening the economy. In their study of public discourses in policy documents on greening the economy in Denmark, Norway, and Sweden, Khan et al. (2021) found that all three countries lack comprehensive national policy strategies on the green economy. Instead of radical approaches, the public discourses in the three countries were based on the optimistic view that economic growth and environmental improvements can occur simultaneously through the use of technological solutions and innovation enabling recycling, sharing, and re-use of resources. Khan et al. (2021) also found an emphasis on state-led transition to sustainable development where technology is a key factor in greening the economy. The main role of the state is to regulate the economy and promote technical change and innovation in green technologies in close cooperation with the social partners and relevant stakeholders, while the welfare state is important in securing social welfare and human well-being (Khan et al., 2021).

11.3.5 The Case in View of Our Framework

Our case study highlights how Norway has safeguarded its economic prosperity based on oil and gas revenues by framing and organizing petroleum policy and climate policy as two separate policy areas. The petroleum sector accounted in 2019 for about 35% of Norway's exports and the petroleum-related employment constituted about 5.8% of total employment (NOU, 2021b, p. 56; Statistics Norway, 2021). Moreover, the petroleum export has made Norway's score on the target *CO_2 emissions embodied in fossil fuel exports (kg/capita)* by far the worst among the Nordic countries and even higher than that of, for example, the United States (see Sachs et al., 2022). The disconnection between petroleum policy and the climate policy has been justified with direct and indirect references to core values of Social Democracy such as *solidarity* with workers in the oil industry and *universalism* or the need to maximize employment and welfare (see Normann & Tellmann, 2021; Houeland et al., 2020; Lahn, 2019; Cox, 2004; Esping-Andersen, 1989). Moreover, the sovereign wealth fund (oil fund) is presented as a manifestation of *solidarity* with future generations whose *decommodification* (freed from the market) is secured by paying all state revenue from oil and gas into the fund to finance their pension (see Bang & Lahn, 2020; Arvin, 2021). This pension solidarity with future generations makes it difficult to criticize the Norwegian government, for compromising the ability of future generations to meet their own needs with its oil and gas production.

Solidarity, universalism, and decommodification are core Social Democratic values that have become embedded in the Nordic/Norwegian model including tripartite regulation of the labor market and policymaking, on the one hand, and the Social Democratic welfare state model providing extensive welfare services and generous welfare benefits, on the other hand (see NOU, 2021a; Mósesdóttir & Ellingsæter, 2019; Cox, 2004; Esping-Andersen, 1989). Hence, the political majority in Norway has wielded discursive and institutional power through references to core Social Democratic values to gain broad support for its implementation of oppositional strategy involving investment in green technologies to prevent cuts in oil production, and for its reactive strategy supporting reduction in oil consumption through economic incentives (see Geels, 2014). At the same time, the country's wealth and the Nordic/Norwegian model underlay Norway's full score on five SDGs within the social and economic dimensions, that is, on zero poverty, gender equality, affordable and clean energy, reduced inequalities, and partnerships for the goals (SDGs 1, 5, 7, 10, and 17). None of the other top-scoring Nordic countries has managed this (see Sachs et.al., 2022; pp. 342–343).

Ambitious global climate goals since 2009 have empowered less powerful actors in their efforts to contain production of oil and gas in Norway. The political majority and the labor movement (LO) have, therefore, come under increased pressure to acknowledge that these two policy areas are interconnected. These internal political pressures and climate urgency have made the political elite more willing to restrain new oil and gas extraction, and to acknowledge that extraction industry may gradually wind down due to falling global demand (see Lahn, 2019; Bang & Lahn, 2020). However, the volume of petroleum exports from Norway continues to be steered by market conditions rather than sustainability concerns.

Governments in Norway, Denmark, and Sweden have been able to integrate the SDGs into prevailing governance structures such that conflicts over policy efforts to attain the goals have been avoided (policy without politics). The compatibility between, on the one hand, the SDGs' ideas of inclusive development and fair distribution and, on the other hand, core Social Democratic values of universalism and solidarity has enabled this integration and made the current transition pathways path-dependent or shaped by prevailing power relations, embedded norms, and institutional arrangement. A similar compatibility underlay the top performance of Denmark, Finland, Norway, and Sweden on the SDGs. However, the governmental commitment and efforts to attain the SDGs have been greater in Denmark, Finland, and Sweden than in Norway (Sachs et al., 2022, Figure 3.6). Hence, Norway's climate leadership ambitions appear to be greater outside of its national border than inside (cf. achievement of SDG 17).

In Norway, the separation of climate policy and petroleum policy, and relatively good performance on the SDGs have meant that policy changes have been incremental or in the form of policy layering (cf. green technologies, carbon taxes, and carbon trading). However, the SDGs facilitating "creative destruction" (e.g., SDGs 6, 9, 12, 13, 14, 15) poses a dilemma for the Norwegian/Nordic model of industrial relations and the welfare state as it was originally not designed to solve the tensions between economic growth and environmental protection. In Norway, employers have, for example, supported the model of extensive tripartite cooperation as it has contributed to industrial growth and moderation of wages, while employees have enjoyed secure working conditions and benefits from growth (Ravn & Øyum, 2018, p. 2). Moreover, economic growth and stable stream of returns from the oil fund's investments are essential for the financial sustainability of the extensive Social Democratic welfare state.

In Norway, there exists a real risk of lock-in situation at the current high level of the SDGs due to the enormous wealth creation of the petroleum industry and lack of technological advances allowing "decoupling" of environmental harm from economic activity. Internal actors do not appear to be able to solve this lock-in problem by phasing out the fossil fuel industry in Norway, although policy framework for retraining and compensation of workers is in place. This became apparent after the parliamentary election in 2021, when smaller parties with ambitious disruptive approaches to climate change fared less well than expected. In the political platform of the current government, climate change is stated as top priority at the same time as the country's petroleum industry "must be developed and not terminated" (Office of the Prime Minister, 2021).

11.4 CONCLUSION

The main assumption of our framework of path dependency and real critical junctures is that ideas underlying the SDGs need to be in line with political values embedded in the institutional context and supported by material conditions for a country to achieve top performance on the goals without a disruptive change. We argue that the SDGs contain ideas of "destructive creation", inclusive development, and fair distribution of opportunities and income. The last two ideas are compatible with Social Democratic values of solidarity and universalism that are to vary degrees entrenched in the institutional context of countries with high scores on the SDGs. Our case study highlights how political actors in Norway have managed to separate fossil fuel production from climate policy with direct and indirect references to institutionally embedded Social Democratic values of solidarity, universalism, and decommodification. This separation and the Nordic/Norwegian model have secured political support for path-dependent pathway to sustainability.

Most transition pathways to sustainable development are initially path-dependent or shaped by institutions such that transition actors only make small adjustments in response to changing global discourse on sustainability. The small adjustments implemented in Norway involve measures to cut petroleum consumption and not its production, and then acknowledging the need to contain new oil and gas extraction. Path-dependent pathways will eventually create a lock-in situation manifested in growing political conflicts arising from persistent lack of progress on the SDGs and no prospect of a technological solution to unsustainable resource extraction, production, and consumption.

Inflexible power relations and socio-technical systems will move the country to a lock-in pathway marked by a real critical juncture or struggle among actors for different societal alternatives. External pressures arising from climate emergency and the domestic lock-in crisis may tilt the political balance of power toward more radical alternatives, paving the way for a move toward a disruptive pathway. The enormous wealth creation of the petroleum industry and lack of technological advances to solve emission problems creates a real risk of lock-in situation in Norway at the current high level of the SDGs. For Norway to continue to realize its ambition to be a climate leader, political support for large-scale "creative destruction" is needed. This political support is lacking today, although the country has wealth (oil fund) to finance its pursuit of a sustainable societal development that will benefit both present and future generations.

11.5 REFERENCES

Acemoglu, D., & Robinson, J. A. (2010). Political losers as a barrier to economic development. *The American Economic Review, 90*(2), 126–130. https://doi.org/10.1257/aer.90.2.126

Arvin, J. (2021, January 15). Norway wants to lead on climate change. But first it must face its legacy of oil and gas. *Vox*. www.vox.com/22227063/norway-oil-gas-climate-change

Asdal, K. (2014). From climate issue to oil issue: Offices of public administration, versions of economics, and the ordinary technologies of politics. *Environment and Planning A: Economy and Space, 46*(9), 2110–2124. https://doi.org/10.1068/a140048p

Bang, G., & Lahn, B. (2020). From oil as welfare to oil as risk? Norwegian petroleum resource governance and climate policy. *Climate Policy, 20*(8), 997–1009. https://doi.org/10.1080/14693062.2019.1692774

Carstensen, M., & Schmidt, V. A. (2016). Power through, over and in ideas: Conceptualizing ideational power in discursive institutionalism. *Journal of European Public Policy, 23*, 318–337. https://doi.org/10.1080/13501763.2015.1115534

Corvellec, H., Stowell, A. F., & Johansson, N. (2021). Critiques of the circular economy. *Journal of Industrial Ecology, 26*(2), 421–432. https://doi.org/10.1111/jiec.13187

Cox, R. (2004). The path-dependency of an idea: Why Scandinavian welfare states remain distinct. *Social Policy and Administration, 38*(2), 204–219. https://doi.org/10.1111/j.1467-9515.2004.00386.x

Dølvik, J. E., & Steen, J. R. (2019). *Drivkrefter og utfordringer for den norske modellen. Framtidens arbeidsliv*. Fafo-notat 2019:15. Fafo. www.fafo.no/images/pub/2019/10305.pdf

Duygan, M., Stauffacher, M., & Meylan, G. (2019). A heuristic for conceptualizing and uncovering the determinants of agency in socio-technical transitions. *Innovation and Societal Transitions, 33*, 13–29. https://doi.org/10.1016/j.eist.2019.02.002

Edmondson, D. L., Kerna, F., & Roggea, K. S. (2019). The co-evolution of policy mixes and socio-technical systems: Towards a conceptual framework of policy mix feedback in sustainability transitions. *Research Policy, 48*(10), 1–14. https://doi.org/10.1016/j.respol.2018.03.010

Esping-Andersen, G. (1989). *The three worlds of welfare capitalism*. Polity Press.

Fosser, P-K. (2020). *The Office of the Auditor General's investigation of the management and review of the national follow-up of the sustainable development goals*. The Office of the Auditor General. www.riksrevisjonen.no/globalassets/reports/en-2020-2021/document-3-3-the-national-follow-up-of-the-sustainable-development-goals.pdf

Fuenfschilling, L., & Binz, C. (2018). Global socio-technical regimes. *Research Policy, 47*(4), 735–749. https://doi.org/10.1016/j.respol.2018.02.003

Geels, F. W. (2014). Regime resistance against low-carbon transitions: Introducing politics and power into the multi-level perspective. *Theory, Culture & Society*, *31*(5), 21–40. https://doi.org/10.1177/0263276414531627

Geisendorf, S., & Pietrulla, F. (2018). The circular economy and circular economic concepts–a literature analysis and redefinition. *Thunderbird International Business Review*, *60*(5), 771–782. https://doi.org/10.1002/tie.21924

Hanger-Kopp, S., Thaler, T., Seebauer, S., Schinko, T., & Clar, C. (2022). Defining and operationalizing path dependency for the development and monitoring of adaptation pathways. *Global Environmental Change*, 72, 1–11. https://doi.org/10.1016/j.gloenvcha.2021.102425

Hickel, J., & Kallis, G. (2020). Is green growth possible? *New Political Economy*, *25*(4), 469–486. https://doi.org/10.1080/13563467.2019.1598964

Houeland, C., Jordhus-Lier, D., & Angell, F. H. (2020, January). Solidarity tested: The case of the Norwegian confederation of trade unions (LO-Norway) and its contradictory climate change policies. *Area*, *53*, 413–421. https://doi.org/10.1111/area.12608

Howlett, M., & Rayner, J. (2013). Patching vs packaging in policy formulation: Assessing policy portfolio design. *Politics and Governance*, *1*(2), 170–182. https://doi.org/10.17645/pag.v1i2.95

ILO. (n.d.). *Frequently asked questions on just transition*. Nordic Council of Minister. www.ilo.org/global/topics/green-jobs/WCMS_824102/lang-en/index.htm

Kalt, T. (2021). Jobs vs. Climate justice? Contentious narratives of labor and climate movements in the coal transition in Germany. *Environmental Politics*, *30*(7), 1135–1154. https://doi.org/10.1080/09644016.2021.1892979

Kalt, T. (2022, January 19). Agents of transition or defenders of the status quo? Trade union strategies in green transitions. *Journal of Industrial Relations*, 1–23. https://doi.org/10.1177/00221856211051794

Kern, F. (2011). Ideas, institutions, and interests: Explaining policy divergence in fostering 'system innovations' towards sustainability. *Environment and Planning C: Government and Policy*, *29*(6), 1116–1134. https://doi.org/10.1068/c1142

Khan, J., Johansson, B., & Hildingsson, R. (2021). Strategies for greening the economy in three Nordic countries. *Environmental Policy and Governance*, *31*(6), 592–604. https://doi.org/10.1002/eet.1967

Kivimaa, P., & Kern, F. (2016). Creative destruction or mere niche support? Innovation policy mixes for sustainability transitions. *Research Policy*, *45*(1), 205–217. https://doi.org/10.1016/j.respol.2015.09.008

Koch, M. (2020). The state in the transformation to a sustainable postgrowth economy. *Environmental Politics*, *29*(1), 115–133. https://doi.org/10.1080/09644016.2019.1684738

Köhler, J., Geels, F. W., Kern, F. et al. (2019). An agenda for sustainability transitions research: State of the art and future directions. *Environmental Innovation and Societal Transitions*, *31*(2019), 1–32. https://doi.org/10.1016/j.eist.2019.01.004

Korhonen, J., Honkasalo, A., & Seppälä, J. (2018, January). Circular economy: The concept and its limitations. *Ecological Economics*, *143*, 37–46. https://doi.org/10.1016/j.ecolecon.2017.06.041

Kuzemko, C., Lockwood, M., Mitchell, C., & Hoggett, R. (2016, February). Governing for sustainable energy system change: Politics, contexts and contingency. *Energy Research & Social Science*, *12*, 96–105. http://dx.doi.org/10.1016/j.erss.2015.12.022

Lahn, B. (2019). *Norwegian petroleum policy in a changing climate*. (CICERO report 2019:10). CICERO Center for International Climate Research. https://pub.cicero.oslo.no/cicero-xmlui/handle/11250/2607906

Lidskog, R., & Elander, I. (2012). Ecological modernization in practice? The case of sustainable development in Sweden. *Journal of Environmental Policy & Planning*, *14*(4), 411–427. https://doi.org/10.1080/1523908X.2012.737234

Lockwood, M. (2015). The political dynamics of green transformations: The roles of policy feedback and institutional context. In I. Scoones, M. Lech, & P. P. Newell (Eds.), *Pathways to sustainability. The politics of green transformation* (pp. 86–101). Routledge.

Markard, J., Raven, R., & Truffer, B. (2012). Sustainability transitions: An emerging field of research and its prospects. *Research Policy, 41*(6), 955–967. https://doi.org/10.1016/j.respol.2012.02.013

Mies, A., & Gold, S. (2021, October). Mapping the social dimension of the circular economy. *Journal of Cleaner Production, 321*, 1–17. https://doi.org/10.1016/j.jclepro.2021.128960

Mósesdóttir, L., & Ellingsæter, A. L. (2019). Ideational struggles over women's part-time work in Norway: Destabilizing the gender contract. *Economic and Industrial Democracy, 40*(4), 1018–1038. https://doi.org/10.1177/0143831X16681483

Mósesdóttir, L., & Jonsson, I. (2020). Theorizing transformative innovation: The role of agency in real critical junctures. In I. Bernhard, U. Gråsjö, & C. Karlsson (Eds.), *Diversity, innovation and clusters: Spatial perspectives* (pp. 104–124). Edward Elgar Publishing. https://doi.org/10.4337/9781789902587.00010

Normann, H. E., & Tellmann, S. M. (2021, September). Trade unions' interpretation of a just transition in a fossil fuel economy. *Environmental Innovation and Societal Transitions*, 40, 421–434. https://doi.org/10.1016/j.eist.2021.09.007

NOU. (2021a). *Den norske modellen og fremtidens arbeidsliv, Utredning om tilknytningsformer og virksomhetsorganisering.* NOU 2021: 9. www.regjeringen.no/contentassets/7b8fb44f93a44029 81ed7f279b345dbe/no/pdfs/nou202120210009000dddpdfs.pdf

NOU. (2021b). *Norge mot 2025.* NOU 2021: 4. www.regjeringen.no/contentassets/612755ca2628423 29ae0a7968e66351f/no/pdfs/nou202120210004000dddpdfs.pdf

Office of the Prime Minister. (2021, October 14). *Hurdalsplattformen 2021–2025.* www.regjeringen.no/contentassets/cb0adb6c6fee428caa81bd5b339501b0/no/pdfs/hurdalsplattformen.pdf

Ravn, J. E., & Øyum, L. (2018). Towards 'multi-collar' unionism: Cases of trespassing professionals in Norwegian industrial relations. *Economic and Industrial Democracy, 41*(4), 887–909. https://doi.org/10.1177/0143831X17743794

Sabato, S., & Fronteddu, B. (2020). *A socially just transition through the European Green Deal?* Working Paper 2020.08, ETUI. www.etui.org/sites/default/files/2020-09/A%20socially%20 just%20transition%20through%20the%20European%20Green%20Deal-2020-web.pdf

Sachs, J. D., Lafortune, G., Kroll, G., Fuller, G., & Woelm, F. (2022). *Sustainable development report 2022. From crisis to sustainable development: The SDGs as roadmap to 2030 and beyond.* Cambridge University Press. https://s3.amazonaws.com/sustainabledevelopment.report/2022/2022-sustainable-development-report.pdf

Schmidt, V. A. (2008). Discursive institutionalism: The explanatory power of ideas and discourse. *Annual Review of Political Science, 11*, 303–326. https://doi.org/10.1146/annurev.polisci.11.060606.135342

Schumpeter, J. A. (1942). *Capitalism, socialism and democracy.* Harper & Row.

Statistics Norway. (2021). *Ringvirkninger av petroleumsnæringen i norsk økonomi. Basert på endelige nasjonalregnskapstall for 2019.* Statistics Norway. www.ssb.no/nasjonalregnskap-og-konjunkturer/konjunkturer/artikler/ringvirkninger-av-petroleumsnaeringen-i-norsk-oko nomi/_/attachment/inline/e194b68a-c7c1-4ebd-abe8-e65c2568d4fb:932c6de2bb912d11d9f7f 3b92b93afe2139e0b02/RAPP2021-35.pdf

Svenning, L., Solem, T. P. P., & Nordhagen, G. G. (2022, June 3). Olje-spørsmålet ble løst med kompromiss på LO-kongressen. Her er vedtaket. *Fribevegelse.* https://frifagbevegelse.no/nyheter/oljesporsmalet-ble-lost-med-kompromiss-pa-lokongressen-her-er-vedtaket-6.158.879872.cbd0bfa567?utm_source=newsletter&utm_medium=email&utm_campaign=FF040622loko ngressen

Tellmann, S. M. (2012). The constrained influence of discourses: The case of Norwegian climate policy. *Environmental Politics, 21*(5), 734–752. https://doi.org/10.1080/09644016.2012.692936

Tjernshaugen, A. (2011, March). The growth of political support for CO_2 capture and storage in Norway. *Environmental Politics*, *20*, 227–245. https://doi.org/10.1080/09644016.2011.551029
Turnheim, B., & Geels, F. W. (2012, November). Regime destabilisation as the flipside of energy transitions: Lessons from the history of the British coal industry (1913–1997). *Energy Policy*, *50*, 35–49. https://doi.org/10.1016/j.enpol.2012.04.060
Tvinnereim, E., & Ivarsflaten, E. (2016, September). Fossil fuels, employment, and support for climate policies. *Energy Policy*, *96*, 364–371. https://doi.org/10.1016/j.enpol.2016.05.052
UN. (2019). *The future is now. Science for achieving sustainable development*. UN. www.un.org/sustainabledevelopment/development-agenda-retired/
von Platz, J. (2020). Democratic equality and the justification of welfare-state capitalism. *Ethics*, *131*(1), 4–33. https://doi.org/10.1086/709981
Werbeloff, L., Brown, R. R., & Loorbach, D. (2016). Pathways of system transformation: Strategic agency to support regime change. *Environmental Science & Policy*, *66*, 119–128. http://dx.doi.org/10.1016/j.envsci.2016.08.010
Xinxin Wang, K. L. (2021, December). Just transition: A conceptual review. *Energy Research & Social Science*, *82*, 1–11. https://doi.org/10.1016/j.erss.2021.102291
Zhao, W., Yin, C., Hua, T., Meadows, M. E., Li, Y., Liu, Y., Cherubini, F., Pereira, P., & Fu, B. (2022). Achieving the sustainable development goals in the post-pandemic era. *Humanities and Social Sciences Communication*, *9*(258). https://doi.org/10.1057/s41599-022-01283-5

CHAPTER 12

Capitalism, Sustainability, and Democracy

Harald Borgebund

CONTENTS

12.1 INTRODUCTION

This chapter is inspired by Joseph Schumpeter's classical analysis of capitalism, socialism, and democracy (1942). I ask the same questions as Schumpeter: (1) Can capitalism survive? (2) Can the UN's sustainability goals be realized? and (3) Are democracy and sustainability compatible? My analysis will emphasize the role of technology in Schumpeter's analysis of capitalism. In addition, I replace socialism with sustainability.

My motivation for going back to Schumpeter's analysis is because the world in 1942 during WWII confronted some of the most challenging situations faced by humanity in the 20th century (if not by all human history). Similarly, today the world is confronting situations of the same magnitude as then, given the threat of climate change. Schumpeter emphasized capitalism's creativity and innovative role and by replacing socialism with sustainability I want to connect capitalism and sustainability to some of the UN's Sustainable Development Goals (SDGs) such as goals 8 and 9 on economic growth and innovation, and also 12 and 13 on responsible production and climate action (see Chapter 2). Answering the previous three questions contributes to a tentative answer to the question of technological progress and innovation can contribute toward realizing a sustainable society.

The structure of this chapter follows the order of the three questions asked earlier. First, I discuss if capitalism can survive. Then, I discuss if sustainability can work before I in the third and final section, ask if democracy and sustainability are compatible.

DOI: 10.1201/9781003325086-12

12.2 CAN CAPITALISM SURVIVE?

Schumpeter argued that capitalism's failures would not cause capitalism's breakdown. Instead, "its very success undermines the social institutions which protect it" (Schumpeter, 1942, p. 61). Socialism would inevitably replace capitalism. However, 80 years later, capitalism is still a dominant economic system, and socialism has not replaced capitalism. Socialism has been abandoned by most of its adherents over the last three decades. Those societies still adhering to a strict socialist ideology are repressive, and authoritarian societies disregard their people and the environment. Although socialism in its strict Soviet style has largely been abandoned, that does not mean that socialist ideas have been abandoned. After WWII, welfare state capitalism transformed Western societies by making capitalism more "social" by alleviating poverty, providing public healthcare, and offering educational opportunities for large groups of society. In this sense, capitalism survived by adapting to a new situation by including some socialist principles (Esping-Andersen, 1990).

Today, capitalism is still a contested concept (Hall & Soskice, 2001). There are many reasons for questioning capitalism, but my analysis will emphasize two reasons linked to the UN's sustainability goals. First, increasing economic inequalities have led many critics to argue that capitalism should be abandoned for more equality (Piketty, 2014, 2020). Second, capitalism is blamed for the purportedly ruthless exploitation of natural resources. Pollution and biodiversity losses are two important examples of how capitalism exploits the environment. The threat of climate change encapsulates many if not most of the ways capitalism detriments the environment. What increasing economic inequality and climate change share is that, according to its critics, both show that capitalism is inadequate to solve the challenges confronting the world in the 21st century (Barry, 2005 and Meyer, 2001). Because of these inadequacies, Schumpeter's question if capitalism can survive is once again an important question to ask. Furthermore, Schumpeter emphasized technological progress and new modes of production as important and defining features of capitalism making his theory an interesting starting point for an analysis of the role of technological solutions to sustainability.

In trying to understand better whether capitalism can survive or not, a good starting point is to analyze some of the points that shaped Schumpeter's understanding of capitalism found in three defining features: (1) mass production and mass consumption, (2) creative destruction, and (3) technological progress.

1. Mass production and mass consumption define capitalism through producing affordable products for ordinary citizens. Mass production aims to lower production costs and make mass consumption possible through affordable prices for most consumers. Industrial capitalism in its modern form is impossible without mass production and consumption.

2. Creative destruction is, for Schumpeter, the essence of capitalism. Through new modes of production and new products, old forms of production are constantly being replaced by new forms of production. Innovation gives rise to new forms of

production, improving existing products or new products replacing existing products. Creative destruction is fundamental to capitalism because "the fundamental impulse that sets and keeps the capitalist engine in motion comes from the new consumers' goods, the new methods of production or transportation, the new markets, the new forms of industrial organization that capitalist enterprise creates. . . This process of Creative Destruction is the essential fact about capitalism" (Schumpeter, 1942, p. 83). According to Schumpeter, constant change and innovation are thus the defining features of capitalism as an economic system.

3. Regarding technological progress Schumpeter claim that "all the features and achievements of modern civilization are, directly or indirectly, the products of the capitalist process" (1942, p. 125). Importantly, the modern world is the product of capitalism, and capitalism has been the main engine in developing modernity. Capitalism set in motion the overthrow of feudalism created the "mental attitude of modern science" (Schumpeter, 1942, p. 124) and was responsible for creating the new social classes. Hence, capitalism was the primary force, while other social and cultural changes were secondary and set in motion by capitalism. Schumpeter's argument here is controversial and does not consider the many significant social and cultural changes that came along with the Reformation and the development of the modern state system, to mention two crucial changes occurring alongside the development of capitalism (Skinner, 1978). I will not discuss these controversial aspects of Schumpeter's view, but note that Schumpeter's interpretation of the development of capitalism must be viewed with caution.

Taken together, these three features are defining features of capitalism as Schumpeter understood capitalism. Capitalism is subject to various and often competing understandings and although Schumpeter's analysis has been influential, it is only one of many possible understandings of capitalism. In contrast, Jürgen Kocka (2016, p. 20), in his history of capitalism, emphasizes decentralization, commodification, and accumulation as defining features of capitalism. Schumpeter's understanding of capitalism does not pay much attention to the role of decentralization and accumulation although the oligopolistic and monopolistic features of capitalism receive some attention from Schumpeter. Thus, Schumpeter's definition of capitalism emphasizes some of the important features of capitalism but can be said to downplay some of the elements highlighted by other theorists of capitalism.

Schumpeter argued that analyses of capitalism must be based on two criteria: (1) capitalism must be judged over time "as it unfolds through decades or centuries" (Schumpeter, 1942, p. 83) and (2) the process of capitalist production is organic. Even if one might worry about oligopolistic and monopolistic tendencies, such worries are a distraction from "how capitalism administers existing structures, whereas the relevant problem is how it creates and destroys them" (Schumpeter, 1942, p. 84). Thus, capitalism is an engine for change and revolution rather than conserving the distribution of present social and economic circumstances.

Now I want to use the three defining features earlier and the two criteria to analyze if capitalism can survive. My analysis will be limited to commenting on the implications of the defining features and the two criteria in a contemporary context. (1) Mass production and consumption are today just as important as during the development of capitalism as some of the biggest and most influential corporations worldwide produce commodities for the masses. Examples range from entertainment such as Netflix to oil and gas producers producing fuel. What these companies share is that they depend primarily on the masses, not on the wealthy. (2) Creative destruction has been observable in various sectors, from how mobile phones and smartphones replaced landlines in most homes to how cleaner energy modes possibly overtake the role of oil and gas in the future. (3) Profit is the motivation for innovation that makes technological progress possible, and Schumpeter believed profit and technological progress were indistinguishable (1942, p. 110). Even if Schumpeter's view on this point is controversial, it is hard to distinguish if profit or technological progress were two distinct processes in developing the first smartphone. These three features offer an economic system that can deliver transformative change for the masses and, in the process, create technological progress through inventing new and better products or modes of production.

Judging capitalism over decades or centuries, as Schumpeter suggested, has the advantage of contributing to revealing the long-term consequences of capitalism. Often judgments of capitalism's performance and consequences are judged at a particular time that is merely a snapshot distorting our perspective. Unfortunately, Schumpeter's analysis made precisely this mistake by assuming what capitalism looked like at a certain point in time and extrapolating what he found into the future. Schumpeter assumed that socialism would replace capitalism and that such a take-over was inevitable given capitalism's internal structure and logic. Instead, capitalism is perhaps more potent in 2022 than in 1942. The unprecedented growth after WWII gave rise to higher living standards and reduced poverty fueled by the expansive welfare states. Schumpeter underestimated how dynamic capitalism could be and how adaptable capitalism could be to changing social circumstances. Because of this adaptability, capitalism should not be underestimated under the current pressure that capitalism is facing. The answer to the question if capitalism can survive is yes, and capitalism can survive. That does not amount to saying that capitalism will survive. My point is mere that in the same way capitalism did survive after WWII capitalism can survive and may survive in the future.

Assuming capitalism stands in the way of developing a sustainable future, disregard the innovative power of capitalism highlighted by Schumpeter and the need for innovation to achieve sustainable societies. As sustainability is understood by the UN sustainability goals and, in this book, capitalism is needed to engineer the necessary innovations for creating a sustainable society, the UN sustainability goals ask for innovations to create a sustainable society, and capitalism may offer a practical framework for creating the needed innovation. The following section will elaborate on this claim and show how the sustainability goals may require capitalism to reach the established goals.

12.3 CAN SUSTAINABILITY WORK?

I answer this question by first arguing that technological progress and innovation can contribute toward a sustainable society through capitalism. There are several objections to such an argument, and I will address two such objections. The second part of my argument states that although capitalism can contribute, concerted collective action is also necessary to reach a sustainable society. My analysis is limited to emphasizing a conceptual framework for achieving sustainability. Furthermore, it suggests what kind of obstacles might stand in the way of a sustainable future and which direction might be favorable to move toward the goal of sustainability. Before starting my analysis, one more point must be emphasized. Sustainability is not politically controversial in the same way as socialism. Socialism defined itself in opposition to liberalism and capitalism, while sustainability is compatible with different political frameworks.

Sustainability is nonetheless a contested concept, and there are differing understandings of what sustainability should mean in theory and practice. Despite these differing understandings of sustainability, the concept itself is not controversial in the same way as socialism. I will not address these disagreements in any detail.

By sustainability, I follow the framework established earlier in this book. I base my analysis on the view that sustainability is primarily to further human development within certain limits. Such limits are environmental but also economic, social, and political (see Chapter 2). UN's 17 sustainability goals exemplify how such limits can be operationalized. Sustainability such understood implies an instrumental and anthropocentric view of sustainability. Such notions are perhaps influential and, to some extent, mainstream understandings of sustainability, but controversial and not beyond criticism. I largely overstep the criticisms of this chapter's instrumental and anthropocentric understanding of sustainability. However, I acknowledge that a somewhat different understanding of sustainability would likely reach different conclusions. The most relevant of the SDGs for my analysis are 8, 9, 12, and 13. Goal 8 is about decent work and economic growth, while SDG 9 is about industry, innovation, and infrastructure. SDG 12 deals with responsible consumption and production, while SDG 13 is perhaps one of the most important and is about climate action. These goals concern human development and well-being and presuppose economic growth. However, economic growth is restricted by the constraints of climate change and the other goals earlier. My analysis emphasizes both the role of capitalism and sustainability and is in line with the SDG which also emphasizes the role of markets, economic growth, and environmental issues important for a sustainable society.

My argument in favor of capitalism as part of the answer to creating a sustainable society over time is that the creative destruction of capitalism replaces current technologies with new forms of technology and production. An example regarding sustainability is that when it comes to the dominant position of oil and gas in most economies oil and gas will fade as these technologies will be displaced by new modes of production. In this way, the modern world can sit back and let the creative destruction of capitalism do its work. Because of the organic nature of capitalism, innovation will constantly offer new technological solutions. Because this process will take decades or centuries, climate change might cause irreversible damage to the planet.

SDG 8 asks for economic growth and SDG 9 asks for innovation. These two goals are not necessarily dependent on capitalism, as economic growth and innovation might be possible without capitalism. Given capitalism's prevalence in the modern world, capitalism is an obvious candidate for achieving economic growth and innovation. SDG 12 is about responsible production and consumption, and SDG 13 is about reducing climate change. These two goals may be undermined by capitalism as there is nothing inherent in capitalism that will reduce emission of climate gases, although the innovative power of capitalism has potential to contribute to reduced emissions. Thus, the sustainability goals appear contradictory. The SDGs exemplify some of the dilemmas confronted by trying to reach a sustainable society through capitalism. On the one hand, sustainability as understood by the UN is compatible or even supportive of capitalism, but on the other hand reduction of climate gas emissions may be exacerbated by capitalism.

In this chapter, I will discuss two objections related to sustainability. First is the objection that capitalism focuses on short-term profitability and excludes future generations. Capitalism focuses on current shareholders and not the broader society or future generations. A second objection is that relying on capitalism to contribute to creating a sustainable society is too uncertain. Capitalism may contribute, but it may also contribute to environmental degradation and make it harder to reach a sustainable society. For example, relying on capitalism to combat climate change is like driving a car while braking and accelerating simultaneously. Innovation might reduce waste and pollution while increasing inequalities within and among countries (goal number 10). Thus, offering progress on one goal while moving in the opposite direction on other goals. What this breaking and accelerating will lead to is uncertainty about trusting capitalism's ability to contribute to a sustainable society.

Future generations are mostly excluded from capitalism. Although investment horizons can be long-term, few investment decisions have a perspective longer than two or three decades (Williamson, 2018). I will not discuss whether we have obligations to future generations or not but assume that a sustainable society means considering the interests of future generations. One response is to argue that future generations will be better off because of economic growth and technological progress, even with climate change. They will have more resources and technology to resolve the challenges caused by economic growth and capitalism. Hence, capitalism contributes to future generations if economic growth makes societies wealthier. Economic growth has made it possible to devote more resources to welfare and healthcare, increasing life expectancy in many developed countries. Environmental degradation is a price to be paid and is worth it if it enhances opportunities for both present and future generations. Despite the temptation to endorse this argument, the problem is that it does not consider future generations *per se*. Instead, future generations' increased amount of resources is a side effect of capitalism. Thus, capitalism is seemingly unconcerned by future generations, which is a structural feature of capitalism and not something that can be easily changed.

Uncertainty is both the strength and weakness of capitalism. Creative destruction can be surprising and unpredictable. Predicting economic changes and how technology may change society is difficult, if not impossible, at least with any accuracy. Trusting in capitalism's possible contribution to sustainability is based on uncertainty, although there are reasons for thinking that techno-optimism can contribute to resolving climate change among other issues John Danaher has argued that a modest version of techno-optimism "that does not assume that technology will save humanity by itself, nor that technology is sufficient for the good to prevail, is defensible" (Danaher, 2022, p. 54). Given this feature of capitalism, the concern should be twofold.

On the one hand, the creative and innovative features of capitalism should be promoted to unleash the potential that lies in capitalism. On the other hand, collective action is warranted to reduce uncertainty and promote sustainability goals specifically. The two objections discussed previously point toward weaknesses with capitalism that must be countered with other measures. This takes us to the second part of my argument, which states that collective action should counter the lack of concern for future generations and the uncertainty of creative destruction unleashed by capitalism. Since Garrett Hardin (1968) pointed out the problem with the tragedy of the commons, the limits of capitalism have been clear. Elinor Ostrom stated that

> the tragedy of the commons, prisoner's dilemma, and the logic of collective action are closely related concepts in the models that have defined the accepted way of viewing many problems that individuals face when attempting to achieve collective benefits. At the heart of these models is the free-rider problem.
>
> (Olstrom, 1990, p. 6)

Individuals or individual countries are incentivized to let the other participants take the costs of reducing carbon emissions while avoiding emissions themselves. The result is that the issue of climate change has not been resolved and emissions are not being reduced, and binding collective action has not been possible.

At the end of WWII, few would have anticipated the development of the modern welfare state that followed the war in most Western countries. It required coordination and cooperation between previously antagonistic actors such as labor and capital. Disagreements were overcome and exemplified the potential of collective action to change society and reach goals previously out of reach. Something along the lines of the compromises that created the modern welfare state is needed to achieve a sustainable society. The welfare state shows that it is possible for a society to come together and establish institutions few had thought would be possible for a society to create. A world war allowed an entirely new direction, and the opportunity was taken. Today climate change may be an existential threat, but the circumstances are perhaps not as extreme as the situation was after WWII. There are limits to the example of the development of the welfare state in the West in the way that the welfare state could be realized within the constraints of the nation-state, while a sustainable society cannot be created by one society alone at least in so far as climate change is part of a sustainable society. Furthermore, the main point of this example is not to argue in favor of the greatness of a welfare state but merely to draw attention to the potential of collective action to radically change a society.

This concludes the argument of this section where I have argued that capitalism engenders creativity and innovation that can contribute toward a sustainable society and that collective action is necessary to reduce the uncertainty and lack of concern for future generations in capitalism. I have also suggested that collective action is possible through the example of the development of the welfare state after WWII. Just that something is possible does not mean that it will happen. Democratic politics focus on short-term gains just as much as capitalism (or perhaps even more) and question the realism in creating collective action of the magnitude necessary to create a sustainable society. The following section extends the analysis to discuss if sustainability and democracy are compatible.

12.4 SUSTAINABILITY AND DEMOCRACY

I argue that sustainability is compatible with democracy in principle, but in practice, sustainability becomes merely one of many issues competing for attention. There is nothing in principle with democracy as a political system that prevents a polity from developing a sustainable society. In practice though, several obstacles are standing in the way of developing a sustainable society (De-Shalit, 2000). I limit my discussion to three: (1) short-term focus on electoral democracy, (2) competition between different political issues, and (3) knowledge and competence of the electorate and politicians.

For example, developing a sustainable society and reducing carbon emissions can be said to be the right thing to do. Nonetheless, democracy may be unable to deliver the "right" result. Paradoxically democracy as a legitimate form of government might sanction policies that undermine the long-term viability of the same society. One of this chapter's main conclusions is that more attention should be devoted to the weaknesses of democracy.

Schumpeter was aware of the weaknesses of democracy, and his bleak and realistic view of modern democracy has been widely criticized for being minimalistic and pessimistic. Despite the criticisms of the bleak description of modern democracy, Schumpeter's description still paints a realistic picture of how democracy functions. For Schumpeter, "democracy is a political *method*, that is to say, a specific type of institutional arrangement for arriving at political – legislative and administrative – decisions and hence incapable of being an end in itself" (1942, p. 242, original italics). Here, two things are essential. Democracy is seen mainly as a method of decision-making. It is a means of making decisions and is not valuable in itself apart from the benefits it represents for the electorate.

Schumpeter defines the democratic method as "that institutional arrangement for arriving at political decisions in which individuals acquire the power to decide by means of a competitive struggle for the people's vote" (1942, p. 269). Democracy is the political institution where individuals compete over the power to rule society. It emphasizes political leadership and the competition over political power as the central elements of democracy. The will of the people or self-government plays no vital role. In this view, the essence of democracy is competition among political elites for power. For a society to be democratic real, although not perfect, competition between political parties or political interests must be present. Furthermore, democracy means that the political elite rules the people. Voters' primary role is electing a government and playing a limited role.

Schumpeter emphasizes that the focus on winning elections means politicians are tempted to prioritize short-term gains over what is most beneficial over the long term. Career politicians are concerned about extending their time in government and offering policies that will provide support from a majority or large enough proportion to be re-elected. Some policies associated with sustainability are related to short-term gains, such as improving access to healthcare and education. Reducing inequalities and poverty may also have some short-term impacts. Thus, on the one hand, some sustainability goals are compatible with the incentive of emphasizing short-term gains.

On the other hand, other aspects of sustainability, such as exploitation of natural resources and the interests of future generations, are tempting to discount. Postponing difficult decisions to avoid short-term pains for present generations is a common strategy among politicians to avoid reducing carbon emissions. On the problematic issues of future generations and climate change, democracy offers incentives to make the "wrong" decisions. Democracy is, therefore, a hindrance to sustainability on this point.

Competing interests among political parties and organizations is a second point challenging the compatibility between sustainability and democracy. Interest groups often dominating role in democratic politics have been the subject of much discussion among democratic theorists building on Schumpeter's democratic theory (Dahl, 1956; Lindblom, 1977). These theorists have interpreted group interests as an integral aspect of democratic politics because political freedom means that people and organizations can organize and work for political change according to their preferences and interests. Most democracies have a broad set of organizations and groups advocating specific policies beneficial to themselves. In an open society, such rights are essential and foundational for the democratic process. Critics have argued that wealthy individuals and corporate interests can gain unduly political influence through donations and public campaigns by accessing abundant financial resources (Cohen & Rogers, 1983). Interest groups are often concerned with narrow interests. If some interests become influential, the interests of the rest of society may be undermined. Sustainability can, for example, be undermined by corporate interests arguing for the importance of jobs and tax income when the government is threatening to regulate or tax polluting activities. Environmental groups often have access to more limited financial resources making it difficult to challenge wealthier corporate interests. In addition, competing interest groups mean that sustainability is only one of many issues on the political agenda. Voters and politicians are confronted by competing interests and must prioritize which interests are most important, and in that struggle, sustainability might not win. Competing interests thus stand in the way of sustainability to be realized, as politics is often about the power to get the world to bend to your will.

Furthermore, the electorate often lacks knowledge about politics and important issues in the political debate. Schumpeter claimed that

> the typical citizen drops down to a lower level of mental performance as soon as he enters the political field. He argues and analyzes in a way that he would readily recognize as infantile within the sphere of his real interests.
>
> (1942, p. 262)

This characterization of the electorate points out that many citizens are uninterested or unknowledgeable about politics. If the electorate were more knowledgeable and better informed, then they would hold the politicians to account and elect politicians who would, for example, prioritize sustainability rather than short-term goals. Much research has shown that the electorate lacks the basic skills and knowledge to make informed political decisions (Achen & Bartels, 2016). Educating the electorate through democratic education or in other ways might contribute to a more informed electorate. Still, there are doubts about whether democratic education can make an impact that will change voters' priorities (Brennan, 2016). Because the electorate lacks knowledge about the basics of politics and the issues at hand, sustainability risks are not being prioritized by the voters. Hence, democracy is a potential barrier to sustainability.

Democracy is mainly unchallenged as a legitimate form of government in contemporary political theory. The discussion in this section has pointed out three features of democracy, making it potentially more challenging to realize a sustainable society. Based on the discussion here, I want to conclude this section by arguing that more attention should be devoted to democracy's weaknesses. By acknowledging the flaws of democracy, it is possible to address them and work toward improving democracy. As Schumpeter describes democracy, democracy and sustainability are only partly compatible. A democratic government, therefore, does not necessarily support a sustainable society.

12.5 CONCLUSIONS

I started this chapter by asking three questions and have answered them through the theoretical framework of Schumpeter. Today the world confronts perhaps equally difficult situations as during WWII. I have argued that capitalism can support and undermine sustainability. The innovation and power in creative destruction might offer technological progress supporting sustainability. The exploitation of non-renewable resources and the environment may undermine sustainability. Capitalism may or may not contribute toward a sustainable society, but it is difficult to identify alternative economic frameworks that might promote creativity and innovation in the way that capitalism does. Thus, the world is left with a fickle and unreliable economic system in want of something better, which means that capitalism to some extent can be justified and maybe even a necessary tool in achieving a sustainable society.

In addition, sustainability requires collective action to be realized. The aftermath of WWII showed the potential of collective action regarding the development of the welfare state in many Western societies, which means that transformative collective action within a democratic context might be possible. Still, democracy contains features undermining sustainability. Taken together, asking these three questions put us in a position to answer the overarching question of this anthology related to technological solutions to the issue of sustainability. Capitalism's flexibility and innovative features can contribute to technological progress beneficial to realizing sustainability by inventing newer modes of production and ways to reduce carbon emissions. In this way, capitalism may contribute to realizing several of the UN's goals for sustainable development. However, historically capitalism has led to environmental degradation just as well as technological progress and innovation.

Nonetheless, there is a tension between the UN's goals for sustainable development by emphasizing innovation and technological progress and simultaneously advocating a reduction in carbon emissions. Capitalism may reconcile some of the tensions between the UN's goals for sustainable development, although that is far from assured. Collective action is also necessary to move toward a sustainable society, and I have argued that such collective action is possible but far from assured. In addition, the weakness of democracy complicates the kind of collective action that is necessary for a sustainable democracy. Is the potential to undermine is far from a simple or obvious answer to realizing a sustainable society as technological confront structural and institutional challenges making it difficult to create sustainable societies. To sum up, this chapter concludes that capitalism's creativity, developing institutions for collective action, and addressing the weaknesses of democracy are necessary tasks for sustainability to be realized.

12.6 REFERENCES

Achen, C. H., & Bartels, L. M. (2016). *Democracy for realists – why elections do not produce responsive governments.* Princeton University Press.

Barry, J. (2005). *Environmental and social theory.* Routledge.

Brennan, J. (2016). *Against democracy.* Princeton University Press.

Cohen, J., & Rogers, J. (1983). *On democracy – Toward a transformation of American Society.* Penguin Books.

Dahl, R. A. (1956). *A preface to a democratic theory.* Chicago University Press.

Danaher, J. (2022). Techno-optimism: An analysis, an evaluation and a modest defence. *Philosophy & Technology, 35*, 54. https://doi.org/10.1007/s13347-022-0550-2

De-Shalit, A. (2000). *The environment: Between theory and practice.* Oxford University Press.

Esping-Andersen, G. (1990). *The three worlds of welfare state capitalism.* polity.

Hall, P. A., & Soskice, D. (Eds.). (2001). *Varieties of capitalism – the institutional foundations of comparative advantage.* Oxford University Press.

Hardin, G. (1968). The tragedy of the commons. *Science, 162*(3859), 1243–1248.

Kocka, J. (2016). *Capitalism – A short history.* Princeton University Press.

Lindblom, C. (1977). *Politics and markets.* Basic Books.

Meyer, J. M. (2001). *Political nature – environmentalism and the interpretation of western thought.* MIT Press.

Olstrom, E. (1990). *Governing the commons.* Cambridge University Press.

Piketty, T. (2014). *Capital in the twenty-first century.* The Belknap Press of Harvard University Press.

Piketty, T. (2020). *Capital and ideology.* The Belknap Press of Harvard University Press.

Schumpeter, J. A. (1942). *Capitalism, socialism and democracy.* Harper Perennial.

Skinner, Q. (1978). *The foundations of modern political thought – volume 2 the age of reformation.* Cambridge University Press.

Williamson, S. D. (2018). *Macroeconomics.* Pearson.

CHAPTER 13

Nudging Policy or Crowding It Out? Green Nudges as Ideational Technologies

Stuart Mills and Richard Whittle

CONTENTS

13.1 INTRODUCTION

Few would deny that new ideas are of great importance to society, and – as the challenges of various sustainability crises begin to take hold – are as vital as they ever have been. A curious, yet highly disruptive, policy approach of the past 15 years has been *nudging*. Nudging, based on insights from behavioral economics, is the notion that small changes in how existing systems and incentive structures are framed to the public can have a significant and predictable impact on behaviors. Nudges are said to *encourage*, rather than *coerce*, individuals toward choosing better outcomes for themselves, and in this sense *may* promote welfare and freedom (Thaler & Sunstein, 2008). Politically, nudging has been seized upon as a means of achieving substantial policy ends by modest policy means (Chater & Loewenstein, 2022).

Besides the slew of curiosities such an idea throws up for political scientists and economists as regards the role of the state in society, understanding nudging is important within sustainable development given the apparent benefits of the approach given the apparently small costs. The aforementioned sustainability challenges – climate crises, environmental breakdown, ecological collapse, oceanic pollution, etc. – are typically understood to require *substantial* policy responses (e.g., Klein, 2014; IPCC, 2022; Wainwright & Mann, 2020; Wallace-Wells, 2019). Given this, one might wonder – as some (Hagman et al., 2019;

DOI: 10.1201/9781003325086-13

Harford, 2022; Loewenstein & Chater, 2017; Chater and Loewenstein, 2022; Maki, 2019) have begun to – whether an idea like nudging *distracts from*, or *crowds out*, others, more substantive policy solutions. Such concern is not dissimilar to those expressed by others in this book regarding other "low cost, high reward" technologies such as geoengineering (see Chapter 4).

As we will show through an examination of the history of nudging and the political–economic forces which surrounded the idea's ascendence, this notion of nudging "crowding out", another more substantive policy, is a reasonable one (see also Chapter 14 on the politics of overpromising). For some initial evidence, consider a recent policy strategy document produced by the Intergovernmental Panel on Climate Change (IPCC).

The IPCC (2022) report on climate breakdown mitigation features a prominent section on *choice architecture* (read *nudging*), alongside various calls for reducing fossil fuel dependence and investment in net-zero technologies. So-called green nudging, as an *idea*, remains an established tool in the policymaker's toolkit, even when directly juxtaposed against the scale of the sustainability changes which must presently be overcome. In this chapter, we will explain why and explore what the consequences of this might be.

The structure of this chapter is as follows. *First*, we will briefly outline our philosophical and conceptual approach, arguing that *ideas* can be understood as a *special kind of technology*, in the Heideggerian sense of the word, as they are able to transform (a) the world in which they exist and (b) those who use them. But they are *special* insofar as ideas can (c) transform how *other technologies* are used. Ideas themselves are understood as important factors in politics and economics and are treated as such in various areas of political economy, which we also draw greatly upon. *Second*, we turn our attention to nudging and behavioral insights, focusing on the epicenter of the "Nudge-Revolution", the United Kingdom (UK). We explain how the *idea* of nudging became popular because of specific political–economic conditions, and how this *idea* was a useful political technology for changing how the technologies of the state (e.g., tax and spend) were to be used. *Finally*, with this context established, we turn our attention to green nudges, and their suitability within sustainability policy and the sustainable development goals (SDGs). We argue that nudging *as a policy* is an inadequate response to sustainability challenges such as the climate crisis, but that nudging as an *ideational technology* still has political capital within the political economy of a neoliberal state such as the United Kingdom. Our perspective thus aligns with those of others (e.g., see Chapters 12, 17, and 18) in this book, who are also skeptical of the limits of technology within existing economic systems.

13.2 IDEAS AS A SPECIAL, TRANSFORMATIVE TECHNOLOGY

Ideas and narratives have an especially crucial role within the political economy of technology. Heidegger (2010) has been very influential in this space, defining technology as a *transformative tool*. Here, the tool serves the purpose of transforming, say, a piece of wood into a chair, but *simultaneously* transforms the person using the tool, say from a *person* into a *carpenter*. In this sense, the tool becomes an extension of the person who uses it, as it enables the transformation of the individual and, in turn, the individual's capacity to transform the world around them. Such mutual transformation also accounts for

Heidegger's (2010) notion of technology becoming *invisible*, or Latour's (1999) notion of "blackboxing". For instance, while it is *the tool* which transforms the wood into the chair, the person (now *carpenter*) comes to be seen as the transforming entity (e.g., "the carpenter built the chair", rather than, "the person used the tool to build the chair").

Classical political economy, specifically Marx (2013), may also contribute to this discussion. Marx likewise regards technology as transformative of both nature and the labor which use it, but further argues that the relationship between technology and humanity is one which is *socially determined.* As earlier, we may say that the person who uses a tool to build a chair is a carpenter, but if they are a *hobbyist*, their relationship with and understanding of the tool is likely different from that of a *worker.* For the former, the tool becomes a means of self-expression, relaxation, and perhaps even a means of expressing one's power over nature. For the latter, the tool becomes a means of survival through laboring, and perhaps a symbol of their *need to labor* (Marx, 1973).

Ideas and narratives become vital to the political economy of technology if one accepts the previous philosophical perspectives on (a) technology as a means of transforming both the user and the user's environment, and (b) that the user's relationship with technology is socially determined. This is because *how* the user of technology comes to *understand* the technology they use, and how they *imagine* technology *can* be used, determines the role of technology in society (see Chapter 15; Spencer, 2017). A worthwhile example here is that of the 19th English *Luddite* movement, a collection of textile workers whose objection to technology was not predicated on a dislike or distrust of technology *per se*, but in opposition to how technologies were being used by factory owners to transform textile production in England, at the *expense* of artisan knowledge (Linton, 1992). This objection is an objection against the transformation of the *user*, based on conflicting ideas about the role of textile technology (i.e., for the factory owner, textile technologies were a means of boosting profits; for the artisan, they were a means of disempowering them and destroying knowledge).

To an extent, the argument that ideas are crucial to understanding how technologies are used, and are conceived of as being usable, is a self-referential one. This is because ideas *themselves* can be understood as *technologies* insofar as they demonstrate this doubly transformative character (Feyerabend, 2010; Kuhn, 2012). For example, the carpenter who uses their tool as a weapon is transformed, in their social role, into a soldier, or an assailant. But the *idea* that the tool *may be a weapon* is also transformative *of the tool itself.* Equally, for the Luddite, the *idea* that a textile machine could destroy artisan knowledge renders the machine a threat; but under different economic conditions, with different social relations, the textile machine may come to be seen as a means of *liberating* the Luddite, or even as a way of *discovering new knowledge* (Braverman, 1974; Marx, 1973). Thus, the political economy of technology is a deeply interconnected study, where technology can change political–economic ideas, and where political–economic ideas can shape the uses and imaginaries of technology.

We conceive of ideas as a *special kind of technology.* In the first instance, ideas are technologies insofar as they transform those who use them (e.g., a politician becomes a policymaker). Yet, in the second instance, ideas are a special kind of technology insofar as they *transform other technologies* (e.g., the state can tax, and can spend, but *should* the state tax, *or* spend?; see also Ellul's (1964) notion of the *technique*, to which parallels to our perspective on ideas could be drawn). Yet which ideas emerge and can be expressed (which is to say, which ideational technologies are used) depend on the political economy at the time. For instance, if there is no state, the *idea* that the state can tax is irrelevant.

Worthwhile ideas emerge constantly, yet only some of these ideas become popular and influential. Understanding why, and under what conditions, ideas become influential within society (particularly ideas concerning policy and the economy) are objectives central to the *political economy* study of ideas (Mukand & Rodrik, 2018). This study is a complex one, in part because it is difficult to determine what an idea is, and in part because it is difficult to attribute the causality of an action or event to an idea (Blyth, 1997). Nevertheless, several prominent perspectives on how ideas function in political economy can be found in the literature. A relatively brief overview is sufficient for our purposes here.

The central question of the political economy of ideas is how ideas shape the political life of economic concepts (Hay, 2004). For instance, central to Gramsci's notion of *hegemony* is that of idea acceptance; that those who succeed economically and politically within society do so only through establishing hegemonic control of the ideas which pervade society (Gramsci, 1992). Susan Strange's work in international political economy expresses a similar perspective, though more in the vein of Hay (2004), suggesting that economists (rather than *political economists*) are limited in their understanding of the economy because their models are often divorced from the structures of power, geopolitics, and culture which determine how those models are understood, and ultimately wielded (Brown, 1999). Ironically, the economist Milton Friedman has argued candidly that the economic ideas which gain political support are not those which are necessarily the best, but those which are "lying around", at a time when other ideas begin to fail (Friedman, 2002; Klein, 2007). For Kuhn (2012), such failure is the result of "anomalous" observations – unexplained by existing ideas – which build up into a crisis. New ideas – which can explain these anomalies – then proceed to overtake the old in an act of *paradigm shift*. For Blyth (2002, 2009, 2013), what matters is often not anomalies, but uncertainties, and the ideas which come to the fore are those which are best disposed to vanquish uncertainty, at least in the short term (see also Hay, 2004).

With our conception of *ideas as technologies* now established, and some theories of how ideas come to be used or ignored offered, let us now turn our attention to an especially pertinent idea of modern policymaking: *Nudging*.

13.3 THE RISE OF "NUDGE" IN THE UNITED KINGDOM

The rise of "nudging" and behavioral insights within the United Kingdom occurred during a 10-year period from 2004 to 2014. Crucial to this timeline is the Global Financial Crisis (GFC), a crisis of economics and finance which would have profound effects on policymaking in the United Kingdom and establish the political–economic conditions for the ascent of behavioral insights. Broadly, the impact of the GFC on nudging can be split into two occurrences, which we will discuss in turn. *First*, the crisis led to a "crisis of legitimacy" in mainstream economics and policymaking, creating the intellectual space for a new idea – behavioral economics and nudging – to rise to prominence. *Second*, the GFC led the UK Conservative party (also known as the Tory party) under David Cameron to adopt a macroeconomic program of fiscal austerity, which was complemented by the "microeconomic" program of nudging.

The GFC precipitated an intellectual crisis in economics and policymaking (Colander et al., 2009). Prior to the GFC, the economics mainstream (typically called *neoclassical* economics) was dominant academically and politically. The dominance of this economic paradigm came from its successful track record (Lucas, 2003), securing 14 consecutive years of economic growth in the United Kingdom, for example (Carney, 2021). Yet this paradigm also failed to foresee the crisis and failed to respond to the crisis once it occurred, in part because of a belief in the infallibility of the paradigm, and in part because the paradigm itself "positively denied it [the GFC] could happen" (Allen, 2009; Asensio, 2014; Kirman, 2009; Stiglitz, 2010). Such denial came from a fundamental axiom of *neoclassical* economic thought – *rationality*. The *neoclassical* school, and its followers in government, held that economies consisted of rational, representative agents: Individuals with perfect information and ruthless utility maximization. Under such an assumption, the GFC could not happen; when the GFC *did happen*, the *neoclassical* school struggled to mount a defence as its legitimacy collapsed (Brown & Spencer, 2014; Green & Hay, 2015; Lux & Westerhoff, 2009; Solow, 2008; Stiglitz, 2010; Streeck, 2014). New ideas were now required.

Behavioral economics was one such idea. Tversky and Kahneman's (1974) "heuristics and biases" program is widely regarded as the origin of a modern "behavioral" school in economics and policymaking (Thaler, 2015). This program had, since the 1970s, sought to cultivate a niche within economics (Heukelom, 2012, p. 810; Kahneman, 2011), and succeeded in part by retaining a close alignment with the *neoclassical* school, unlike previous "behavioral" turns – such as Herbert Simon's (1955) work – which outright attacked the paradigm (Heukelom, 2012, p. 817). Behavioral economics (later insights) abandons the rationality assumption (e.g., people do not have perfect information, and do not always maximize utility), but only on a *descriptive* level. For instance, the school holds that people rarely succeed in "maximizing utility" but does not question the logic that people *want* to maximize utility.

This slight change – which is in the spirit of economics' own long-standing criticisms of the rationality assumption (Schumpeter, 1940; Marshall, 1980) – positioned behavioral economics as *inoffensively dissimilar* to the *neoclassical* school. Yet, its specific departure from the *neoclassical* school – the abandonment of rationality – was the ideal solution to the "anomaly" the GFC had demonstrated in the *neoclassical* school. The GFC, which preceded the crisis of legitimacy, seems to have been necessary to shift the political economy in favor of behavioral ideas (Halpern, 2015, p. 46; Osborne & Thaler, 2010, para. 2–3). For instance, the eventual head of the Behavioural Insights Team (BIT), David Halpern, would in 2015 explain how earlier efforts to integrate behavioral ideas into policymaking (Halpern et al., 2004) had been met with "heavy press attack, leading to the Downing Street Press Office to distance itself from [the policy]" (Halpern, 2015, pp. 35–36). Similarly, reflecting on the success of *Nudge* (2008), Richard Thaler (*Nudge* co-author and former BIT advisor) revealed that he and his co-author, Cass Sunstein, found the book's success to be a "big surprise" which far exceeded their expectations (Thaler, 2021).

Second, following the GFC, UK politics became dominated by the idea of fiscal austerity – that cutting the size of the state could produce an outsized economic benefit (Reinhart &

Rogoff, 2009). In the 2010 General Election, the UK Conservative Party ran on this macroeconomic platform, and the subsequent Tory-led Coalition Government (2010–15) pursued this program with zeal.

Austerity represented a unique opportunity for nudging and behavioral insights. The Conservative Government had also committed to a slew of social initiatives. Yet, the austerity regime demanded reduced spending and reduced taxation (this formula is sometimes called *British* Austerity, as "textbook" austerity is a reduction in state spending and an *increase* in taxation; the UK Government in 2010 did not pursue the latter component), as well as calls for reduced "red tape", which also discouraged the use of legislation (Halpern, 2015). With these traditional ideas of how the state could be used now being politically rejected, the idea of nudging rose to the fore as an alternative (Eaglesham, 2008). Promising that the state could influence behavior without mandating actions (e.g., legislation) or incentivizing activities (e.g., tax and spend), nudging became a "microeconomic" complement of macroeconomic program of fiscal austerity, and represented what George Osborne (UK Chancellor of the Exchequer, 2010–16) and Richard Thaler would call, "a fundamentally conservative approach", to government (Osborne & Thaler, 2010, para. 8).

The link between nudging and austerity in the United Kingdom is well evidenced in various materials from the time. David Halpern (2015, p. 13) is explicit when discussing how the *idea* of austerity had transformed the *technologies of the state*, and how the *idea* of nudging could further this transformation:

> Britain, like many other countries, was in the grip of austerity. Most departments faced major cuts. . . At the same time, the new government was pushing them to deregulate, cut red tape and avoid legislation. In effect, the two main tools that most departments had relied on for the past 50 years – spending money and legislation – had been put in a box labelled "do not touch. . . Here in front of them was a tool [nudging] they could use.

Osborne and Thaler (2010, para. 8) championed this narrative too:

> Evidence from behavioural economics and social psychology can't only help us meet our goals more effectively, it can also help us to achieve them more cheaply, and without intrusive and burdensome regulations. This is therefore a fundamentally conservative approach, which can help us to reduce government spending and get the deficit down.

The political complement of the *idea* of nudging, following the *idea* of austerity, was not lost on journalists at the time either, as Eaglesham (2008, para. 6) writing in the *Financial Times* notes:

> The theory appears to offer an answer to one of Mr Cameron's central dilemmas – how to achieve his social goals, such as tackling poverty and obesity, without resorting to. . . centrist state measures.

In sum, the rise of nudging in the United Kingdom is a story of an *ideational technology*. *First*, the GFC created an ideational crisis in economics and policymaking which saw austerity economics and nudging rise to prominence. *Second*, in doing so, the uses of the state were transformed. Austerity transformed how traditional functions of the state – tax, spend, and legislation – could be used, namely, arguing that these functions *could not be used*. Nudging transformed the state through this policy gap with an alternative idea of how the state could be used – to nudge, rather than to mandate or incentivize. As such, changing political–economic conditions (the GFC) changed which ideational technologies were used (i.e., which ideas became influential), which in turn transformed how existing technologies (e.g., the state) were *to be* used.

13.4 BEYOND STATISTICAL SIGNIFICANCE

The growing importance of sustainability in daily life, and of various environmental crises regarding the question of the *continuance* of life, should be understood as an important shift in the political–economic landscape of most states. The SDGs are a particularly powerful statement to contextualize the sustainability challenges facing modern society, and our commentary has close implications with those concerning more "nudge-friendly" goals such as SDG-12 (sustainable consumption) as well as "harder" goals such as SDGs 13, 14, and 15, all of which concern broad environmental action.

Given this shifting landscape, it is interesting to look at nudging, as an ideational technology, to see how, and for what *political* ends, the idea is being deployed. Behavioral scientists are nothing if not ambitious. The broad topic of the environment (which should be taken to include *global warming, climate breakdown, ecological breakdown, sustainability*, and more) has been a steadfast feature of the nudging agenda in this regard. Thaler and Sunstein (2021) include a dedicated chapter to this topic in *Nudge: The Final Edition*, titled "Saving the Planet".

Yet, there remains debate about the veracity of green nudges to "save the planet". Some critics worry that, *at best*, green nudges will distract from more effective policy, and "may give the impression that sufficient progress is being made, reducing the need for other, more aggressive climate policies" (Maki, 2019, p. 439; see also Hagman et al., 2019 and Chapter 14). At worse, such policies may be used as purposeful *distractions* from more effective – but politically inconvenient – policies (Harford, 2022; see also Chater & Loewenstein, 2022 and Chapter 4). Others have been more optimistic about the role of green nudges in climate policy. Fischhoff (2021) has argued that it is plain folly to regard climate policy as *not* having a significant behavioral component (e.g., SDG-12), while van der Linden et al. (2021, p. 431) have expressed their concern that, "traditional policy tools, such as relying on local market mechanisms to raise the price of carbon and drive technological change, may be insufficient, or they may operate on too slow of a timeframe to meet key adaption and mitigation targets" (for a similarly critical perspective on capitalist mechanisms to promote sustainability, see Chapter 17). Based on these concerns, van der Linden et al. (2021) argue alternative policy approaches – namely, *behavioral science and nudging* – should be utilized.

Our broad approach in this chapter has not been to question whether nudges work. Instead, we have focused on *why* nudging has been used in policymaking, *irrespective* of the policy effectiveness of nudging. We continue this approach here.

Advocates of "behavioral climate policy", such as Fischhoff (2021) and van der Linden et al. (2021) have argued that climate policy must be understood as behavioral because most carbon emissions can be attributed to consumption behaviors, broadly defined. This frames climate policy largely as a matter of individual choice, and climate mitigation largely as a matter of changing individual choices, which creates the intellectual space for green nudges to be a viable policy response. For instance, Sparkman et al. (2021, p. 433) argue – specifically concerning climate policy – that social norm interventions could be utilized because people have a "tendency to conform to trends".

While we have no objection to this argument as it is *broadly conceived*, we note that contained within the "consumptive behaviors" that such nudges would target are many consumables which are *economically* driven, rather than behaviorally. For instance, a worker *must drive to work* if there is no public transport alternative, or if the alternative is unaffordable to them. This "behavior" is not driven by a "tendency to conform", but by the infrastructural and economic conditions facing the individual. From this perspective, one may conclude that "traditional policy tools" have been insufficient insofar as they have not been used for transport investment (van der Linden et al., 2021). But this does not necessarily mean that green nudges *are sufficient*; neither does it mean they are *insufficient*. Rather, this is merely a *critique* of how "traditional policy tools" have been used, and *could* be used differently, for instance, investing in public transport and subsidizing its usage.

Once more, we see the importance of ideas. Such a perspective comes from seeing "traditional policy tools", as *ideational* technologies with political lives, rather than *neutral* (see also Chapter 15). Implicitly, authors such as van der Linden et al. (2021) assume the latter. For instance, they argue behavioral insights should be used to educate legislators and the public about the threat of climate change, framing the problem as one of individual knowledge, and implicitly downplaying the role of political and economic interests which others (Harford, 2022; Loewenstein & Chater, 2017; Chater and Loewenstein, 2022) have emphasized (this is not an unusual tendency in more typical discussions of technology and neutrality; see Spencer, 2017). Loewenstein and Chater (2017) argue that this is a common problem of behavioral policy, as all problems are – from the outset – viewed as *behavioral* problems, and thus all solutions – *implicitly* – are regarded as also being behavioral.

A further example of such thinking may be seen in a recent exchange of criticism in *Nature* regarding the effectiveness of pro-environmental interventions. Nisa et al. (2019) undertake a meta-analysis of behavioral interventions designed to change household environmental behaviors and conclude that while the overall effect of these interventions is *positive*, it is so negligibly positive that, for the authors, prioritizing such interventions would be a policy mistake. In a response, van der Linden and Goldberg (2020) replicate the results and argue that Nisa et al. (2019) likely underestimate the average effect size by a factor of two. For van der Linden and Goldberg (2020) this is an important result. In their response, however, Nisa et al. (2020) note that even doubling a negligible effect size does not change the validity of the original conclusions (that such interventions do not substantially help), among a slew of other criticism regarding the replication procedure.

Ultimately, Nisa et al. (2020, p. 2) are quite cutting when they write, "A thoughtful debate beyond statistical significance is long overdue to make psychological and behavioral science more relevant to intervention and policy-making". This statement could, and we would argue *should*, be seen as an acknowledgment that nudges, green or otherwise, have a *political dimension* as much as they have an academic one. The notion that "statistically significant" is *sufficient* for policy ends (either to solve the policy problem or to satisfy some political agenda) is an example of viewing nudging as a neutral technology, rather than one deployed for political–economic purposes (or perhaps failing to see the political life inherent in *all* ideational technologies). Of course, opponents to this view may argue that irrespective of *how* or *why* green nudges are used, evidence of a positive effect suggests using them would be beneficial (van der Linden & Goldberg, 2020). Yet, as we have argued, nudging is an *ideational technology*, which transforms how *other technologies are conceived to be used*. If, through nudging, other more impactful policies are deemed *politically* unnecessary, then the net result may be that *less progress* is made, compared to the counterfactual (Chater & Loewenstein, 2022). In the economics parlance, this is a variation on the notion of *crowding out* (see also Chapter 4).

An interesting example to consider in this debate thus far is the use of green license plates in the United Kingdom. Introduced in December 2020, electric cars registered in the United Kingdom are given a green-colored license plate (UK Department for Transport, 2020). This initiative follows policies already introduced in countries such as Norway and Canada (Helm, 2018) and was explicitly marketed as a "behavioral" policy backed by the BIT (Costa et al., 2018). The rationale was that these license plates would make electric cars more salient, and thus nudge people into purchasing an electric car as they come to be seen as normal. The UK's Transport Secretary described the policy, following the COVID-19 pandemic, as an opportunity to "build back greener" (Shapps, 2020, para. 6). Such a policy is an excellent example to draw upon, for three reasons.

First, while it is too early to determine whether this policy has indeed nudged electric car sales, we can likely already conclude the effect has been rather small. As electric vehicle registration data from the UK Department of Transport (2022) shows, the third quarter of 2020 (the quarter *before* the license plate nudge was introduced) was *already* the highest quarter on record for the registering of hybrid electric (615,000), plug-in electric (182,000), and battery electric (152,000) cars. This is to say, even prior to the nudge, there was an upward trend in electric car consumption; even if this nudge proves to be "statistically significant", it is unlikely to be the driver of electric car sales, pardon the pun.

Second, the nudge is insufficient as part of a mass, pro-environmental policy shift, despite its framing in 2018 (Helm, 2018). Electric cars do not in themselves address the question of energy generation, with a growing *absolute* usage of electric cars potentially *continuing* to perpetuate environmental harm as this coincides with growing energy demand (Marx, 2022). Furthermore, the various extraction and production processes in the manufacture of lithium-ion batteries – key to modern electric vehicles – remain extremely environmentally damaging, as well as politically contentious given the associated resource conflicts which engulf lithium-rich countries such as Bolivia (Agusdinata et al., 2018). This is all to say that even if this nudge has a *substantial impact on behavior* (i.e., increasing electric car usage), it may be a *deleterious* and *harmful* policy when considered outside of its isolation.

Third, the policy may be understood as a resurrection of the political life of nudging in the austerity era of UK politics. In the previous section, we demonstrated how nudging was seen as a means of the government appearing to do *something*, while embarking on a policy program (austerity) designed to reduce the capacities of the state to do *anything* (Eaglesham, 2008). One may make the same argument regarding green license plates; the same government which introduced green license plates has – at the time of writing – greenlit new oil and gas developments in the UK's North Sea (Tidman, 2022) and scraped subsidies for electric car infrastructure (Kollewe, 2022). Just as nudging in the austerity era allowed the government to promote a socially progressive program while reducing economic investment in such programs, so too might one see green license plates as a political tool for bolstering environmental credentials (e.g., seeming to tackle SDG 12), while not *challenging* more meaningful environmental policy (e.g., fossil fuel development projects covered by SDGs 13–15).

13.5 CONCLUSION

In this chapter, we have argued that ideas can function as technologies, insofar as they transform *those who use them*, and insofar as they transform how *other technologies are conceived to be useable*. Ideas are also political–economic creatures, with the political economy – and the social relations contained within – shaping which ideas proliferate, and which falter. We have focused on a policy idea which has been very popular in the past decade, and which continues to influence environmental policy: Nudging.

By considering the political–economic conditions in which the notion of "Nudge", came to be popular, we have demonstrated the notion of an ideational technology. Nudging was able to transform the academic economic paradigm, which was beleaguered by the Global Financial Crisis, and subsequently, transform conceptions of how the state (itself a technology) could be used to influence behavior. Nudging also transformed those who deployed the idea, being of political usage to politicians in the United Kingdom who supported economic austerity, but still wished to pursue various social programs.

Green nudges now exist as a set of policy prescriptions for pro-environmental ends. In this chapter, we have explored some of these ideas, as well as the criticism of them. While all environmental policies may have some "behavioral" component to them (by virtue of almost always involving *humans*), the role of behavioral policy is often less obvious than one may derive from this perspective. Once more, *how* technologies of the state are used, and are *perceived to be useable*, is political–economic in nature. The "failure" of traditional policy tools is not necessarily a failure of the tools *per se*, but a demonstration of how the tools of been chosen to be used; the flipside of this argument is that the "failure" of a traditional tool does not necessitate a behavioral tool be used instead. To fall into such a trap is to risk allowing nudges to crowd out policy.

13.6 REFERENCES

Agusdinata, D. B., Liu, W., Eakin, H., & Romero, H. (2018). Socio-environmental impacts of lithium mineral extraction: Towards a research agenda. *Environmental Research Letters, 13*, e. 123001.

Allen, D. (2009). *The institutional revolution: Measurement and the economic emergence of the modern world.* University of Chicago Press.

Asensio, A. (2014). "The Achilles" heel of the mainstream explanations of the crisis and a post Keynesian alternative. *Journal of Post Keynesian Economics, 36*(2), 355–379.

Blyth, M. M. (1997). "Any more bright ideas?" The ideational turn of comparative political economy. *Comparative Politics, 29*(2), 229–250.

Blyth, M. M. (2002). *Great transformations: Economic ideas and institutional change in the twentieth century.* Cambridge University Press

Blyth, M. M. (2009). Coping with the Black Swan: The unsettling world of Nassim Taleb. *Critical Review, 21*(4), 447–465.

Blyth, M. M. (2013). *Austerity: The history of a dangerous idea.* Oxford University Press.

Braverman, H. (1974). *Labor and monopoly capital: The degradation of work in the twentieth century.* Monthly Review.

Brown, A., & Spencer, D. (2014). Understanding the global financial crisis: Sociology, political economy, and heterodox economics. *Sociology, 48*(5), 938–953.

Brown, C. (1999). Susan strange: A critical appreciation. *Review of International Studies, 25*(3), 531–535.

Carney, M. (2021). *From credit crisis to resilience | Reith lecture.* Retrieved March 1, 2022, from www.youtube.com/watch?v=CTZKraxwQ2s

Chater, N., & Loewenstein, G. (2022). The i-frame and the s-frame: How focusing on individual-level solutions has led behavioral public policy astray. *SSRN.* Retrieved June 30, 2022, from https://papers.ssrn.com/sol3/papers.cfm?abstract_id=4046264

Colander, D., Föllmer, H., Haas, A., Goldberg, M. D., Juselius, K., Kirman, A., Lux, T., & Sloth, B. (2009). *The financial crisis and the systemic failure of academic economics.* University of Copenhagen Department of Economics Discussion Paper No. 09–03

Costa, E., Reiner, C., & Fitzhugh, E. (2018). *How new number plates could green Britain's roads.* Behavioural Insights Team. Retrieved June 28, 2022, from www.bi.team/blogs/how-new-number-plates-could-green-britains-roads/

Eaglesham, J. (2008). *Nudging not nnaying to achieve social goals.* Retrieved January 31, 2022, from https://www.ft.com/content/0e6f1088-6280-11dd-9a1e-000077b07658

Ellul, J. (1964). *The technological society.* Vintage Books.

Feyerabend, P. (2010). *Against method.* Verso Books.

Fischhoff, B. (2021). Making behavioral science integral to climate science and action. *Behavioural Public Policy, 5*(4), 439–453.

Friedman, M. (2002). *Capitalism and freedom.* University of Chicago Press.

Gramsci, A. (1992). *Prison notebooks.* Columbia University Press.

Green, J., & Hay, C. (2015). Towards a new political economy of the crisis: Getting what went wrong right. *New Political Economy, 20*(3), 331–341.

Hagman, D., Ho, E. H., & Loewenstein, G. (2019). Nudging out support for a carbon tax. *Nature Climate Change, 9*, 484–489.

Halpern, D. (2015). *Inside the nudge unit: How small changes can make a big difference.* W. H. Allen.

Halpern, D., Bates, C., Mulgan, G., Aldridge, S., Beales, G., & Heathfield, A. (2004). *Personal responsibility and changing behaviour: The state of knowledge and its implications for public policy.* The National Archives. Retrieved January 24, 2022, from https://webarchive.nationalarchives.gov.uk/ukgwa/+/http:/www.cabinetoffice.gov.uk/media/cabinetoffice/strategy/assets/pr2.pdf

Harford, T. (2022). What nudge theory got wrong. *The Financial Times.* Retrieved June 30, 2022, from www.ft.com/content/a23e808b-e293-4cc0-b077-9168cff135e4

Hay, C. (2004). Ideas, interests and institutions in the comparative political economy of great transformations. *Review of International Political Economy, 11*(1), 204–226.
Heidegger, M. (2010). *Being and time*. State University of New York Press.
Helm, T. (2018). Green number plates 'could boost sales of electric cars' in the UK. *The Guardian*. Retrieved June 28, 2022, from www.theguardian.com/environment/2018/sep/09/green-cars-get-green-number-plates-clean-boost-sales
Heukelom, F. (2012). Three explanations for the Kahneman-Tversky programme of the 1970s. *The European Journal of the History of Economic Thought, 19*(5), 797–828.
IPCC. (2022). *Climate change 2022: Mitigation of climate change*. UN International Panel on Climate Change. Retrieved June 28, 2022, from www.ipcc.ch/report/sixth-assessment-report-working-group-3/
Kahneman, D. (2011). *Thinking, fast and slow*. Penguin.
Kirman, A. (2009). *The economic crisis is a crisis for economic theory*. CESifo Economic Studies Conference on "*What's Wrong with Modern Macroeconomics*?"
Klein, N. (2007). *The shock doctrine*. Penguin Books.
Klein, N. (2014). *This changes everything*. Penguin Books.
Kollewe, J. (2022). Government pulls plug on its remaining UK electric car subsidies. *The Guardian*. Retrieved June 28, 2022, from www.theguardian.com/business/2022/jun/14/government-pulls-plug-on-its-remaining-uk-electric-car-subsidies
Kuhn, T. S. (2012). *The structure of scientific revolutions*. University of Chicago Press.
Latour, B. (1999). *Pandora's hope: Essays on the reality of science studies*. Harvard University Press.
Linton, D. (1992). The Luddites: How did they get that bad reputation? *Labor History, 33*(4), 529–537.
Loewenstein, G., & Chater, N. (2017). Putting nudges in perspective. *Behavioural Public Policy, 1*(1), 26–53.
Lucas, R. E. (2003). Macroeconomic priorities. *American Economic Review, 93*(1), 1–14.
Lux, T., & Westerhoff, F. (2009). Economics crisis. *Nature Physics, 5*, 2–3.
Maki, A. (2019). The potential cost of nudges. *Nature Climate Change, 9*(6), 435–439.
Marshall, A. (1980). *Principles of economics*. Macmillan.
Marx, K. (1973). *Grundrisse*. New Left Review and Penguin Books.
Marx, K. (2013). *Capital*. Wordsworth.
Marx, P. (2022). *Road to nowhere: What silicon valley gets wrong about the future of transportation*. Verso Books.
Mukand, S., & Rodrik, D. (2018). *The political economy of ideas: On ideas versus interests in policymaking*. NBER Working Paper no. 24467. Retrieved May 23, 2022, from https://nber.org/papers/w24467
Nisa, C. F., Bélanger, J. J., Schumpe, B. M., & Faller, D. G. (2019). Meta-analysis of randomized controlled trials testing behavioural interventions to promote household action on climate change. *Nature Communications, 10*(4545), 1–13.
Nisa, C. F., Sasin, E. M., Faller, D. G., Schumpe, B. M., & Bélanger, J. J. (2020). Reply to: Alternative meta-analysis of behavioural interventions to promote action on climate change yields different conclusions. *Nature Communications, 11*(3901), 1–3.
Osborne, G., & Thaler, R. H. (2010). We can make you behave. *The Guardian*. Retrieved January 25, 2022, from www.theguardian.com/commentisfree/2010/jan/28/we-can-make-you-behave
Reinhart, C. M., & Rogoff, K. S. (2009). The aftermath of financial crises. *American Economic Review, 99*(2), 466–472.
Schumpeter, J. A. (1940). The meaning of rationality in the social sciences. *Journal of Institutional and Theoretical Economics, 140*(4), 577–593.
Shapps, G. (2020). *Consultation outcome: Green number plates for ultra-low emission vehicles: Government response*. UK Government Department for Transport. Retrieved June 28, 2022, from www.gov.uk/government/consultations/introduction-of-green-number-plates-for-ultra-low-emission-vehicles/outcome/green-number-plates-for-ultra-low-emission-vehicles-government-response

Simon, H. A. (1955). A behavioral model of rational choice. *The Quarterly Journal of Economics, 69*(1), 99–118.
Solow, R. (2008). The state of macroeconomics. *Journal of Economic Perspectives, 22*(1), 243–246.
Sparkman, G., Howe, L., & Walton, G. (2021). How social norms are often a barrier to addressing climate change but can be part of the solution. *Behavioural Public Policy, 5*(4), 528–555.
Spencer, D. (2017). Work in and beyond the Second Machine Age: The politics of production and digital technologies. *Work, Employment and Society, 31*(1), 142–152.
Stiglitz, J. (2010). Needed: A new economic paradigm. *The Financial Times*. Retrieved May 25, 2022, from www.ft.com/content/d5108f90-abc2-11df-9f02-00144feabdc0
Streeck, W. (2014). *Buying time: The delayed crisis of democratic capitalism*. Verso Books.
Thaler, R. H. (2015). *Misbehaving: The making of behavioural economics*. Penguin.
Thaler, R. H. (2021). *Nudge: The final edition | LSE online event*. London School of Economics and Political Science. Retrieved January 24, 2022, from www.youtube.com/watch?v=FEkfqQAp6wk
Thaler, R. H., & Sunstein, C. R. (2008). *Nudge: Improving decisions about health, wealth, and happiness*. Penguin Books.
Thaler, R. H., & Sunstein, C. R. (2021). *Nudge: The final edition*. Penguin Books.
Tidman, Z. (2022). UK government approves new oil and gas field after finding it 'would not significantly affect environment. *The Independent*. Retrieved June 28, 2022, from www.independent.co.uk/climate-change/news/oil-gas-field-cop26-climate-b2005123.html
Tversky, A., & Kahneman, D. (1974). Judgment under uncertainty: Heuristics and biases. *Science, 185*(4157), 1124–1131.
UK Department for Transport. (2020). *Green number plates get the green light for a zero-emission future*. UK Government Department for Transport. Retrieved June 28, 2022, from www.gov.uk/government/news/green-number-plates-get-the-green-light-for-a-zero-emission-future
UK Department for Transport. (2022). *Vehicle licensing statistics data tables*. UK Government Department for Transport. Retrieved June 28, 2022, from www.gov.uk/government/statistical-data-sets/vehicle-licensing-statistics-data-tables#ultra-low-emissions-vehicles
van der Linden, S., & Goldberg, M. H. (2020). Alternative meta-analysis of behavioral interventions to promote action on climate change yields different conclusions. *Nature Communications, 11*(3915), 1–2.
van der Linden, S., Pearson, A. R., & van Boven, L. (2021). Behavioural climate policy. *Behavioural Public Policy, 5*(4), 430–438.
Wainwright, J., & Mann, G. (2020). *The climate leviathan: A political theory of our planetary future*. Verso Books.
Wallace-Wells, D. (2019). *The uninhabitable earth: A story about the future*. Penguin Books.

CHAPTER 14

The Fallacy of Disruptive Technologies and the Primacy of Politics

Sustainable Development Goals as an Example

Imad Antoine Ibrahim

CONTENTS

14.1 INTRODUCTION

Over the last decade, new technologies have emerged affecting the implementation of the United Nations (UN) Sustainable Development Goals (SDGs). These include Artificial Intelligence (AI), Big Data Analytics, Blockchain, Internet of Things, next-generation robotics, digital agriculture, gene editing, the microbiome, and additive manufacturing, among others (Bringing Ingenuity to Life, n.d). Through them, the international community hopes to ensure the realization of the 17 aims in the areas of science, technology, and innovation (Doyle, 2021; Mohieldin, 2018). Examples include SDG 2 "Zero Hunger", SDG 3 "Good Health and Well-being", SDG 6 "Clean Water and Sanitation for all",

DOI: 10.1201/9781003325086-14

SDG 12 "Responsible Consumption and Production", SDG 16 "Peace, Justice and Strong Institutions", and SDG 17 "Partnerships for the Goals" (UN DESA, n.d.).

Disruptive technologies are game changers in the field of sustainability and all existing sectors as they highlight a paradigm shift in governance (Sousa et al., 2020). This notion is considered part of the theory of "disruptive innovation" established by Christensen in the 1990s. It describes new technologies that can affect and change business models and enable the creation of new markets (Ibrahim et al., 2022). Disruption in this case refers to a variation that leads to the inefficiency of products, services, and processes. This may take place as new technologies cost less, are of a better quality, alter a consumer's behavior, and there may, correspondingly, be new regulations and resource scarcity. Disruptive innovation refers to the introduction of commercial products or services that affect the operations of existing players at various levels, such as industry segments and structures, and the social system. Disruptive technology has the potential of creating disruptive innovation (Millar et al., 2018). The assumption in the literature is that such technologies will significantly improve governance when it comes to sustainable development (Akkucuk, 2021). There is an atmosphere of hope and optimism that has begun resulting in international organizations and states riding the disruptive technology wave (ECOSOC, 2019). Nonetheless, technological progress, if left unchecked, can negatively impact the achievement of the SDGs given the role of Big Tech and the need to regulate an emerging field that remains mysterious to numerous policymakers and regulators (Truby, 2020; see also Chapter 10).

This chapter argues that politics remains the main element impacting the implementation of the 17 goals regardless of whether disruptive technologies are used and despite their well-recorded advantages (Keping, 2018). Even with the technological developments taking place, the political interests of various nations affect whether a specific SDG shall be achieved. This resulted in the emergence of a vast literature on the interplay between politics and governance theoretically and in practice (Eraydin & Frey, 2019) in various fields such as economics (Hickey et al., 2015) and development (Schofield & Caballero, 2015).

To prove the aforementioned claim, the author focuses implicitly and explicitly on several goals, specifically SDGs 2, 3, 6, 12, 16, and 17. This chapter analyzes these objectives through two case studies. The first concerns the use of Big Data for transboundary water resource management. The second addresses the deployment of disruptive technologies in the agricultural sector. This chapter first starts with a brief overview of the interplay between technology and politics. It then examines the two examples. In both cases, a focus on the impact of politics is emphasized.

14.2 TECHNOLOGY AND POLITICS

The interplay between technology and politics has been discussed in the literature from various angles (Papacharissi, 2015; Kurgan, 2013; Street, 1992; Wills, 2008; Winner, 1977). This is because the rules applicable to technology have a political nature establishing boundaries, rights, and obligations in the public and private domains (Sussman, 1997) and impact the governance system in place. This has been the case, for instance, with the rise of the Internet affecting political regimes (Weare, 2006). The relationship between these two notions changes with the emergence of new eras where the state attempts to regulate and use technological developments for political purposes (Schot, 2003). For instance, the government adopts policies concerning the diffusion of renewable energy (Jacobsson & Lauber, 2006). Technology also drives politics related to energy and climate change (Schmidt & Sewerin, 2017). Sometimes, such developments are banned when it does not suit the state's political goals (Meckling & Nahm, 2019). All this leads to technological and political change based on the situation (Street, 1992).

Globally, questions are raised on whether technological progress or politics will shape the course of international governance (Levy, 1975) and as to why certain nations are better than others in the technological and scientific fields (Taylor, 2016). This is mainly because such developments are seen through the lens of power in international relations (McCarthy, 2015), as its impact on politics has already been noticed. However, scholars have argued that there is an overestimation of its effect on nations, cultures, and military innovations (Deutsch, 1959). Others also claim that an exaggeration is taking place concerning the importance of politics in technological advancement (Asdal et al., 2008). This takes place in a context where some see technology as a means to advance society, whereas others are concerned about its negative consequences (Mayer et al., 2014). This has resulted in the emergence of various concepts to highlight this interdependency, such as "Digital Politics" (Karpf, 2017), "Technological Determinism" (Agre, 2002), "Voluntarist Views of Technology" (van der Ploeg, 2003), and "Technopolitics" (Kurban et al., 2017). Some have gone further in their classification. For instance, Jasanoff (2008, p. 745) argued that technology as a "site and object of politics displays itself clearly in four linked yet separate aspects: as risk, as design, as standard, and as ethical constraint". This focus on the classification of technology is understandable given the complete reliance on it in all sectors (Asaro, 2000) and the public misinterpretation and understanding of its role in society (Bromley, 2002), and its occasional perception as a threat (González, 2005; Van Slyke, 2008). This has resulted in the emergence of Science and Technology Studies and Science, Technology, and Society (STS) as new fields tackling politics, governance, and regulatory practices, among other areas of concern (Irwin & Wynne, 1996; Webster, 1991).

Society is currently being ordered and organized in keeping with existing technologies, which has resulted in various political questions with respect to its neutrality and legitimacy as a technical means to achieve social ends (Introna, 2007). This is where politics is important as policymakers and other stakeholders such as industries and civil society organizations are seeking to integrate societal and ethical considerations into the technological sphere. The objectives are to address potential future technological problems, ensure the adaptability of the governance system, and allow citizen participation among other things (van Oudheusden, 2014). Thus, the relationship between technology and politics and the analysis of the interplay between both is expected to gain further importance (Brown, 2015), as technology is not neutral but impacted by politics (Delvenne & Parotte, 2019) and the latter is influenced by technological evidence (May, 2006).

This is why the law plays an important role in technological progress. Politics rely on the legislation that creates and regulates it in the general framework of justice and social order. It is the driving force behind the establishment of legal norms (Cerar, 2009). Politics include decisions taken by a group or part of a group, often in the form of a government concerning governance matters. The law is used to implement the outcome representing the authority of those who govern (Alexander, 2018). Legislations concerning technologies have already developed and are expected to be adopted in the future, reflecting politicians views on this matter (Salmerón-Manzano, 2021; Malby, 2018; Gifford, 2007; Mandel, 2007). It is a compromise that needs to be made to ensure that innovation is not stifled while citizens are still protected (Tranter, 2011). This is complicated as the law cannot catch up with the rapid technological advancements (Griffith, 2019).

14.3 BIG DATA FOR TRANSBOUNDARY WATER GOVERNANCE

This section tackles the impact of politics on the use of big data in the management of shared freshwaters. It shows that despite the promises of technology, interstate politics remain the principal factor affecting the way this resource is governed, thus influencing directly the realization of SDG 6 and indirectly goals 3, 12, 16, and 17.

14.3.1 Background

Water flow in a hydrological cycle, which does not recognize the existence of political boundaries or physical characteristics. Having data and information on this resource's cycle is essential for its long- and short-term governance. These data are used for monitoring, planning, policymaking, and designing infrastructure. Other types also affecting water management include data on policies and regulations, engineering, culture, and the various modes of water use, be it agricultural, industrial, and household consumption, among others. Data on the links between water resources and the ecosystem are equally important (Leb, 2020).

International water conventions and instruments, mainly the Convention on the Law of the Non-navigational Uses of International Watercourses (1997), the Convention on the Protection and Use of Transboundary Watercourses and International Lakes (1992), and the 2008 Draft Articles on the Law of Transboundary Aquifers (Draft Articles, 2008) include explicit provisions related to the regular data and information exchange; the notification of natural and human triggered emergencies and planned measures (Ibrahim et al., 2022). Similar provisions are found in basin water agreements such as the revised Southern African Development Community Water Protocol (SADC, 2000).

The inability, unwillingness, absence, and insufficient exchange of data and information across state borders is one of the most challenging problems that affects the management of transboundary water resources. Nations can have the technical capacity to collect and share water data but may withhold such information to protect their bargaining positions. They might also lack the ability to collect and exchange data given the great costs associated with the process and the lack of institutional capacity (Ibrahim, 2020). Various factors culminate in this situation, such as the absence of compatible needs, mistrust among parties, and the inability to perceive benefits from cooperation (Chenoweth & Feitelson, 2001). New technologies are considered a means to address these issues (Gupta et al., 2020; World Economic Forum, 2018).

In the transboundary water context, data are collected from different institutions and disciplines such as social and hydrological sciences. Its great volume has emerged from various sources like satellites, social media, and monitoring stations providing instant information that must be factored into the decision-making process (Ibrahim et al., 2022). Its veracity is questioned as it is not clear as to what data are valuable in drawing up policies. Big Data through machine learning holds the promise of collecting, storing, managing, processing, and providing added value from a huge amount of variable water data that emerges rapidly. This new information, which is considered credible and timely, can be used for the management of shared water resources (Ibrahim, 2020). The importance of Big Data led to the establishment of projects such as the Big Data Analytics and Transboundary Water Collaboration for Southern Africa initiative to improve regional water management through this disruptive technology (Bocchino & Adkisson, 2020). The opportunities for Big Data in the water sector have led to an increased focus in the literature on this topic (Gohil et al., 2021).

14.3.2 Role of Politics

Water management in the transboundary context is extremely complicated. Each state usually looks to secure the biggest quantity of this resource, which is considered a matter of national security. The importance of politics has resulted in the emergence of a new term called "hydropolitics", based on which scholars have concluded that shared freshwater is simultaneously a source of conflict and cooperation (Bréthaut et al., 2022). Factors that play a role include whether a state has asymmetrical power in comparison to other nations that share the resource and whether the country holds an upstream or downstream position (Mirumachi & Allan, 2007). This is worsened by the overuse and pollution of shared freshwaters. Climate change, the rise in global population, and economic growth have impacted them negatively, too.

The literature shows that water challenges are mainly governance crises (Johns & VanNijnatten, 2021) and a political issue. The effectiveness of existing water conventions and instruments at the international, regional, and basin levels and joint water institutions depends on the political willingness of states to abide by established rules to solve matters such as water pollution (Ibrahim, 2020). Politics has played an important role in the drafting and adoption of international water conventions (McCaffrey, 2008), to the point where many provisions in these treaties have a political character (Eckstein, 2008). Its impact is far more obvious at the basin level such as the Nile river basin, which is shared among several countries, each of which has different interests and needs (Brunnie, 2008).

States may have different political positions and attitudes toward the use of big data, based on whether it will positively or negatively impact their water needs. Various scenarios can unfold as all or few nations or even one may decide to use Big Data. It is also possible that none of them use it. If all states agree to use Big Data, problems may unfold if they disagree over key areas of concern. If some states or even one nation agrees to deploy it, but others do not, more challenges may emerge. Why would any of the other countries accept the results of data processing if they do not participate in the process? What if each state uses its own big datasets and draws out different results? What data should be adopted to avoid bias? Why would nations accept new data that may leave them with less favorable conditions? Should a third party take on data processing to ensure objectivity (Ibrahim, 2020)? These are some of the political challenges facing big data; other technologies like geoengineering are likely to face similar issues (see Chapter 4).

When disruptive technologies are deployed with the promise of enhancing transboundary water management, politics remains the core element in deciding whether actual changes will occur. Suggestions have been made for the adoption of legal mechanisms to reduce the impact of politics and allow the efficient use of Big Data. These include the development of water protocols at the basin level, tackling only the question of big data including provisions pertaining to the integration of water and information from multiple databases and the establishment of water data administration funds or accounts to finance such operations. Globally, incorporating a new provision on disruptive technologies in international water law has been suggested (Ibrahim et al., 2022). Only time will tell whether such suggestions will be considered and will yield the needed outcomes.

Even though technology has great potential for the realization of SDG 6, on clean water and sanitation in the transboundary context, politics has been decisive on whether this objective shall be achieved. This reality indirectly affects the implementation of SDGs 3, 12, 16, and 17 (UN DESA, n.d.). The good management of shared water resources is essential for the health and well-being of the population, especially as water needs to be exploited responsibly through strong and just joint institutions to guarantee peace among states sharing it.

14.4 DISRUPTIVE TECHNOLOGIES IN THE AGRICULTURAL FIELD

This section shows that despite the proven positive influence of disruptive technologies in agriculture, a wide range of concerns pertaining to the relationship between farmers and agritech companies require the political intervention of the state. The aim is to regulate the relationship between both parties and ensure the adequate realization of SDG 2 explicitly and SDGs 3, 8, 9, 10, 12, 15, 16, and 17 implicitly.

14.4.1 Background

Agriculture is a risky business given its

> relatively low operational efficiency and small managerial power due to farm size limitations, a high level of uncertainty because of weather and environmental conditions, and a volatile balance between food supply and demand due to growing and breeding times of crops and livestock.
>
> (Osinga et al., 2022, p. 2)

Thus, investments were and are made to improve the efficacy of operations and reduce uncertainties. The aim is to ensure the production of food with great nutritional value; constant food supply and reduction of environmental harm; and various types of benefits including ecological, social, and economic ones (Osinga et al., 2022). Farming is complex and involves various costs such as those for labor and land. Farmers use expensive machines and equipment, fertilizers, and pesticides, and ensure proper irrigation (Paraforos et al., 2016). The agri-food sector is challenged by population growth and climate change, which results in environmental degradation (land, water, and air) in addition to loss of biodiversity and increase in foodborne diseases (Leader et al., 2020).

New technologies are reshaping the agricultural sector, which has led to the emergence of Agricultural Technology (AgriTech) as a field (Spanaki et al., 2022). These include smart farming technologies that have supported agricultural practices, digital ones that have applied big data and machine learning, and precision agriculture focused on using data collected from satellites and other technologies to improve productivity and reduce costs. These include AI, drones, crop monitoring, farming robotics, autonomous transport, radio frequency identification sensors, tracking, and Machine Learning and Analytics (Leader et al., 2020). Disruptive technologies focus on increasing productivity and decreasing economic and physical burdens, among others. The aim is to facilitate the work of a farmer by tackling difficult, unwanted and tiring work, which will give them time to focus on improving and developing the farm. For instance, AI allows the use of unmanned agricultural machinery, and Big Data leads to a detailed analysis of farming data to understand the ways in which farming practices can be improved (Ryan, 2020).

Globally, efforts have been made for the establishment of an International Digital Council for Food and Agriculture at the 12th Global Forum for Food and Agriculture in January 2020 in Berlin. Its role is to ensure the good use of digital technologies in the agricultural sector and is to be supported by various units and national governments. Additionally, an International Platform for Digital Food and Agriculture under the Food and Agriculture Organization framework was created (FAO, 2020). The International Organization for Standardization (ISO) has issued standards governing the use of digital technologies in agriculture and has established committees to that end (Gasiorowski-Denis, 2017; ISO, 2017). The World Bank, the Organisation for Economic Co-Operation and Development (OECD), and others have also begun addressing this topic (Mattson, 2019; OECD, n.d.).

14.4.2 Role of Politics

Despite the opportunities highlighted earlier due to the use of disruptive technologies in the agricultural sector, various political obstacles have emerged in recent years, requiring state intervention at the domestic level and the international community's attention (Wiseman et al., 2019; OECD, 2020). Innovation has led to the emergence of agritech companies such as Monsanto, Bayer, and John Deere. These are profit-driven, which has resulted in political practices that negatively affect farmers who are the main beneficiaries of their services (Moon, 2019).

Owing to the great costs associated with investing in new technologies and human capital, as well as technical knowhow, few agritech companies are expected to have the lion's share of the market, resulting in monopolies dictating the terms and conditions to farmers who are set to lose the most from this unbalanced situation (Schönfeld et al., 2018; Shastry & Sanjay, 2020). This is already resulting in farmers complete dependency on agritech service providers through strict and often harsh legal contracts that limit their capacity to abandon the contract altogether. Additionally, due to this, they are unable to seek the services of other agritech providers (Ryan, 2020). This is worsened by the fact that the agritech company with whom the farmer has signed a contract has all the data on the farm and the farmer does not know how these data are being used, which implies information asymmetry between both parties. The farmer does not know whether their agricultural data are disclosed to a third party, and whether agritech, as a result of these data, may impose different legal conditions on each farmer on a case-by-case basis. The legal contract signed between both parties does not protect the farmer, as they are usually extremely complicated and long-winded agreements (Ryan, 2020; Carbonell, 2016). Thus, the state has to intervene politically to organize the relationship. This has resulted in new and innovative laws. For example, the European Union adopted the Code of Conduct on Agricultural Data Sharing by Contractual Agreement to regulate data sharing in the field. The code contains detailed provisions on data ownership, access, control, portability, protection, transparency, privacy, and security; and liability and Intellectual Property Rights (European Union Code of Conduct, 2018). The American Farm Bureau Federation (AFBF) adopted the Privacy and Security Principles for Farm Data, which guide companies on education, ownership, collection, access, control, notice, transparency, and consistency, choice, portability, terms and definitions, disclosure, use and sale limitation, data retention and availability, contract termination, unlawful and anti-competitive activities, and liability and security safeguards (Ag Data Transparent, n.d.). These attempts, alongside the international community's efforts through the International Digital Council for Food and Agriculture, the International Platform for Digital Food and Agriculture, and ISO standards, aim to address the political challenges emerging from the use of disruptive technologies in agriculture.

Thus, in the agricultural field and despite the promises of disruptive technology as seen earlier, politics remains essential in deciding whether SDG 2 shall be achieved. It also affects the indirect implementation of SDGs 3, 8, 9, 10, 12, 15, 16, and 17 (UN DESA, n.d.). Good governance in the agricultural sector has impacts on land, health, well-being, work opportunities especially for farmers, and the overall economic growth of the country. It has a great influence on the industries, mainly those investing in this sector and ensuring that equality and responsible consumption and production take place. All this must unfold while ensuring the existence of peaceful and inclusive societies within the general framework of partnerships among many stakeholders.

14.5 CONCLUSION

Politics remain the core element that determines whether disruptive technologies can help improve the implementation of the SDGs (Ibrahim et al., 2022). The question is then how we address the interplay between politics and technology in the sustainability field. One way is to tackle this through regulations (see Chapter 10). As seen earlier, rules are being developed internationally, regionally, nationally, and locally to address this topic and for the realization of the 17 goals (Schönfeld et al., 2018). Scientists should be given a greater role to play in the decision-making process to avoid having politicians with different agendas drive the use of these technologies in the sustainability field (Siddhpura et al., 2020; Jameson, 2014). This will not solve the problem of politics but rather help reduce its impact (Susskind, 2018; McCarthy, 2018; Hilpert, 2016; Jacobsen, 2015).

It is left to the voluntary will of each state to use disruptive technologies efficiently for the fulfillment of the SDGs and report on the application through the follow-up and review mechanism established for tracking progress (High-Level Political Forum on Sustainable Development, 2021; VNR, 2019, 2020). This would happen once a nation sees that the political benefits far outweigh disadvantages at various levels (Ibrahim et al., 2022). The growing literature on the topic has highlighted the increasing interest in the interplay between the SDGs and disruptive technologies by various stakeholders, including states, which will eventually impact the role of politics in this area (Iizuka & Hane, 2021; UNCTAD, 2019; IISD, 2017; ITU, 2021).

14.6 REFERENCES

Ag Data Transparent. (n.d.). *Ag data's core principles: The privacy and security principles for farm data.* www.agdatatransparent.com/principles#:~:text=Farm%20data%20should%20be%20protected, a%20breach%20should%20be%20established.

Agre, P. E. (2002). Real-time politics: The internet and the political process. *The Information Society, 18*(5), 311–331.

Akkucuk, U. (Ed.). (2021). *Disruptive technologies and eco-innovation for sustainable development.* IGI Global.

Alexander, L. (2018). Law and politics: What is their relation? *Harvard Journal of Law & Public Policy, 41*(1), 355–363.

Asaro, P. M. (2000). Transforming society by transforming technology: The science and politics of participatory design. *Accounting, Management and Information Technologies, 10*(4), 257–290.

Asdal, K., Borch, C., & Ingunn Moser, I. (2008). The technologies of politics. *Distinktion: Journal of Social Theory, 9*(1), 5–10.

Bocchino, C., & Adkisson, K. (2020, July 28). Using big data analytics for transboundary water management. *New Security Beat.* www.newsecuritybeat.org/2020/07/big-data-analytics-transboundary-water-management/

Bréthaut, C. et al. (2022). Exploring discursive hydropolitics: A conceptual framework and research agenda. *International Journal of Water Resources Development, 38*(3), 464–479.

Bringing Ingenuity to Life. (n.d.). *Achieving the UN sustainable development goals.* www.paconsulting.com/insights/sustainability/united-nations-global-compact/

Bromley, D. A. (2002). Science, technology, and politics. *Technology in Society, 24*(1/2), 9–26.

Brown, M. B. (2015). Politicizing science: Conceptions of politics in science and technology studies. *Social Studies of Science, 15*(1) 3–30.

Brunnie, J. (2008). Law and politics in the Nile basin. *Proceedings of the ASIL Annual Meeting, 102,* 359–363.

Carbonell, I. M. (2016). The ethics of big data in big agriculture. *Internet Policy Review, 5*(1), 1–13.

Cerar, M. (2009). The relationship between law and politics. *Annual Survey of International & Comparative Law, 15*(1), 19–23.

Chenoweth, J. L., & Feitelson, E. (2001). Analysis of factors influencing data and information exchange in international river basins: Can such exchanges be used to build confidence in cooperative management? *Water International, 26*(4), 499–512.

Convention on the Protection and Use of Transboundary Watercourses and International Lakes, March 17, 1992. https://unece.org/fileadmin/DAM/env/water/pdf/watercon.pdf

Draft articles on the Law of Transboundary Aquifers. Adopted in 2008. https://legal.un.org/ilc/texts/instruments/english/draft_articles/8_5_2008.pdf

Delvenne, P., & Parotte, C. (2019). Breaking the myth of neutrality: Technology assessment has politics, technology assessment as politics. *Technological Forecasting and Social Change, 139*, 64–72.

Deutsch, K. W. (1959). The impact of science and technology on international politics. *Quantity and Quality, 88*(4), 669–685.

Doyle, M. (2021, June 30). The role of sustainable technologies in achieving the SDGs. *Academy Insights*. https://theworldwecreate.net/insights/role-of-sustainable-technologies-in-achieving-sdgs

Eckstein, G. (2008). Examples of the political character of international water law. *Proceedings of the ASIL Annual Meeting, 102*, 364–366.

ECOSOC, United Nations Economic and Social Council, Commission on Science and Technology for Development. (2019, May 13–17). *The impact of rapid technological change on sustainable development*. Report of the Secretary-General; Twenty-second session; Geneva; Item 3 (a) of the provisional agenda; Distr.: General, March 4, 2019.

Eraydin, A., & Frey, K. (Eds.). (2019). *Politics and conflict in governance and planning: Theory and practice*. Routledge.

European Union Code of Conduct on agricultural data sharing by contractual agreement adopted in 2018.

FAO, Food and Agriculture Organization of the United Nations. (2020). *Hundred and sixty-fourth session; 6–10 July; international platform for digital food and agriculture*. Annex: Realizing the Potential of Digitalization to Improve the Agri-Food System: Proposing a New International Digital Council for Food and Agriculture: A Concept Note.

Gasiorowski-Denis, E. (2017, May 9). *The future of farming*. International Organization for Standardization. www.iso.org/news/Ref2183.htm

Gifford, D. J. (2007). Law and technology: Interactions and relationships. *Minnesota Journal of Law Science & Technology, 8*(2), 571–587.

Gohil, J. et al. (2021). Advent of big data technology in environment and water management sector. *Environmental Science and Pollution Research, 28*, 64084–64102.

González, W. J. (Ed.). (2005). *Science, technology and society: A philosophical perspective*. Netbiblo.

Griffith, J. (2019, April 12). A losing game: The law is struggling to keep up with technology. *Journal of High Technology Law at Suffolk University Law School*. https://sites.suffolk.edu/jhtl/2019/04/12/a-losing-game-the-law-is-struggling-to-keep-up-with-technology/

Gupta, A. D. et al. (2020). Smart water technology for efficient water resource management: A review. *Energies, 13*, 1–23.

Hickey, S., Sen, K., & Bukenya, B. (Eds.). (2015). *The politics of inclusive development: Interrogating the evidence*. Oxford University Press.

High-Level Political Forum on Sustainable Development. (2021). *Voluntary national reviews at the 2021 High-level political forum on sustainable development*. Secretariat Background Note.

Hilpert, U. (Ed.). (2016). *Routledge handbook of politics and technology*. Routledge.

Ibrahim, I. A. (2020). Legal implications of the use of big data in the transboundary water context. *Water Resources Management, 34*, 1139–1153.

Ibrahim, I. A., Brown, R. D., & Jon Truby, T. (2022). *Big data analytics and its impact on basin water agreements and international water law: A study of the Ramotswa aquifer*. Brill.

Iizuka, M., & Hane, G. (2021). Towards attaining the SDGs: Cases of disruptive and inclusive innovations. *Innovation and Development, 11*(2/3), 343–364.

International Institute for Sustainable Development. (2017, October 11). *Why innovation is critical to achieving the sustainable development goals*. www.iisd.org/articles/insight/why-innovation-critical-achieving-sustainable-development-goals

ISO. (2017). Smart farming. *ISO Focus, 12*.

International Telecommunication Union. (2021). *Digital technologies to achieve the UN SDGs*. www.itu.int/en/mediacentre/backgrounders/Pages/icts-to-achieve-the-united-nations-sustainable-development-goals.aspx

Introna, L. D. (2007). Maintaining the reversibility of foldings: Making the ethics (politics) of information technology visible. *Ethics and Information Technology*, *9*, 11–25.

Irwin, A., & Wynne, B. (1996). *Misunderstanding science? The public reconstruction of science and technology*. Cambridge University Press.

Jacobsen, K. L. (2015). *The politics of humanitarian technology: Good intentions, unintended consequences and insecurity*. Routledge.

Jacobsson, S., & Lauber, V. (2006). The politics and policy of energy system transformation–explaining the German diffusion of renewable energy technology. *Energy Policy*, *34*(3), 256–276.

Jameson, J. L. (2014). Disruptive innovation as a driver of science and medicine. *The Journal of Clinical Investigation*, *124*(7), 2822–2826.

Jasanoff, S. (2008). Technology as a site and object of politics. In R. E. Goodin & C. Tilly (Eds.), *The Oxford handbook of contextual political analysis* (pp. 745–764). Oxford University Press.

Johns, C., & VanNijnatten, D. (2021). Using indicators to assess transboundary water governance in the Great Lakes and Rio Grande-Bravo regions. *Environmental and Sustainability Indicators*, *10*, 1–13.

Karpf, D. (2017). Digital politics after Trump. *Annals of the International Communication Association*, *41*(2), 198–207.

Keping, Y. (2018). Governance and good governance: A new framework for political analysis. *Fudan Journal of the Humanities and Social Sciences*, *11*, 1–8.

Kurban, C., Peña-López, I., & Haberer, M. (2017). What is technopolitics? A conceptual schema for understanding politics in the digital age. *Revista de Internet, Derecho y Política*, *24*(24), 3–20.

Kurgan, L. (2013). *Close up at a distance: Mapping, technology, and politics*. Zone Books.

Leader, J. et al. (2020). *Disruptive technologies in the agri-food sector*. Ontario Ministry of Agriculture, Food and Rural Affairs (OMAFRA).

Leb, C. (2020). *Data innovations for transboundary freshwater resources management: Are obligations related to information exchange still needed?* Brill.

Levy, S. A. (1975). INTELSAT: Technology, politics and the transformation of a regime. *International Organizations*, *29*(3), 655–680.

Malby, S. (2018). Strengthening the rule of law through technology. *Commonwealth Law Bulletin*, *43*(3/4), 307–317.

Mandel, G. N. (2007). History lessons for a general theory of law and technology. *Minnesota Journal of Law Science & Technology*, *8*(2), 551–570.

Mattson, S. (2019, June 13). *CIAT, World Bank and partners announce digital agriculture country profiles initiative*. https://blog.ciat.cgiar.org/ciat-world-bank-and-partners-announce-digital-agriculture-country-profiles-initiative/

May, C. (2006). Mobilising modern facts: Health technology assessment and the politics of evidence. *Sociology of Health & Illness*, *28*(5), 513–532.

Mayer, M., Carpes, M., & Knoblich, R. (2014). The global politics of science and technology: An introduction. In M. Mayer, M. Carpes, & R. Knoblich (Eds.), *The global politics of science and technology: Concepts from international relations and other disciplines* (pp. 1–35). Springer.

McCaffrey, S. C. (2008). Introduction: Politics and sovereignty over transboundary groundwater. *Proceedings of the ASIL Annual Meeting*, *102*, 353–355.

McCarthy, D. R. (2015). *Power, information technology, and international relations theory: The power and politics of US foreign policy and the internet*. Springer.

McCarthy, D. R. (Ed.). (2018). *Technology and world politics: An introduction*. Routledge.

Meckling, J., & Nahm, J. (2019). The politics of technology bans: Industrial policy competition and green goals for the auto industry. *Energy Policy*, *126*, 470–479.

Millar, C., Lockett, M., & Ladd, T. (2018). Disruption: Technology, innovation and society. *Technological Forecasting and Social Change, 129*, 254–260.

Mirumachi, N., & Allan, J. A. (2007). Revisiting transboundary water governance: Power, conflict cooperation and the political economy. In *Proceedings from CAIWA international conference on adaptive and integrated water management: Coping with scarcity* (Vol. 1215). Basel, Switzerland.

Mohieldin, M. (2018, May 10). Leveraging technology to achieve the sustainable development goals. *World Bank Blogs*. https://blogs.worldbank.org/voices/leveraging-technology-achieve-sustainable-development-goals

Moon, E. (2019). *Big ag monopolies have stifled small farmers. 2020 democrats want to break them up.* https://psmag.com/social-justice/big-ag-monopolies-have-stifled-small-farmers-2020-democrats-want-to-break-them-up

Organization for Economic Co-Operation and Development. (2020). *Issues around data governance in the digital transformation of agriculture – The farmers perspective*. Trade and Agriculture Directorate Committee for Agriculture; Working Party on Agricultural Policies and Markets.

Organisation for Economic Co-operation and Development. (n.d.). *New technologies and digitalisation are transforming agriculture and offering new opportunities to improve policy.* www.oecd.org/agriculture/topics/technology-and-digital-agriculture/

Osinga, S. A. et al. (2022). Big data in agriculture: Between opportunity and solution. *Agricultural Systems, 195*, 1–12.

Papacharissi, Z. (2015). *Affective publics: Sentiment, technology, and politics.* Oxford University Press.

Paraforos, D. S. et al. (2016). A farm management information system using future internet technologies. *IFAC-PapersOnLine, 49*(16), 324–329.

Ryan, M. (2020). Agricultural big data analytics and the ethics of power. *Journal of Agricultural and Environmental Ethics, 33*, 49–69.

SADC – Southern African Development Community. (2000). *Revised protocol on shared watercourses.* https://www.sadc.int/document/revised-protocol-shared-watercourses-2000-english

Salmerón-Manzano, E. (2021). Laws and emerging technologies. *Laws, 10*(2), 2–3.

Schmidt, T. S., & Sewerin, S. (2017). Technology as a driver of climate and energy politics. *Nature Energy, 2*(17084), 1–3.

Schofield, N., & Caballero, G. (Eds.). (2015). *The political economy of governance: Institutions, political performance and elections.* Springer.

Schönfeld, M. V., Heil, R., & Bittner, L. (2018). Big data on a farm–Smart farming. In T. Hoeren & B. Kolany-Raiser (Eds.), *Big data in context: Legal, social and technological insights* (pp. 109–120). Springer.

Schot, J. (2003). The contested rise of a modernist technology. In T. J. Misa, P. Brey, & A. Feenberg (Eds.), *Modernity and technology* (pp. 257–278). The MIT Press.

Shastry, K. S., & Sanjay, H. A. (2020). Data analysis and prediction using big data analytics in agriculture. In K. Pattnaik, R. Kumar, & S. Pal. (Eds.), *Internet of things and analytics for agriculture* (pp. 201–224). Springer.

Siddhpura, A. et al. (2020). Current state of research in application of disruptive technologies in engineering education. *Procedia Computer Science, 172*, 494, 494–501.

Sousa, M. J., Melé, P. M., & Gómez, J. M. (2020). Technology, governance, and a sustainability model for small and medium-sized towns in Europe. *Sustainability, 12*(3), 1–15.

Spanaki, K. et al. (2022). Disruptive technologies in agricultural operations: A systematic review of AI-driven Agritech research. *Annals of Operations Research, 308*, 491–524.

Street, J. (1992). *Politics and technology.* The Macmillan Press.

Susskind, J. (2018). *Future politics: Living together in a world transformed by tech.* Oxford University Press.

Sussman, G. (1997). *Communication, technology, and politics in the information age.* Sage Publications.

Taylor, M. Z. (2016). *The politics of innovation: Why some countries are better than others at science and technology.* Oxford University Press.

Tranter, K. (2011). The laws of technology and the technology of law. *Griffith Law Review, 20*(4), 753–762.

Truby, J. (2020). Governing artificial intelligence to benefit the UN sustainable development goals. *Sustainable Development, 28*(4), 946–959.

UN DESA. (n.d.). *Do you know all 17 SDGs?* United Nations Department of Economic and Social Affairs. https://sdgs.un.org/goals

United Nations Conference on Trade and Development. (2019, May 13). *Deploying disruptive innovations for the SDGs.* https://unctad.org/news/deploying-disruptive-innovations-sdgs

United Nations Convention on the Law of the Non-navigational Uses of International Watercourses, May 21, 1997, https://legal.un.org/ilc/texts/instruments/english/conventions/8_3_1997.pdf

van der Ploeg, I. (2003). Biometrics and privacy: A note on the politics of theorizing technology. *Information, Communication & Society, 6*(1), 85–104.

van Oudheusden, M. (2014). Where are the politics in responsible innovation? European governance, technology assessments, and beyond. *Journal of Responsible Innovation, 1*(1), 67–86.

Van Slyke, C. (2008). *Information communication technologies: Concepts, methodologies, tools, and applications.* IGI Global.

Voluntary National Reviews. (2019). *Submitted to the 2019 high-level political forum for sustainable development: A comparative analysis.* https://sdghelpdesk.unescap.org/e-library/voluntary-national-reviews-submitted-2019-high-level-political-forum-sustainable

Voluntary National Reviews. (2020). *A snapshot of trends in SDG reporting.* https://sdghelpdesk.unescap.org/e-library/2020-voluntary-national-reviews-snapshot-trends-sdg-reporting

Weare, C. (2006). The internet and democracy: The causal links between technology and politics. *International Journal of Public Administration, 25*(5), 659–691.

Webster, A. (1991). *Science, technology and society: New directions.* Palgrave Macmillan.

Wills, D. (2008). *Dorsality: Thinking back through technology and politics.* University of Minnesota Press.

Winner, L. (1977). *Autonomous technology: Technics-out-of-control as a theme in political thought.* MIT Press.

Wiseman, L. et al. (2019). Farmers and their data: An examination of farmers reluctance to share their data through the lens of the laws impacting smart farming. *NJAS-Wageningen Journal of Life Sciences, 90/91*, 1–10.

World Economic Forum. (2018). Harnessing the Fourth Industrial revolution for water. *Fourth Industrial Revolution for the Earth Series.* https://www3.weforum.org/docs/WEF_WR129_Harnessing_4IR_Water_Online.pdf

CHAPTER 15

Technology and the Distribution of Power

Faridun Sattarov

CONTENTS

15.1 INTRODUCTION

The notion of techno-solutionism has emerged to capture a rather modern phenomenon – attempts of using engineering and technology to solve complex social problems, which are themselves often the result of past technological interventions (see Chapter 2; Morozov, 2013; see also Stilgoe, 2020). As the other chapters of this book have amply shown, while technological ingenuity has been the driving force for many great innovations, technological solutions can create more problems than they solve. One such problem concerns the impact of technology on the distribution and exercise of power in society. Technological fixes can generate inequalities of power, which can have significant implications for society, be they moral, social, economic, or political.

Inequalities of power, in their turn, can impact socially sustainable development and constrain the achievement of the Sustainable Development Goals (SDGs). The SDGs were introduced by the United Nations in the 2015 publication titled *Transforming our World: The 2030 Agenda for Sustainable Development* (United Nations, 2015). The 17 goals (see Chapter 2) comprise a wide array of aims, emphasizing people, planet, prosperity, peace, and partnership (United Nations, 2015). Of these 17 SDGs, at least 6 can directly be impacted by inequalities of power: "Reduced inequalities" (SDG 10), "Gender equality" (SDG 5), "Decent work and economic growth" (SDG 8), "No poverty" (SDG 1), "Quality education" (SDG 4), and "Peace, justice and strong institutions" (SDG 16).

DOI: 10.1201/9781003325086-15

This chapter aims to further our understanding of the relation between technology and power from the perspective of equality and sustainable development. Specifically, it aims to show how recent advances in AI technology can affect power relations and lead to unequal distribution of power in society, thereby impeding the attainment of SDGs. In doing so, the discussion here aims to address the following three questions:

1. What is technological power?

   The notion of technological power has been the subject of great many discussions. However, in relating technology to political and social power, most such discussions fail to make explicit their conception of power, instead relying on an implicit understanding of power which can sometimes lead to confusion given that authors and their readers may have different, and even conflicting, interpretations of what power means (Sattarov, 2019). For this reason, this chapter aims first to disentangle different senses of power and consider how these different senses of power can be applied to technology.

2. How does technology lead to inequality of power?

   Traditional economic thought was dominated by the view that advances in technology lead to less inequality as more and more people begin to take advantage of the resulting opportunities (see Chapters 12 and 17). However, this chapter presents several counterexamples in the area of AI, to show that advances in technology do not necessarily level off inequality; on the contrary, they can create new and further reproduce existing inequalities. This chapter discusses how bias and discrimination embedded in such practices lead to inequality by giving rise to digital segregation and algorithmic exclusion of marginalized and vulnerable people (see Chapter 5).

3. Why does equality of power matter for sustainable development?

   Traditional liberal philosophy tends to hold that inequality and excesses of power in society can be tolerated so long as the rights and liberties of the powerless are protected. However, drawing on republicanism, Marxism, and identity politics, this chapter argues that failure to address asymmetries and disparities of power is symptomatic of a rather "head in the clouds" conception of justice, which is unsuitable to "here and now" circumstances of disadvantaged and unfortunate persons and groups. This chapter further highlights the importance of equality in the distribution of technological power in the context of sustainable development.

15.2 FROM HUMAN TO NONHUMAN POWER

That technology plays a central part in structuring power relations in society is nothing new and has been the subject of many discussions in a number of areas and disciplines including critical theory, the philosophy and ethics of technology, Science and Technology Studies (STS), sociology and history of technology, cyborg anthropology, etc. (Sattarov, 2019). It has also been an underpinning element of several influential social and philosophical theories and approaches, such as historical materialism (Marx; Fuchs), actor–network theory (Latour; Callon), postphenomenology (Ihde; Verbeek), and social construction of technology (Bijker; Woolgar).

Power is defined variously, for example, as one's "present means . . . to obtain some future apparent good" (Hobbes, 1839, p. 74); as "the probability that one actor within a social relationship will be in a position to carry out his own will despite resistance" (Weber, 1947, p. 152); as "the human ability not just to act but to act in concert" (Arendt, 1970, p. 44); or, formulaically, as "A has power over B to the extent that he can get B to do something that B would not otherwise do" (Dahl, 1957, pp. 202–203). Despite this diversity of notions of power, those who speak of "technological power" often forget to clarify what they mean by "power". To avoid ambiguity in our discussion of technology and power, we need a pluralist approach to power that first disentangles different notions of power and then applies them to technology. This pluralist approach to technological power has been developed in an earlier work (Sattarov, 2019), which, however, does not address the question of how technological power affects equality, especially in the context of sustainable development. This chapter aims to fill this gap.

A review of literature on power (e.g., Clegg, 1989; Haugaard & Clegg, 2009; Haugaard, 2010; Allen, 2016) can show that there are (at least) four main notions of power: (1) episodic, (2) dispositional, (3) systemic, and (4) constitutive. The episodic notion refers to a person being subject to the power of another. The dispositional notion denotes a person having certain abilities, capacities, or capabilities. The systemic notion views power as the property of entire social systems that structure the actions and possibilities of individuals or groups. Finally, the constitutive notion regards power as something that acts on people from within, as something that constitutes their very selves.

When viewed from the episodic perspective, technology can have power over people, by affecting their moods, decisions, and actions. For example, AI-enabled news and social media websites can affect the behavior of their readers by tweaking the frequency of appearance of good and bad news on their newsfeed (see, e.g., Kramer et al., 2014). When viewed from the dispositional perspective, technology can empower people by giving them new, or enhancing their existing, abilities, or capabilities. For instance, AI-enabled online translation services can enhance the ability of a human interpreter to translate text and speech in amounts larger than what the human can do on their own (see, e.g., Kushner, 2013). When viewed from the systemic perspective, technology can permeate social systems and institutions by creating and recreating asymmetrical relations of power (see Chapter 17; Sætra, 2021a). For example, AI-enabled applications for sorting and profiling people can lead to an exclusion of certain individuals and groups in the distribution of social goods and resources through algorithmic bias and discrimination (see, e.g., Barocas & Selbst, 2016). Finally, when viewed from the constitutive perspective, technology can become implicated in the myriad of power relations which shape, constitute, and produce modern selves. For instance, information technologies have become an integral part of the "surveillance society" (Lyon, 2001) which shapes and produces the "postliberal" (Cohen, 2012) selves.

The episodic and dispositional views of power are in accord with methodological individualism, as they focus on individuals holding and exercising power. On the other hand, the systemic and constitutive views are in agreement with holism, as they attend to higher-level social and political institutions and systems. Similarly, the four views of power correspond to different levels of analysis: Micro, meso, and macro (see Chapter 2; see also Sætra, 2022). The three levels of analysis, and the corresponding senses of power, can show how technological power may have different effects on different levels. What is good on one level may not be good on other levels. For example, a person is refused a bank loan, say, for funding a house extension, based on an automated assessment of her creditworthiness. On the micro-level, the algorithm and the bank have exercised power over the credit applicant (episodic power), thereby negatively affecting her options in funding her home extension (dispositional power). On the meso-level, we can imagine the bank(s) benefitting from the automation of routine credit assessments (systemic power). Finally, on the macro-level, mass automation of credit assessment across banks produces a class of uncreditworthy "quantified selves" in society (constitutive power).

15.3 SOME ARE MORE EQUAL THAN OTHERS

The four senses of power, introduced earlier, give rise to different concerns about equality of power. The episodic sense, to begin with, readily raises the issue of inequality, as it views power as an asymmetrical relation between two or more agents, in which one agent (with more power) exercises power over another agent (with less power). Such asymmetrical relations are often described in terms of zero-sum (or negative-sum) games, meaning that within episodic power relations an increase in the power of the power wielder corresponds to a decrease in the power of the power endurer. This can also mean that the more power an algorithm (and someone who controls it) has in deciding whether or not you get a bank loan, the less power you have in deciding about your welfare, housing, health, etc., as you are left with fewer options and opportunities. Asymmetrical power relations can arise through delegated seduction (e.g., targeted advertising), coercion (e.g., surveillance systems), force (e.g., lethal autonomous weapons), manipulation (e.g., algorithmic nudging and recommendation), persuasion (e.g., persuasive technology), and authority (AI decision-making). The infamous experiment to produce emotional contagion among users by Facebook (Kramer et al., 2014) and the manipulation of voters by Cambridge Analytica (Berghel, 2018) are thus instances of asymmetrical power relations between corporate and individual users, marked by disparity and inequality of power. Yet another example of asymmetrical power in the area of AI technology is who controls the technology: Training large AI systems necessitates access to large amounts of data and costs significant sums of money, which raises the possibility of AI technology being concentrated in the hands of governments or large corporations (Crawford, 2021), resulting in a disparity of power between state/corporation and people.

While the episodic sense considers how agents exercise power over other agents, the dispositional sense studies how agents become empowered and disempowered through acquisition or loss of abilities, potential or capabilities necessary for self-determination, and leading a good or successful life. There is a wealth of literature arguing that technology has become a primary means of empowerment (e.g., Johnstone, 2007; Brey, 2008, p. 87). The issue of inequality arises as we consider *differential* empowerment and disempowerment through technology. For example, technology can make the rich richer and the poor poorer, because the former has access to new technology, while the latter has not. This is an example of the problem of the "digital divide", characterized by the disparity between those with access to information technology and those without such access (van Dijk, 2020; see also Chapter 8). A differential empowerment and disempowerment through technology can occur as technology impacts the overall value of one's skills and abilities. Classic studies by Winner (1980) and Noble (1984) show how mechanical automation led to deskilling among farm and factory workers, respectively. Today, it is algorithmic automation that presents a similar challenge. Brynjolfsson and McAfee (2014) argue that advances in AI beget greater inequality by eliminating routine jobs through automation and favoring a small elite of highly skilled individuals, thus hollowing out the middle class.

Unlike the episodic and dispositional senses of power, the systemic sense puts on a broader lens and considers how power relations between individuals and possibilities for empowerment and disempowerment are structurally and systemically conditioned. In this view, technology can affect the power structure of an entire social system, by favoring some people to the detriment of others. There have been numerous discussions of how sorting and profiling algorithms used for recidivism prediction and social welfare management leads to exclusion and marginalization of certain individuals and groups. Of critical concern here is the potential for bias and error in algorithmic decision-making systems. For example, Hillman (2019) writes how a number of US states have been experimenting with the use of AI in assessing the risk of recidivism among criminal defendants. He argues that current AI technology poses serious risks of bias and error. Indeed, a study by Larson and colleagues (2016) found that a recidivism assessment algorithm used in the state of Wisconsin attributed a higher risk of recidivism to black defendants than to white defendants. If the algorithm is trained on data from past risk assessments performed by humans, it can simply reproduce and even increase existing social bias (see also Chapter 9). Moreover, as argued by Benjamin (2019), technologies designed for policing have expanded into other areas including hospitals, schools, banks, social services, etc., and recreate existing social hierarchies and engender new systems for social control. Considering such cases, the issue is not that a particular AI technology has a specific consequence for some individuals (which can be explained in terms of episodic or dispositional power); rather, the main issue is that AI technology contributes to the reproduction of existing social, economic, and political asymmetries and hierarchies in society as a whole (which is what the systemic view of power aims to describe).

Reproduction of existing societal hierarchies can also be in non-economic, symbolic, and cultural forms. Noble (2018) argues that a combination of factors to do with the private interests of corporations, the monopoly status of some internet search engines, and the nonneutral nature of algorithms has led to "algorithmic oppression", whereby search engines reinforce racism and sexism, especially with regard to Black women. Noble writes that "the everyday racism" found on the web is bad in itself; however, "it is entirely different with the corporate platform vis-à-vis an algorithmically crafted web search that offers up racism and sexism as the first results". What her examples suggest is that discrimination found offline is reproduced and further reinforced online through algorithm-powered internet search engines, portraying women of color as having less moral, social, and political value, and sustaining racist and sexist stereotypes and cliches that continue to haunt these women even online.

However, the effects of technology do not stop with the reproduction of inequalities through exclusion and marginalization. According to the constitutive sense of power, technology becomes implicated in power relations, where the exclusion and marginalization of certain individuals and groups can become internalized and normalized – as how they see themselves or how others see them. Simply put, technology becomes implicated in the production and shaping of subjects, selves, and bodies (see Chapter 5). To illustrate the point, we can further develop our earlier example concerning automation and deskilling. As automation increases, routine jobs become eliminated, resulting in deskilling and unemployment of those who used to perform such jobs. Although automation in this case may spare jobs requiring higher levels of skill or creativity (at least in the short term), even highly skilled and well-paid workers cannot escape the negative implications of automation. As Azmanova writes, under "precarity capitalism", "economic and social insecurity has become a core feature of our societies" (2020, p. 105). And as Moore argues, "psychological changes arising from precarity contribute to the formation of anxious selves who have internalised the imperative to perform, a two-part subjectification of workers as observing, entrepreneurial subjects and observed, objectified labouring bodies" (2018, p. 21). Furthermore, surveillance and exploitation of employees in what Zuboff (2019) calls "surveillance capitalism", and current predictions about potential automation of non-routine, creative, jobs through AI based on so-called "foundational models" (Bommasani et al., 2021), can add to and exacerbate existing fears of being replaced by machines. Some jobs are more precarious than others, and some selves are more quantified than others (Lupton, 2016), but the effects of both precarity and quantification go into the very selves of individuals, affecting, shaping, and constituting them.

For some time, economic thought was dominated by the view that advances in technology lead to less inequality as more and more people begin to take advantage of the resulting opportunities (see, e.g., Kuznets, 1955). As we have seen, there are numerous counterexamples which we need to address and overcome, before the long arc of history bends toward equality and justice (see Chapter 17). Importantly, our discussion of how AI technology affects equality of power shows that of the 17 top-level Sustainable Development Goals, at least 6 can directly be impacted by inequalities of power: "Reduced inequalities" (SDG 10), "Gender equality" (SDG 5), "Decent work and economic growth" (SDG 8), "No poverty" (SDG 1), "Quality education" (SDG 4), and "Peace, justice and strong institutions" (SDG 16).

15.4 EQUALITY NOW!

Our preceding discussion has focused on how technology, especially AI, can negatively affect equality in the distribution of power in society. However, one could question our preoccupation with equality, as it is one of the lesser-regarded concepts in the traditional liberal philosophy. Introductory texts on political philosophy rarely dedicate a chapter to the concept (Coeckelbergh, 2022). Its unpopularity can be explained in terms of its instrumentality: Equality, unlike justice or liberty, is valued as a means to something else, rather than as an end in itself. For example, equality is of instrumental value to democracy in the form of equal distribution of voting rights among citizens, while economic inequality is a problem for democracy: The rich may use their economic and financial resources to unduly influence the electoral processes to the detriment of the poor (Frankfurt, 2015, p. 6).

To articulate the importance of equality in the distribution of technological power, the discussion proceeds with a critique of traditional liberal philosophy. Specifically, it briefly outlines three different criticisms of traditional liberal philosophy, based on republicanism, Marxism, and politics of recognition. What this eclectic trio of political theories shares in common, as we shall see later, is that they disagree with traditional liberal philosophy, by highlighting different ways in which traditional liberal philosophy, including modern liberal democracies, tend to overlook inequality and injustice suffered by marginalized people, groups, and classes. Each of the outlined criticisms sheds light on a significant aspect of the problem of inequality in the distribution of power as impacted by AI. Importantly, they direct us toward a recognition of equality as an important principle of social and political justice, in line with socially sustainable development, and SDG 10 in particular.

A republican critique of liberalism can be found in the works of Philip Pettit (1997, 2012). At the core of Pettit's argument is the idea of "freedom as non-domination", which he offers as a third option to "freedom as non-interference" and "freedom as self-determination" as famously introduced by Isaiah Berlin (see also Sætra, 2021b, pp. 23–28). According to freedom as non-domination, humans are free insofar as they are not subject to the domination of others, exposed and vulnerable to their will, whims, and desires. Pettit argues that many liberals, including Berlin, have long espoused freedom as non-interference, according to which the powerful in society can be tolerated insofar as they do not interfere in the affairs of the powerless. However, Pettit argues, freedom as non-interference largely focuses on *actual* interference and overlooks *potential* interference. Were a powerful despot to choose not to exercise his or her power, those subject to this power would still have to live with a constant worry that this can change at any moment, which Pettit reasonably finds objectionable. Hence, Pettit maintains that the republican conception of freedom as non-domination better serves the institutions of a constitutional democracy than its liberal counterpart. On this view, then, being subject to the power of automated decision-making systems constrains one's freedom (as non-domination) as long as there remains a potential for arbitrariness in such systems (Sætra, 2021b; Creel & Hellman, 2022; Barocas et al., 2017, p. 152).

Marxism is similarly critical of traditional liberal philosophy for emphasizing abstract ideals of justice, without dealing with the capitalist structure of society, which actually creates inequality between those who own the means of production, and those who become exploited as a result (see Chapter 13). Automation power by algorithms, in the hands of corporate owners, produces a proletariat of unemployed people and exploits those still employed through "surveillance capitalism" (Zuboff, 2019) under conditions of "precarity capitalism" (Azmanova, 2020). Hence, inequality in the *distribution* of social goods is not the main issue (which liberals aim to remedy by falling back on some conception of equality or justice); rather, the main problem is inequality in the *production* of social goods (that is, in the ownership of the means of production). Put simply, in its quest for justice and fairness, the liberal tradition is "barking up the wrong tree": Instead of trying to fix the symptoms of injustice, we should address the root cause of injustice – capitalism itself (see Chapter 17). On this view, then, algorithms and AI technologies themselves do not lead to inequality; rather they support and reinforce the capitalist system of production and distribution, which actually creates asymmetrical relations of power between the haves and the have-nots.

Another critique of the traditional liberal philosophy is "identity politics" (also known as "politics of difference" or "politics of recognition") consisting of "a wide range of political activity and theorizing found in the shared experiences of injustice of members of certain social groups" (Heyes, 2020). Advocates of identity politics criticize liberal political theories at least on three counts. First, ontologically speaking, liberal theories imagine people as essentially independent and similar individuals, devoid of their identities, differences, and affiliations. Second, historically speaking, political liberalism underestimates the past origins of injustices, especially race, gender, or ethnicity. Third, empirically speaking, liberal–democratic order has fallen short in stopping the exclusion and marginalization of specific social groups. The core idea is therefore to recognize historically excluded and oppressed groups and identities: "universal recognition based on a shared humanity is not enough, particularly on the part of groups that have been discriminated against in the past. Hence modern identity politics revolves around demands for recognition of group identities" (Fukuyama, 2006, p. 9). On this view, then, analyses offered by Noble (2018) and Benjamin (2019) about how algorithms are impacting people of color are specimens of identity politics approach to AI, which raise important questions about identifying and recognizing vulnerable groups at the receiving end of injustices (see also Chapter 5).

Overall, much can be said about the interconnections (and disagreements) between republicanism, Marxism, and identity politics. First, all three criticisms constitute an important part of the discourse concerning the overly universalist and individualist shortcomings of liberalism. Second, all three theories provide a necessary corrective to idealized and ahistorical justifications of liberalism. Finally, they all share a common concern for equality, especially for equality in the distribution of power in society. To clarify, for republicanism, institutions of a constitutional democracy must be designed with the goal of promoting "people's equal enjoyment of freedom as non-domination" (Pettit, 2012, p. 123), where people can pass the "eyeball test", by looking "others in the eye without reason for fear or deference that a power interference might inspire" (p. 103). As for Marxism, equality is important at least insofar as inequality in the ownership of the means of production is the root cause of capitalist injustice. Finally, for politics of recognition, equality figures in claims demanding equal recognition of diverse identities and groups, whether excluded or marginalized.

We may not agree with all three criticisms of liberalism in their entirety. The point is not to offer a comprehensive theoretical framework that could unify all these theoretical perspectives. Rather, the purpose is to combine them in a neat program for policy action to mitigate the negative impact of AI technology and facilitate the achievement of relevant SDGs. The first (i.e., republicanism) explains why inequality in power brought about by AI technology is objectionable, even in liberal–democratic societies, for there is no good life in having to live your life at the mercy of the powerful even if they are not exercising their power just yet. The second (i.e., Marxism) identifies the root cause of the unequal distribution of benefits and burdens resulting from AI technology, which suggests that we ought to pay more attention to those who own and control the technology. Finally, the third (i.e., identity politics) identifies the foremost vulnerable segments of population, on the frontline of AI revolution, which require priority protection. Together they suggest that unequal distribution of power through AI can become a factor in the unsustainability of socio-technical systems (Sætra, 2021a).

The republican call for freedom as non-domination through equality of power in society is in large agreement with several SDGs which aim for equality, inclusivity, justice, and strong institutions. Equalizing power relations in society can ensure inclusive and equitable quality education for all (SDG 4), achieve gender equality and empower all women (SDG 5), reduce inequality within countries (SDG 10), as well as provide access to justice and build inclusive institutions at all levels (SDG 16). To safeguard the achievement of these goals and their relevant targets, it is necessary to address issues of bias, error, and discrimination inherent in AI decision-making systems to be used by diverse institutions which provide and manage health services, public education, social welfare assistance, public security and policing, etc.

Marxist concern with who owns the means of production and who is being exploited as a result is, too, in accord with SDGs which emphasize security, inclusivity and sustainability in areas of employment, innovation, industrialization, and production. Reducing sharp asymmetries of power between, on one hand, employers who develop, produce, and deploy AI technologies, and on the other hand, employees who are subjected to surveillance and quantification, job insecurity, and anxiety can help promote full and productive employment and decent work for all (SDG 8), promote inclusive industrialization (SDG 9), ensure sustainable production and consumption patterns (SDG 12), and indirectly, reduce poverty (SDG 1), fight hunger (SDG 2), and reduce inequality within countries (SDG 10).

Last, but not least, identity politics, with its emphasis on recognition, emancipation, and decolonization of racialized and gendered selves and bodies is in line with a number of SDGs that promote equality, justice, and inclusivity in diverse areas of public life, including housing, health, education, and government. Recognition of diverse identities and groups, especially those who are made vulnerable, or unfairly marginalized and excluded from the distribution of social goods based on their racialized or gendered identities, can be a sure step toward building peaceful and inclusive societies, and provide access to justice for all (SDG 16), achieving gender equality and empowering women and girls (SDG 5), making cities and human settlements inclusive and safe (SDG 11), and, overall, reducing inequality within countries (SDG 10).

These considerations further infuse a healthy level of concern with power and politics into discussions about the goals and targets of sustainable development as they relate to technology and innovation. The UN SDGs are rather wide-ranging. As such, the 17 top-level SDGs are sometimes categorized into broader categories of economy, society, and environment, which are regarded as the "three dimensions of sustainability" (Vinuesa, 2020, p. 2). As argued by Sætra (2021a), what makes such categorization problematic is that politics largely disappears from picture. Hence, there is a real need to emphasize power and politics intricately woven into technological and innovation processes (see Chapter 11). Highlighting the power and political dimension of technology in general, and AI in particular, is necessary not only for our understanding of the effects of technology on specific SDGs but also for our realization of how power and politics underpin our attempts to strengthen and streamline our efforts to achieve the SDGs.

15.5 CONCLUSION

Technological solutions to social problems can create problems of their own. One such problem is the negative impact of technology on the distribution and exercise of power in society. The chapter set out to answer three interrelated questions: (1) What is technological power? (2) How does technology lead to inequality of power? and (3) Why does equality matter in the distribution of power? – all part of the main question addressed by this chapter: How do recent advances in AI technology can impact equality in the distribution of power in society? In addressing these questions, it has been argued that (1) distinguishing between different senses of power and applying them to (AI) technology helps avoid potential ambiguity in discussions of technological and algorithmic power; (2) applying different levels of analysis (which correspond to different senses of power) to technological change can show how AI technology may have different effects on different levels; and (3) the ideal of equality, albeit sometimes overlooked, still figures prominently in discussions of social and political justice, and can have important implications for policy in mitigating the undesirable effects of AI technology.

Equality, indeed, matters, as there is no good life in having to live your life at the mercy of the powerful. Philip Pettit (1997, p. 5) wrote about

> the grievance expressed by the debtor who has to depend on the grace of the moneylender, or the bank official, for avoiding utter destitution and ruin; and by the welfare dependant who finds that they are vulnerable to the caprice of a counter clerk for whether or not their children will receive meal vouchers.

In the past, such grievances may have been only about people. Today, and in future, they are going to be more about machines.

15.6 REFERENCES

Allen, A. (2016). Feminist perspectives on power. In *The Stanford encyclopedia of philosophy* (Fall 2016 ed.). https://plato.stanford.edu/archives/fall2016/entries/feminist-power

Arendt, H. (1970). *On violence*. Penguin.

Azmanova, A. (2020). *Capitalism on edge: How fighting precarity can achieve radical change without crisis or utopia*. Columbia University Press.

Barocas, S., Hardt, M., & Narayanan, A. (2017). Fairness in machine learning: Limitations and opportunities. *Nips Tutorial, 1*, 2.

Barocas, S., & Selbst, A. D. (2016). Big data's disparate impact. *California Law Review, 104*, 671–732.

Benjamin, R. (2019). *Captivating technology: Race, carceral technoscience, and liberatory imagination in everyday life*. Duke University Press.

Berghel, H. (2018). Malice domestic: The Cambridge Analytica dystopia. *Computer, 51*(05), 84–89.
Bommasani, R., Hudson, D. A., Adeli, E., Altman, R., Arora, S., von Arx, S., . . . Liang, P. (2021). On the opportunities and risks of foundation models. *arXiv preprint*, arXiv:2108.07258.
Brey, P. (2008). The technological construction of social power. *Social Epistemology, 22*(1), 71–95.
Brynjolfsson, E., & McAfee, A. (2014). *The second machine age: Work, progress, and prosperity in a time of brilliant technologies*. W.W. Norton & Company.
Clegg, S. (1989). *Frameworks of power*. Sage.
Coeckelbergh, M. (2022). *The political philosophy of AI: An introduction*. Polity.
Cohen, J. (2012). What privacy is for. *Harvard Law Review, 126*(7), 1904–1933.
Crawford, K. (2021). *The atlas of AI: Power, politics, and the planetary costs of artificial intelligence*. Yale University Press.
Creel, K., & Hellman, D. (2022). The algorithmic leviathan: Arbitrariness, fairness, and opportunity in algorithmic decision-making systems. *Canadian Journal of Philosophy*, 1–18. https://doi.org/10.1017/can.2022.3
Dahl, R. A. (1957). The concept of power. *Behavioral Science, 2*(3), 201–215.
Frankfurt, H. (2015). *On inequality*. Princeton University Press.
Fukuyama, F. (2006). Identity, immigration, and liberal democracy. *Journal of Democracy, 17*(2), 5–20.
Haugaard, M. (2010). Power: A 'family resemblance' concept. *European Journal of Cultural Studies, 13*(4), 419–438.
Haugaard, M., & Clegg, S. R. (2009). Introduction: Why power is the central concept of the social sciences. In *The SAGE handbook of power*. Sage.
Heyes, C. (2020). Identity politics. In *Stanford encyclopedia of philosophy* (Fall 2020 ed.). https://plato.stanford.edu/entries/identity-politics/
Hillman, N. L. (2019). The use of artificial intelligence in gauging the risk of recidivism. *Judges Journal, 58*, 36.
Hobbes, T. (1839 [1651]). *Leviathan: The English works of Thomas Hobbes* (Vol. III) (Sir William Molesworth, Ed.). John Bohn.
Johnstone, J. (2007). Technology as empowerment: A capability approach to computer ethics. *Ethics and Information Technology, 9*(1), 73–87.
Kramer, A. D. I., Guillory, J. E., & Hancock, J. T. (2014). Experimental evidence of massive-scale emotional contagion through social networks. *Proceedings of the National Academy of Sciences, 111*(24), 8788–8790.
Kushner, S. (2013). The freelance translation machine: Algorithmic culture and the invisible industry. *New Media & Society, 15*(8), 1241–1258.
Kuznets, S. (1955). Economic growth and income inequality. *The American Economic Review, 45*(1), 1–28.
Larson, J., Mattu, S., Kirchner, L., & Angwin, J. (2016, May 23). How we analyzed the COMPAS recidivism algorithm. *ProPublica*. www.propublica.org/article/how-we-analyzed-the-compas-recidivism-algorithm
Lupton, D. (2016). *The quantified self: A sociology of self-tracking*. Polity.
Lyon, D. (2001). *Surveillance society*. McGraw-Hill Education.
Moore, P. (2018). *The quantified self in precarity: Work, technology and what counts*. Routledge.
Morozov, E. (2013). *To save everything, click here: The folly of technological solutionism*. PublicAffairs.
Noble, D. (1984). *Forces of production: A social history of industrial automation*. Knopf.
Noble, S. U. (2018). *Algorithms of oppression: How search engines reinforce racism*. New York University Press.
Pettit, P. (1997). *Republicanism: A theory of freedom and republicanism*. Oxford University Press.
Pettit, P. (2012). *On the people's terms: A republican theory and model of democracy*. Cambridge University Press.

Sætra, H. S. (2021a). AI in context and the sustainable development goals: Factoring in the unsustainability of the sociotechnical system. *Sustainability, 13*(4), 1738. https://doi.org/10.3390/su13041738
Sætra, H. S. (2021b). *Big data's threat to liberty: Surveillance, nudging, and the curation of information*. Academic Press.
Sætra, H. S. (2022). *AI for the sustainable development goals*. CRC Press.
Sattarov, F. (2019). *Power and technology: A philosophical and ethical analysis*. Rowman & Littlefield.
Stilgoe, J. (2020). *Who's driving innovation? New technologies and the collaborative state*. Palgrave Macmillan.
United Nations. (2015). *Transforming our world: The 2030 agenda for sustainable development*. UN Division for Sustainable Development Goals.
Van Dijk, J. (2020). *The digital divide*. John Wiley & Sons.
Vinuesa, R., Azizpour, H., Leite, I., Balaam, M., Dignum, V., Domisch, S., . . . Fuso Nerini, F. (2020). The role of artificial intelligence in achieving the Sustainable Development Goals. *Nature Communications, 11*(1), 1–10.
Weber, M. (1947). *Max Weber: The theory of social and economic organization*. The Free Press.
Winner, L. (1980). Do artifacts have politics? *Daedalus, 109*(1), 121–136.
Zuboff, S. (2019). *The age of surveillance capitalism: The fight for a human future at the new frontier of power*. Profile Books.

CHAPTER 16

What Does Data Valuation Literature Tell Us About Methods and Dimensions? Implications for City Data Marketplaces

Petter Kvalvik, Mary Sánchez-Gordón, and Ricardo Colomo-Palacios

CONTENTS

16.1 INTRODUCTION

Technological solutions are often proposed as quick and flawless ways to solve complex real-world problems. Not only organizations but also cities are looking for technological solutions to alleviate real-world problems. A stronger need to build sustainable cities that last draws attention to more sustainable and resilient growth and development pathways. In this sense, cities around the world are using smart city solutions to build resilience, increase public safety, and create healthier, more livable urban environments. According to

DOI: 10.1201/9781003325086-16

the European Commission (European Commission, n.d), a city should provide improved and smarter public services through a more interactive and responsive administration that also meets the ever-changing needs of individuals who reside, work, or travel in the city. Moreover, Sustainable Development Goal (SDG) 11, named "sustainable cities and communities", highlights that we need to "make cities inclusive, safe, resilient and sustainable". Smart cities have become a popular concept because they have the potential to create a sustainable and liveable urban future.

For smart cities, technology is a necessary but not sufficient component (Augusto, 2020; Granath et al., 2021). Various data sources throughout the cities produce diverse data at an increasing scale. For instance, data sources are personal devices (smartphones, wearables, laptops), smart home devices (lighting, security, heating), public services (health, administration, waste management, water supply management, emergency preparedness), and smart grid (smart neighborhood, smart charging). Therefore, the data generated by individuals, and private and public organizations are also an essential element (European Commission, 2020a). For instance, collecting and sharing real-time data related to unemployment, worker productivity, or citizens' well-being to target aid interventions to vulnerable groups can help to *promote inclusive and sustainable economic growth* (SGD 8). Moreover, a human rights-based approach to data facilitates the gathering of relevant data stakeholders and the development of communities of practice that improve the quality, relevance, and use of data and statistics in accordance with international human rights norms and principles (UN, 2018). However, data values are difficult to evaluate (J. Liang & Yuan, 2021), and the data valuation process of data for sustainable development (see Chapter 2) arises as a nontrivial challenge that involves participation, transparency, privacy, and accountability (UN, 2018).

As data are at the center of any smart city initiative, the full value and potential of that data should be known, but there are no widely accepted approaches that would allow cities to properly understand the value and potential of data (Wdowin & Diepeveen, 2020). Such data could help to *build effective, accountable, and inclusive institutions* (SDG 16) and *promote inclusive and sustainable industrialization, and foster innovation* (SGD 9). In fact, it is expected that Common European dataspaces, in particular, the Green Deal dataspace, will be an important step in developing the digital ecosystem of the environment, on issues such as climate change, pollution, deforestation, biodiversity, and circular economy (Gronkiewicz-Waltz et al., 2020). Gronkiewicz-Waltz et al. highlight that cities play a pivotal role in achieving climate neutrality by 2050 and present the mission "climate neutral and smart cities" to the European Commission. Figure 16.1 illustrates potential research areas aligned with such a mission.

Data demand drives the rapid development of data trading (J. Liang & Yuan, 2021) which requires data markets to support it (F. Liang et al., 2018). As a result, the development of data marketplaces that enable efficient data trading cannot be overlooked. Although several challenges need to be addressed to do so, the first is to determine the proper price for the data to be traded (F. Liang et al., 2018). In other words, one of the central issues in data trading is data pricing mechanisms (J. Liang & Yuan, 2021). In this sense, an understanding of data valuation methods and dimensions allows us to make the right strategic decisions

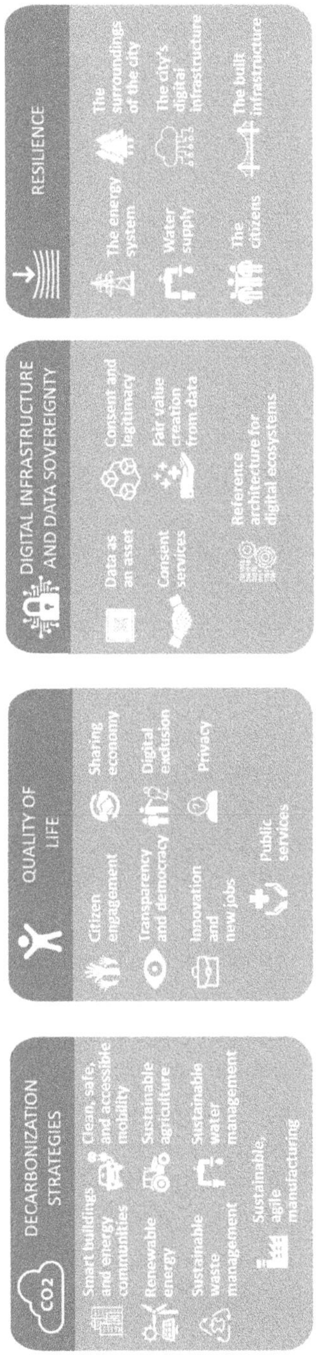

FIGURE 16.1 Research areas aligned with the mission “Climate neutral and smart cities”.

and assess robust investments in data assets, either organically or through acquisitions. At the city level, there is a clear gap in understanding the consequences and impacts of a potential investment in datasets, data platforms, and the usage of algorithms and other analytical tools to extract that potential (Morgenstadt Value of Data, n.d.).

16.2 UNDERSTANDING DATA VALUATION AND CITY DATA MARKETPLACES

Data valuation is an emerging discipline that aims to estimate the information's value as an asset (Leitner-Hanetseder & Lehner, 2022). The value of data could assess according to the level of public interest, sensitivity, and relevance. Moreover, the data value is related to data flow and data categories. Unlike most assets, which lose value as they are utilized, data can potentially be more valuable the more it is used.

In cities, common goods or merit goods such as public infrastructure, public space, air, water, vegetation, traffic flows, public safety, and street lighting are subject to rivalry and hence need to be governed. These goods are increasingly linked in the smart city through another good, whose nature is ambiguous: Data (Morgenstadt Value of Data, n.d). Data are sometimes open (public domain), sometimes restricted, and sometimes only accessible to their owner (private domain) (Gagnon-Turcotte et al., 2021). Publicly funded data lie at the intersection of the public and private domains and could be created, for example, through public–private partnerships. Since data are the basis of public services and urban development in the smart city, this situation exponentially increases the complexity of public value creation, smart city investments, and urban governance.

The city data market today is in the early stages of development. The European strategy for data has launched an initiative to create a single data market that is open to data from all over the world (European Commission, 2020a). However, one of the foremost challenges within these emerging markets is assessing the value of datasets. Current data pricing strategies are often driven by the seller, with little visibility into the cost of collection, cleansing, and packaging to the buyer. The lack of pricing transparency is the result of this asymmetry of information. It is detrimental to both the seller and the buyer, as the seller is unable to price optimally in the market while the buyer is unable to strategically assess pricing options across data service sellers. As the final result from a data value chain is useful insights rather than a tangible product or a service (Leitner-Hanetseder & Lehner, 2022), a more structured data market with a standardized pricing model would improve the transaction experience for all parties.

A city data marketplace would help strengthen the urban economic ecosystem and develop new business models and empower communities through sharing of data (Bass et al., 2020). As a result, while city data marketplace technology has potential for good, it may also have negative implications. In this sense, according to Sætra (2021), privacy concerns are not adequately addressed by any of the SDGs, unless they result in unfair and discriminatory consequences and behaviors. It means that SDGs require a complementary framework like human rights to unveil the negative impacts related to a loss of privacy and autonomy.

16.3 RESEARCH METHOD

On the basis of our research objective, we formulate the following research question: *What methods and dimensions constitute the current body of literature on the valuation of data?* To answer this question, we conducted a multivocal literature review (MLR) which is a systematic review of academic literature and grey literature. The searches are conducted in academic databases for the first kind of literature, while Google Search was used to search grey literature sources such as white papers, blogs, and posts (Garousi & Mäntylä, 2016). It is noteworthy that there are secondary studies on topics such as business data sharing through data Marketplace (Abbas et al., 2021) but our study is the first MLR on the implications of data valuation that discusses the preliminary findings in the context of city data marketplaces. The search process is based on the study protocol depicted in Figure 16.2.

In this MLR, the first author conducted the study selection process, the second author reviewed the process and checked the outcomes, and the third author supported the resolution of disagreements. The main keywords were "data marketplace" and "data valuation", but we also included terms that are synonymous with them. The final search string used Boolean operators to combine all the terms as follows.

> ("data marketplace" OR "data market" OR "data trading" OR "data markets" OR "data space" OR "data spaces") AND
>
> ("data valuation" OR "pricing model" OR "valuing data" OR "data pricing")

The search was conducted on five academic databases, IEEE, ACM, ScienceDirect, Springer, and Wiley. Although the search period was unlimited, it was executed by March 2022. Given that Google Search retrieved about 197,000 results for grey literature, we limited the search space by applying a relevance ranking approach (Google's PageRank algorithm) (Garousi & Mäntylä, 2016). It means that we went to the $(n + 1)^{th}$ page only if the sources on the nth page still appeared relevant. As a result, the first few pages were searched but we went further when necessary (101 sources). In the initial search, we found 1,232 sources.

The sources were screened by reading the titles, abstracts, and keywords in the academic databases, while they were screened based on the title and meta-text provided by Google Search. The inclusion and exclusion criteria were as follows:

- Sources that discuss data valuation and data marketplace.
- Sources that discuss methods and dimensions related to the topic.
- Sources that are accessible online in full text.

When a study was excluded, the following criteria were applied:

- Sources that are not presented in English.
- Sources that are duplicated or extended in other sources.

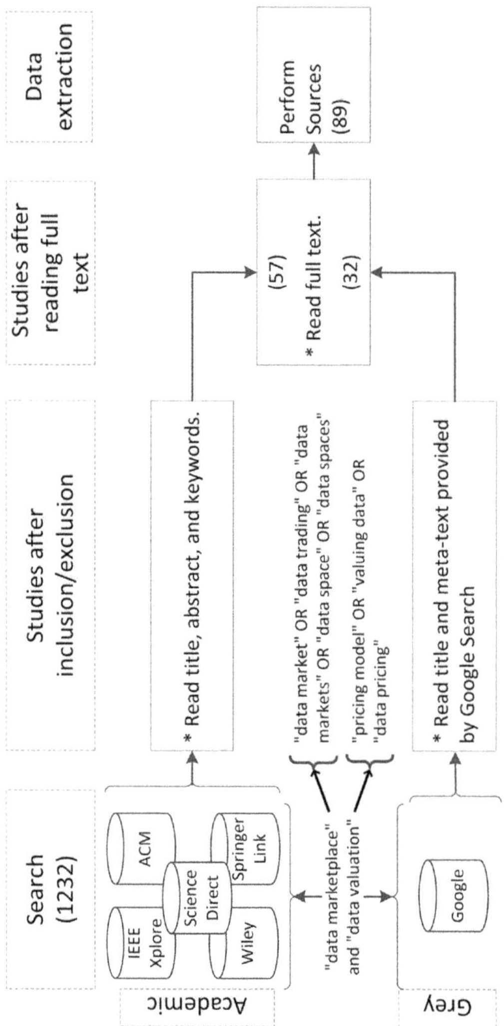

FIGURE 16.2 An overview of the search process.

When a source did not meet the inclusion criteria, it was excluded from the rest of the selection process. In case of doubt, we took the source to full-text reading in the next phase. The full text was then reviewed, and the inclusion and exclusion criteria were applied again to ensure that the source contains relevant information for this MLR. As a result, a set of 89 sources was selected as primary studies. Then, we systematically identified, classified, and analyzed the pricing mechanisms proposed or used by the authors of each selected study.

16.3.1 Data Valuation: Methods and Dimensions

In the literature, we find 12 main categories of pricing mechanisms for a data marketplace as shown in Table 16.1.

TABLE 16.1 Summary of pricing mechanisms

	Pricing mechanisms	Description
1	Auction	The privacy cost of selling the data is calculated. Then, it is sent to a buyer that calculates a utility score related to the data (Gonçalves et al., 2021)
2	Bundling	Package Pricing Strategy is an improved usage-based pricing strategy, for example, vendors provide a data package plan with a fixed price (F. Liang et al., 2018). Subscriptions consider period (Yang et al., 2019)
3	Compensation	Individuals provide their personal data to a data market based on their privacy preferences and are compensated appropriately (Yang & Xing, 2019). The data market charges a price to the data consumer, and the data consumer can acquire data products based on their willingness to pay.
4	Customers' willingness to pay (WTP)	From a consumer behavior view, WTP refers to the price that a customer is willing to pay to buy a specific data product or service (Yang et al., 2019). It could be related to data quality dimensions, for example, the sensitivity level of privacy data (Yang & Xing, 2019)
5	Exponential method	This method can be tailored to suit various auction scenarios (non-numeric problems) while ensuring output quality (Zhang et al., 2021)
6	Flat pricing	A vendor simply considers selling each digital commodity once. As time is the only parameter, this mechanism lacks flexibility and diversity for consumers (F. Liang et al., 2018)
7	Freemium	Basic products or limited services are provided to consumers for free (F. Liang et al., 2018)
8	Laplace method	This method can be tailored to suit numeric problems, but it adds uncertainty which results in sub-optimality (Zhang et al., 2021)
9	Privacy pricing	This method introduces noise into the data or to some components of the data release process to protect users' privacy while releasing data (Nguyen et al., 2020)
10	Query pricing	Query pricing on incomplete data considers factors such as query quality, data completeness, and data contribution (Miao et al., 2020)
11	Subscription	Subscription determines the entire price of certain data items in advance by analyzing consumer and vendor behavior. Although it overlooks the difference among items, it is usually multi-stepped and consumers can select the step they require and then pay for it (Li et al., 2022)
12	Two-part Tariff	Two-part Tariff is a combination of package pricing and flat pricing strategies (F. Liang et al., 2018)

The data value is defined by different dimensions, and the most evident is data quality. However, another important dimension is data complexity which is related to the volume of the data (size based on the number of entries in a dataset and/or the number of variables), the structure of data (e.g., relationships between elements), the heterogeneity of data in terms of diverse values and structure (variety), and the level of granularity (e.g., aggregated data). Moreover, data valuation is impacted by data perishability (Ricart, 2020), a measure of the time since data collection which entails time dependency and devaluation over time.

The data valuation has main implications for city data marketplaces regardless of the pricing mechanism. In the next section, findings are grouped into three main implications: (i) digital data partnership, (ii) use cases, and (iii) data protection, data ownership, and data sovereignty.

16.4 IMPLICATIONS FOR CITY DATA MARKETPLACES

Establishing data markets is critical for trading data in the market in a secure, safe, and fairway (F. Liang et al., 2018). A data market implies data sharing and the creation of digital data partnerships (Gagnon-Turcotte et al., 2021). By reviewing the literature, personal information, privacy, and data ownership have been identified as barriers to participation in digital data partnerships. In fact, preliminary results of this MLR reveal that a data market also needs to protect the privacy of both data commodities and personal information while providing an optimal experience for both vendors and consumers such as citizens, and governmental and non-governmental organizations in a smart city.

16.4.1 Digital Data Partnership

Smart cities entail multi-stakeholder data sharing initiatives that require the development of a data use culture (Gagnon-Turcotte et al., 2021). In Europe, a political agreement was reached on the Data Governance Act (DGA) in April 2022 (European Commission, 2020b). This first legislative building block for establishing a stable and fair data-driven economy aims to set up the right conditions for trustful data sharing in line with European values and fundamental rights, strengthen the digital single market's governance mechanisms, and establish a framework to facilitate general and sector-specific data sharing. However, such initiatives require data from multiple industries, platforms, and a culture that depends on these industries – sector and size – technology choices, and stakeholders' attitudes toward data. Just because data, and the technology behind it, can help to reduce inequality, improve health and education, and spur economic growth does not mean it meets the criteria for privacy, security/safety, autonomy, justice, control, and power.

In a competitive market, the costs of creating and maintaining data might constitute another barrier to participation. In the context of smart cities, new delivery systems could include public service providers and commercial partners (Löfgren & Webster, 2020). This raises concerns about the control, ownership, and access to data, and whether the value is kept by society via government agencies or privatized by commercial interests. Public–private partnerships are expected to become more widespread due to the private sector collects a large portion of data with the greatest potential for public good. The challenge

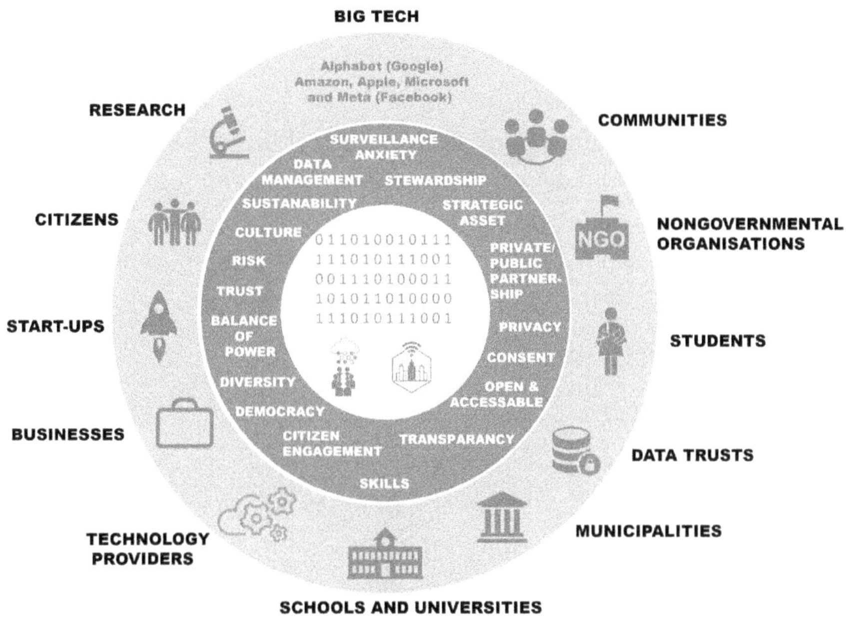

FIGURE 16.3 Overview of stakeholders and data sharing.

will be ensuring that clear frameworks are in place to establish roles and expectations for all parties and that such partnerships are sustainable over time. Figure 16.3 shows an overview of data sharing and potential stakeholders identified from our MLR.

In the context of public services, value is not only multidimensional but also complex and difficult to define and realize by both service providers and users (Löfgren & Webster, 2020). For instance, crime, drugs, and unemployment are challenging issues that require multi-agency responses to realize better services that benefit both citizens and society. In practice, finding policy and service solutions that consider the critical and active agency of citizens requires collaboration.

16.4.2 Use Cases

A use case is a way to use data to achieve specified benefits for the stakeholders involved (Gagnon-Turcotte et al., 2021). However, there are many ways in which data can bring value to different stakeholders. In the context of smart cities, a challenge is to build a shared vision, especially when there are many stakeholders with various interests participating and a large diversity of data that can be shared. Therefore, it includes commitments to increase transparency. Focusing on innovation is more challenging because it implies finding new ways to generate value from previously unshared data. Figure 16.4 illustrates potential data-driven innovation areas reported in the literature, for example, mobility services, sharing economy, intelligent buildings, and public services.

It is not an easy task to determine data value through use cases because the value of a data source is determined not only by its potential intended use but also by whether it could be replaced with another data source. There are also serious concerns about "unintended uses

FIGURE 16.4 Data-driven innovation.

of data" in the real world and the implications of other factors such as data combination, re-selling, and new (at times unknown) uses. Combining multiple datasets, in particular, may result in the re-identification of individuals or groups of individuals, exposing them to potential harm. Therefore, there is a need for anticipation in design decisions considering what might arise in the future, for example, new forms of discrimination may also emerge. Unintended uses could arise for a large variety of reasons and their potential negative consequences go beyond the individual, into economic, social, and political areas (Wigan & Clarke, 2013). In this context, new balances are needed to deal with the resulting power shifts.

Negligent data use can exacerbate existing inequality and harm citizens and communities, for example, the presence of discriminatory and skewed interpretations, privacy invasion, and loss of individual autonomy and agency (Gagnon-Turcotte et al., 2021). Citizens face real consequences as a result of such harm, which can be financial, physical, or emotional. Moreover, when identifiable data is disclosed in delicate situations, it can lead to exclusion, violence, or discrimination.

16.4.3 Data Protection, Data Ownership, and Data Sovereignty

Data protection arises as a critical element for ensuring data ownership (F. Liang et al., 2018). To prevent data misuse or mishandling, proper data protection measures must be implemented. Moreover, the rights to data protection and privacy must be balanced with the right to access information. Data sovereignty, on the other hand, regulates how data should be secured and governed. It reinforces the rights of data owners and gives them control over their data. However, to respect data sovereignty, we must assess individual rights against collective benefits.

In 2020, the European data strategy was published (European Commission, 2020a), envisioning that data will be shared and used to reflect European values, put people first in developing technology, and defend and promote European values and rights in the digital world. Likewise, General Data Protection Regulation (GDPR) established standards for protecting privacy and data ownership (Gagnon-Turcotte et al., 2021). GDPR also covered sovereignty, but each real-world implementation is unique due to its context. In the same vein, the European Commission proposed a legislative building block called the Data Act (European Commission, 2022), which concerns the actual rights regarding access to and use of data. The proposed Data Act includes measures to allow users of connected devices to gain access to data generated by them, rebalance negotiation power for small-medium sized enterprises (SMEs) by preventing abuse of contractual imbalances in data sharing contracts, means for public sector bodies to access and use data held by the private sector that is necessary for exceptional circumstances and new rules allowing customers to effectively switch between different cloud data-processing services providers and putting in place safeguards against the unlawful data transfer. However, there are cases in which the ownership of data is unclear making data confidentiality and security more complex (Gagnon-Turcotte et al., 2021). Identifying who owns the data could require not only the interpretation of different laws but also an understanding of individual attitudes about data.

As the volumes of data grow, and the Internet of Things takes hold, universal surveillance is becoming a practicable proposition. In this situation, concerns extend beyond these raised by data breaches. Policymakers and governments are failing to recognize the social costs associated with the growing scale of data and its detailed intensity (Wigan & Clarke, 2013). In Europe, the "right to be forgotten" gives individuals the legal right to have their data erased. This right is especially important in countries where underrepresented groups suffer discrimination, for example, indigenous peoples are among the poorest socioeconomic groups.

16.5 CONCLUSIONS

As data become an essential component in smart cities, how quantifying the value of data for sustainable development is an increasingly important topic of study with policy, economic, and technological implications. In this chapter, we present a set of main implications in terms of promises and pitfalls for city data marketplaces based on the preliminary findings of our review.

City data marketplaces are a kind of technology that may enable more efficient, agile, and evidence-based decision-making and better monitor progress on the SDGs in a way that is both inclusive and fair if used responsibly. However, there is also a risk of amplified bias and inequality, and fundamental elements of human rights must be safeguarded to realize the opportunities presented by data marketplaces: Participation, transparency, privacy, and accountability are required, and the rights of individuals against the benefits of the collective must be assessed. In the future city data market, it is essential to develop a holistic valuation method for judging data quality based on its inherent properties for pricing purposes.

The value of data is very context-dependent and requires a depth analysis. Thus, although this study may be seen as a starting point and it provides good insights into this topic, we pose the following open question: To what extent city data marketplaces can or cannot provide insight to solve the challenges related to inclusion, safety, resilience, and sustainability?

16.6 REFERENCES

Abbas, A. E., Agahari, W., van de Ven, M., Zuiderwijk, A., & de Reuver, M. (2021). Business data sharing through data marketplaces: A systematic literature review. *Journal of Theoretical and Applied Electronic Commerce Research*, *16*(7), 3321–3339. https://doi.org/10.3390/jtaer16070180

Augusto, J. C. (2020). Smart cities: State of the art and future challenges. In J. C. Augusto (Ed.), *Handbook of smart cities* (pp. 1–12). Springer International Publishing. https://doi.org/10.1007/978-3-030-15145-4_95-1

Bass, T., Shuderland, E., & Symons, T. (2020). *Common knowledge: Citizen-led data governance for better cities*. European Commission. https://decodeproject.eu/publications/common-knowledge-citizen-led-data-governance-better-cities

European Commission. (2020a). *A European strategy for data* (COM(2020) 66 final). European Commission. https://eur-lex.europa.eu/legal-content/EN/TXT/PDF/?uri=CELEX:52020DC0066&from=EN

European Commission. (2020b). *Data Governance Act (Regulation of the European Parliament and of the Council on European data governance)*. https://eur-lex.europa.eu/legal-content/EN/TXT/PDF/?uri=CELEX:52020PC0767&from=EN

European Commission. (2022). *Data Act (regulation of the European parliament and of the council on harmonised rules on fair access to and use of data)*. https://ec.europa.eu/info/law/better-regulation/have-your-say/initiatives/13045-Data-Act-&-amended-rules-on-the-legal-protection-of-databases_en

European Commission. (n.d.). *Smart cities* [Text]. European Commission – European Commission. https://ec.europa.eu/info/eu-regional-and-urban-development/topics/cities-and-urban-development/city-initiatives/smart-cities_en

Gagnon-Turcotte, S., Sculthorp, M., & Coutts, S. (2021). *Digital data partnerships: Building the foundations for collaborative data governance in the public interest* (p. 96). Synapse C and the Montréal Urban Innovation Lab. https://assets.ctfassets.net/e4wa7sgik5wa/6mV2HLHbhKbU2sgtXSTMQX/da0ede46238b1809d60b5ba65732fb2b/Digital_Data_Partnerships_Report-EN.pdf

Garousi, V., & Mäntylä, M. V. (2016). A systematic literature review of literature reviews in software testing. *Information and Software Technology, 80*, 195–216. https://doi.org/10.1016/j.infsof.2016.09.002

Gonçalves, C., Pinson, P., & Bessa, R. J. (2021). Towards data markets in renewable energy forecasting. *IEEE Transactions on Sustainable Energy, 12*(1), 533–542. https://doi.org/10.1109/TSTE.2020.3009615

Granath, M., Axelsson, K., & Melin, U. (2021). Reflection note: Smart City Research in a Societal Context. A Scandinavian perspective and beyond? *Scandinavian Journal of Information Systems, 33*(1), 5–16.

Gronkiewicz-Waltz, H., Larsson, A., Boni, A. L., Krogh Andersen, K., Ferrao, P., Forest, E., Jordan, R., Lenz, B., Lumbreras, J., Nicolaides, C., Reiter, J., Russ, M., Sulling, A., Termont, D., & Vassilakou, M. (2020). *Proposed mission: 100 climate-neutral cities by 2030 – by and for the citizens: Report of the Mission Board for climate-neutral and smart cities.* European Commission. https://op.europa.eu/en/publication-detail/-/publication/bc7e46c2-fed6-11ea-b44f-01aa75ed71a1/language-en/format-PDF/source-160480388

Leitner-Hanetseder, S., & Lehner, O. M. (2022). AI-powered information and Big Data: Current regulations and ways forward in IFRS reporting. *Journal of Applied Accounting Research, ahead-of-print* (ahead-of-print). https://doi.org/10.1108/JAAR-01-2022-0022

Li, B., Wu, M., Li, Z., & Sun, Y. (2022). A pricing model for subscriptions in data transactions. *Connection Science, 34*(1), 529–550. https://doi.org/10.1080/09540091.2021.2024146

Liang, F., Yu, W., An, D., Yang, Q., Fu, X., & Zhao, W. (2018). A survey on big data market: Pricing, trading and protection. *IEEE Access, 6*, 15132–15154. https://doi.org/10.1109/ACCESS.2018.2806881

Liang, J., & Yuan, C. (2021). Data price determinants based on a hedonic pricing model. *Big Data Research, 25*, 100249. https://doi.org/10.1016/j.bdr.2021.100249

Löfgren, K., & Webster, C. W. R. (2020). The value of Big Data in government: The case of 'smart cities'. *Big Data & Society, 7*(1), 2053951720912775. https://doi.org/10.1177/2053951720912775

Miao, X., Gao, Y., Chen, L., Peng, H., Yin, J., & Li, Q. (2020). Towards query pricing on incomplete data. *IEEE Transactions on Knowledge and Data Engineering*, 1–1. https://doi.org/10.1109/TKDE.2020.3026031

Morgenstadt value of data. (n.d.). Fraunhofer IAO. www.morgenstadt.de/content/dam/morgenstadt/de/documents/value-of-data.pdf

Nguyen, K., Krumm, J., & Shahabi, C. (2020). Spatial privacy pricing: The interplay between privacy, utility and price in geo-marketplaces. *Proceedings of the 28th International Conference on Advances in Geographic Information Systems*, 263–272. https://doi.org/10.1145/3397536.3422213

Ricart, G. (2020). Invited paper: Perishable data in smart communities. *2020 IEEE 28th International Conference on Network Protocols (ICNP)*, 1–6. https://doi.org/10.1109/ICNP49622.2020.9259385

Sætra, H. S. (2021). AI in context and the sustainable development goals: Factoring in the unsustainability of the sociotechnical system. *Sustainability, 13*(4), 1738. https://doi.org/10.3390/su13041738

UN. (2018). *A human rights based approach to data.* United Nations. www.ohchr.org/sites/default/files/Documents/Issues/HRIndicators/GuidanceNoteonApproachtoData.pdf

Wdowin, J., & Diepeveen, S. (2020). *The value of data*. Bennett Institute for Public Policy. www.bennettinstitute.cam.ac.uk/media/uploads/files/Value_of_data_literature_review_26_February.pdf

Wigan, M. R., & Clarke, R. (2013). Big data's big unintended consequences. *Computer, 46*(6), 46–53. https://doi.org/10.1109/MC.2013.195

Yang, J., & Xing, C. (2019). Personal data market optimization pricing model based on privacy level. *Information, 10*(4), 123. https://doi.org/10.3390/info10040123

Yang, J., Zhao, C., & Xing, C. (2019). Big data market optimization pricing model based on data quality. *Complexity, 2019*, e5964068. https://doi.org/10.1155/2019/5964068

Zhang, L., Zhu, T., Xiong, P., Zhou, W., & Yu, P. S. (2021). More than privacy: Adopting differential privacy in game-theoretic mechanism design. *ACM Computing Surveys, 54*(7), 136:1–136:37. https://doi.org/10.1145/3460771

CHAPTER 17

Techno-solutionism Facing Post-liberal Oligarchy

Ivar Jonsson and Lilja Mósesdóttir

CONTENTS

17.1 INTRODUCTION

> *Inequalities also tend to become entrenched through the efforts of those at the very top to secure and perpetuate their positions through various channels, such as having a greater say in the political process or weakening anti-trust and other regulatory efforts that are aimed at curbing monopoly power and improving market efficiency.*
>
> (United Nations, 2019, p. xxiv)

Humans are by nature active social beings, who struggle with solving remnant unsolved problems of the past, while framing the opportunities of their future communities and societies. Looming in the background are institutionalized power structures, ideologies, and worldviews that largely set the principles when we are planning our socioeconomic futures. The colors with which we as individuals paint our future societies depend to a large extent on our reflections on these aforementioned institutionalized structures as well as on our socialization, habitus, and virtues. Our focus in this chapter is on social and economic planning, and we compare it with techno-solutionism.[1] First, we need to argue for what we consider obvious, and then we proceed to arguing for what may not be obvious for many of

DOI: 10.1201/9781003325086-17

those who read this chapter. What we take to be obvious for our reader is that the positivist ideal type of "techno-solutionism" cannot be transposed into being an ideal prescript for planning sustainable societies and economies. The main idea behind techno-solutionism is that by refining and implementing frontier physical and/or social technology, various problems facing humans can be solved effectively. The epistemics (i.e., the ontology and epistemology) of societal planning confronts us with far too high level of complexity and uncertainty. What we presume is not obvious for our reader is our claim that planning for sustainable transition is impossible in capitalism. The market system cannot deliver sustainable future without a disruptive change that would compromise its fundamental principles of private property, maximization of profits, and regimes of economic growth. Let us now turn to the problem of epistemics.

Sartre's concept of "bad faith" comes to mind when we look at how scientists approach acute and apparently destructive problems that threaten human societies. Although molded by our past, we are always bound to choose and act, but, as Sartre (1972) stressed, we may also choose not to act "in bad faith", as we do when we excuse our not-acting by simply referring to external circumstances claiming that they are outside our control. Sciences tend in practice to be institutionalized as "bad faith" that scientists use as an excuse for their passive approach to their objects of study. As Sartre puts it, "I train myself in order to be a scientist, etc. Thus, from the outset, the three big categories of concrete human existence appear to us in their original relation: to do, to have, to be" (Sartre, 1972, p. 576). To this view, we would add that the practice most scientists do, the knowledge they have, and what they are, boils down to their choice to become "scientists" and do the necessary training to have the necessary skills to make "scientific" knowledge. But this training requires that they accept to generate objectified images of reality resulting in the reified knowledge that enables them to sell their labor as "neutral" professionals of de-contextualized details.

The process of neutralizing scientists and the reification of scientific knowledge (Lukács, 1971 [1923], pp. 6–10) has taken long time to develop but has escalated fast in recent decades. Science as a reified production process is the outcome of a development in which the aims of academic research have been subsumed under the dominant production process of commodity production. Science has become commodified. The main characteristic of this commodity is narrowly defined units of specific knowledge bits for narrowly specified markets and hierarchies of customers. The reified producers, the scientists, are now excused from actively engaging in politics on societal scale that seek to counteract existential threats to Humanity such as environmental problems arising from climate change due to constantly unsatisfied need for population growth, waste-based consumption, and profits. The scientists' contribution is limited to their professions pregiven ontological reductions of fragmented reality view.

The production process that scientists find themselves in is part of the wider system of social relations of commodity production. The basic social relation of the production is a property relation that undermines what scientists have and can be. Furthermore, as their labor power is a commodity in the same vein as the product they create, scientists are estranged from the aims of their work and their reified product because they do not own themselves the products and the means of production. The virtue of their work and the value of their creation are reduced to the shifting exchange value on the market.

It is in this context of the practice of science that we will approach the concept of "techno-solutionism" by the way of meta-critical analysis. Our point of departure is the Sustainable Development Goals (SDGs) that focus on securing human well-being and social welfare (e.g., the SDGs 1, 2, 3, 4, 5, 6, 7, 9, 10, 16). We concentrate on the human dimension of the various SDGs because it marks the contexts of critical junctures during which scientists are being challenged to address their role as actors of change. This task requires that the scientists reflect on the moral and political nature of their work. Rather than accepting science from the point of view of "bad faith", their role should be that of reflective scientists who actively take part in transforming the relations between science and nature.[2] The nature of science needs to be redefined so that the science of nature would include the human dimension. Hence, science must consider how various types of societies and their relations with nature affect the state of her well-being and vice versa.[3] This appears clearly in the phenomena of climate change.

Our analysis in this chapter focuses on the following research question: Does the scientific community and economic policy regimes have adequate knowledgebase for building effective societal planning that would make the human dimension of the SDGs attainable? We seek to answer this question by scrutinizing the weaknesses of social and economic science in terms of philosophy of science, on the one hand, and by analyzing a contemporary case of socioeconomic planning technique of post-liberal oligarchism, i.e., the World Economic Forum (WEF), on the other hand. The WEF highlights the role of innovation policies and social steering of technological change to save capitalism from socioeconomic and political crises.

17.2 META-CRITIQUE OF "TECHNO-SOLUTIONISM"

"Techno-solutionism" is impregnated with a connotation that the framing of problem-solving is rational. If we manage to bring together certain quantity of particular elements in certain circumstances in time and space, the outcome will with high level of probability eliminate the problem to be solved. Moreover, the most valuable type of solution is the one that solves the same kind of problem in the most cost-effective way. Hence, the search for solution is reduced to the methods of experimental laboratory science and mathematical analysis. It goes without saying that techno-solutionism as we have summarized it earlier, adheres to positivist philosophy of science that a priori beliefs in the fourth main rule of positivism, that is, the essential unity of the scientific method, indeed, the method of physical science and all knowledge must be reduced to that of physics (Kolakowski, 1972, pp. 17–18). It suffices to say that this requirement is empirically unprovable and hence metaphysical, even by positivist standards. Moreover, positivism emphasizes three other rules upon which techno-solutionism must also be based if it is to be considered scientific, that is, phenomenalism, nominalism, and the rule of rejecting value judgments and normative statements (Kolakowski, 1972, pp. 11–16).

It is particularly this last-mentioned rule that appears to limit the usefulness of the reified practice of contemporary science. The growing existential crisis of humanity that is caused by climate change requires actions and policies that take moral and ethical stances. Climate change threatens animal and human live in general as well as social harmony. Moreover, world's biomass and biodiversity are shrinking adding to the present moral challenges of humankind (Diaz et.al., 2019; Sætra, 2020). The present aims and organization of the world economy do neither secure morally acceptable distribution of world production nor global income distribution that is equal enough to satisfy the basic material needs of the growing world population. Indeed, the world economy is dominated by multinational oligopoly corporations that have undermined market competition. Ironically, market competition was used to morally legitimize the expansion of private property in the heydays of an economic system coined as "capitalism" (Amin, 2014, pp. 3–5).[4]

It appears from the discussion earlier that the existential challenges to humanity require transformation of science in which value judgments and moral principles are critically accepted in advance of the reified production process of contemporary science. In this respect, Habermas' (1987, p. 308) categorization of science and his idea of critically oriented sciences with emancipatory cognitive interests is valuable. Critical self-reflection of the sciences and their knowledge and by analyzing their base in language and power structures at a given point in history and society, scientists can transform knowledge of existing social structures and argue how they could have been organized in the past, and how they may be re-organized in the future. However, as Habermas claims, history shows how violence deforms and suppresses repeated attempts at dialog and closes off the path to unconstrained communication and mankind's evolution toward autonomy and responsibility (Habermas, 1987, p. 315).

Indeed, positivist science, its praise of objectivism and naïve "techno-solutionism", if you like, bears the mark of suppressed dialog that reproduces its limited and reified knowledge.

> But the praise of objectivism has its limits. Husserl's critique was right to attack it, if not with the right means. As soon as the objectivist illusion is turned into an affirmative Weltanschauung, methodologically unconscious necessity is perverted to the dubious virtue of a scientistic profession of faith.
>
> (Habermas, 1987, p. 315)

17.3 LIMITS OF OBJECTIVISM AND PREDICTIVE SCIENCE

Objectivism or the idea that knowledge can be objective in the sense that it is value-neutral and has validity beyond individuals and social settings is highly contested in contemporary philosophy of science. As Gadamer (1977, pp. 74–81) claims, we can only build our knowledge of reality by approaching it via language and formulate our view of it via symbolic expressions and language. Moreover, as languages, including the languages of sciences, and the meaning of words and symbols develop through time and via social settings, knowledge cannot stay static through time and social settings. The same goes for different types of knowledge, such as scientific knowledge. Put differently, our knowledge of the world is created in a dialogical, hermeneutic praxis, and changes through time.[5] Furthermore, as the objects of science and research are defined by means of the different languages of sciences, the level of predictability of hypotheses based on inquiry of objects of study in the past tends to diminish in future studies of the same kind of objects of study. Hence, the technical cognitive interest is satisfied with diminishing returns. In terms of "techno-solutionism", the predicted and proclaimed advances of solutions will require special arguments for their usefulness as time goes by. The same goes for cases of different cultures and societies.

Innovation of new technology often has unforeseen and unintended consequences for external disparate areas or phenomena. Technologies that are invented and developed to solve particular and sometimes urgent problems often create other problems in the process. Well-known examples are chemicals like DDT causing widespread cancer problems; combustion engines and burning fossil fuels leading to greenhouse gases, global warming, increasingly stormy weather, and increasing risk of new ice age in the very long term. This might happen due to melting glaciers, less salt in the oceans, and changes in the world system of circulation of the ocean stream (Broecker, 2010; see also Starr et al., 2021).

In recent decades, investment in research and innovation in the field of digital technology and "Artificial Intelligence" (AI) has been enormous. This investment has been led by private corporations and the state, while scientists and engineers employed by these actors have executed the research under the guiding rule of corporations and state institutions. There are many cases of unintended consequences and serious malfunctions that are claimed to be unforeseen. As an example, autopilots have been put into new models of cars that have caused fatal accidents on the roads (Boudette & Chokshi, 2021). The technology in question is put on the general market without sufficient research and public control. Quite often, no one is deemed ethically responsible for such malfunctioning death traps that use public roads or public spaces as sites for experiments.[6]

In some cases, even the categorization "unforeseen" and "unintended consequences" of innovation and technological change is quite hollow, to say the least. Seen from the narrow and reified point of view of inventors it is to be expected that they can hardly know or foresee in advance future consequences of their inventions. Indeed, they (un)intentionally fail to take responsibility for the way in which their inventions are misused. Cases at hand are platforms on the Internet and its numerous examples of intentional misinformation, spread of unreliable knowledge by lay people, fake-news practices and bullying campaigns against individuals, politician, stakeholders, and organizations. As Gatehouse (2022) has documented, dedicated, and organized groups of bloggers copy political conspiracy theories from platforms such as QAnon and mainstream on popular platforms such as Reddit, YouTube, Facebook, and Twitter (Rothchild, 2021). Such conspiracy theories that are generated in the "virtual world" start having a life of their own in "the virtual world" with real-world consequences. QAnon is an example of such activity that produces popular misinformation and conspiracy theories that had substantial impact on the presidential elections in the United States in 2016 in which D. Trump had the upper hand against H. Clinton.

Activities such as these are increasingly felt to undermine democratic societies leading many governments to seek ways to implement policies to "regulate the wild-west" of the Internet. An example at hand is European Union's General Data Protection Regulation (GDPR). The British Parliament has responded in similar vein (The House of Lords and

The House of Commons, 2021). In the last instance, it is humans as actors that have to set the aims and rules for technological change and decide the moral limits of experimenting with AI. Acting differently is an act of bad faith.

Fosch-Villaronga and Sætra (2021) argue that the spheres of science, ethics, and politics should not be conflated so that reactions to AI's implications would lead to adequate course of action. They conclude that "the different functions of science, ethics, and politics must be respected to ensure AI development serves the interests of society". Their conclusion helps somewhat in tackling the problem of "bad faith" and reification of contemporary science. However, it suffers from a very high level of abstraction, and it does not consider the uneven distribution of power that capitalism and its mode of production generates. Hence, their conclusion is unlikely to contribute to satisfactory analysis of how the proposed adequate course of actions is to be realized in terms of disruptive transition of contemporary societies.

Let's elaborate further on this last-mentioned point. How important it is to emphasize disruptive transition from the point of view of uneven distribution of power becomes clear in light of the fate of the scientists' policy recommendations in the UN's Sustainable Development Goals Progress Report (2022). As a part of the work on the progress report, the UN Intergovernmental Panel on Climate Change (IPCC) released in August 2021 the Physical Science Basis report of Working Group II of its Sixth Assessment Report (AR6) (Foster & Clark, 2022, p. 16 and Foster, 2022, pp. 23–24). It claimed that even in the best-case scenario prognosis, the worldwide climate change situation is so dire that it requires nothing less than a rapidly escalating transformation of the entire global system of production and consumption. There was a consensus among the scientists who wrote the report that if such a drastic structural policy will not be implemented, the world will surpass a 1.5°C increase in global average temperature after 2040 and enter vicious circle of existential crisis.

A "Summary for Policymakers" of Working group II was leaked to the French news agency Agence-France Presse in June 2021 (Foster & Clark, 2022). This summary called for transformational change at all levels: Individuals, communities, businesses, institutions, and governments. Such a transformation would have required coordinated action, massive public mobilization, political leadership and commitment, and urgent decision-making to change the global economy. The central role given by the scientists to government action in transformative change constituted a threat to the hegemony of corporations. As Foster & Clark put it: "Unfortunately, such action has been consistently thwarted by capital and global political leaders, who managed to remove the statement from the final published Working Group II report, where it is nowhere to be found" (Ibid., p. 16).

This brings us to the role of actors in shaping transformative change. We will now look at how the aims of innovation and technological change are socialized in a dynamic way, that is, how these aims are limited and steered in regimes of innovation and technological change (RITC).

17.3.1 The Role of Actors in Disruptive Change

The RITCs have become increasingly important for economic growth and societal development of contemporary national states and international semi-sovereign organizations

(ISSO), such as the European union. Moreover, they have become a major issue for policy formation that aims at institutionalizing research and development (R&D) in various countries and ISSOs. It goes without saying that techno-solutionism as a taken-for-granted ideology by scientists, public officials, politicians, and various stakeholders melts into the R&D policies and the institutionalized RITCs by actors of structuration. Our critical approach to the RITCs and the R&D as social constructions is shared by critical realists such as Archer (1995) as well as sociologist Giddens (1984) who emphasize the role of human actors in societal development. Let us now look closer at the role of agency in demystifying, de-reifying, and de-commodifying science and technology.

Whether institutions and social change are determined by structure or agency is an old question in social sciences. According to structuralist scholars, the change of social relations or their reproduction is a result of structural necessity irrespective of the intentions or reflections of individuals. Structuralist view of this kind is opposed by constructivist approaches that emphasize the role of intentions of individuals, their ethics and ideology, and processes in which meaning is created in symbolic interaction.

We argue that social change and the development of institutions can be reduced neither to structural necessity nor to individual intentionality. Social contexts give birth to individuals' intentions, their ideas and morals which they "interiorize" (Jonsson, 2015, p. 141). At the same time, social structures and relations are generated by intentional actions of individuals and groups while the resulting structures are partly unintentional. This results from the fact that intentions are never fully defined and they change in the moment of the interactive practice of agents of change and, furthermore, actors' information of contexts of change is never perfect. Moreover, social contexts of change are constituted by societal structures that constrain the alternatives of change. However, while these structures determine the alternatives of change, the determination is neither exclusively bound to one, and only one alternative, nor an infinite number of voluntary choices. As a consequence, we claim that social change is characterized by morphogenetic (Archer, 1995, pp. 157–158) processes of "structuration"[7] generated by way of active social construction and relational politics.

Earlier, we have argued that science has emerged as if it were a production process that produces knowledge estranged from its producer. Furthermore, the scope of the products, that is, the knowledge, is increasingly limited to fit professions determined by actors of power (Mósesdóttir & Jonsson, 2020) and goals generated in the various RITCs mentioned earlier. These actors belong to societal power blocks that share balance of power and build coalitions for economic and science policy formation that reflect their national and international economic interests.

By organizing science according to professions and narrowing their scope of study as part of production processes, the steering of science has become more effective than ever before. The so-called stakeholders have now a greater say when various research programs are defined and carried out such that the contribution of scientists in particularized professions have become easier to predict. Consequently, the opportunity to invest in and steer multidisciplinary research projects at ever larger scale has become much easier for the power blocks and political elites than it used to be. This last-mentioned point is what the advocates of the thesis of "the finalization phase of science" argue (Schroyer, 1984, pp. 716–718). As sciences reach the stage of increased maturity, research can aim either at post-paradigm theoretical sophistication or, alternatively, to "functionalization" of the object of study, that is, the object of study is externally defined according to practical goals that are generated in interactive relations of science, society, nature, and politics. Unfortunately, as goes for Habermas's abstract idea of "emancipatory cognitive interests", the idea of extra-scientific determination of "functional finalization" of the objects of study is equally hollow. In both cases, what is missing is a substantial dynamic theory, or at least proper conceptualization of real power structures and hegemonic politics.

We presume that professionalization, the narrower scope of sciences, their "finalization", and canalization via the RITCs are not "Manna" that falls from heaven. Investment in science and research as well as policy formation in this field is actively constructed in coalition building that favors some stakeholders at the cost of others.

Hence, shifting balances of power and stakeholder coalitions undermine how secure investment in research can be. The same goes for predictability of steering scientific production: One cannot presume that science and research will necessarily lead to ever more useful and predictable knowledge. Hence, it is very difficult to estimate who will, when and how, generate science and research policies, how and when policies will emerge that aim at solving the increasing existential crises of humankind. This is detrimental for realizing the SDGs. As we discussed earlier, the answers are written on the wall, that is, in the limits of objectivism and the credence in predictive science. Let us concretize this thesis of ours in terms of the political economy of contemporary oligarchic capitalism.[8]

17.3.2 Social and Economic Planning as Technique

The realm of social and economic planning is increasingly dominated by a global industry that already in 2007 counted over 250,000 expert advisors who sell "techno-solutionist" advice to politicians and the state bureaucracy (Røvik, 2007, p. 64).[9] The goods sold by the "techno-solutionists" are developmental strategies that would deliver high level of success measured in terms of improved predictability.

As Sætra discusses in Chapter 2, the concept of technology may be understood in a broad sense. The concept encompasses what we make, but also what we do. But what we make and do is affected by why we decide to make and do particular things. Why we do things is often because techniques as means to make and do are fetishized due to scientists' bad faith. Planning techniques, the way we plan things, are socially pregiven and a result of skills learned in specialized occupational education.

However, promised success of planning and its techniques is questionable. The levels of predictability depend on pregiven societal structures that are actively created in particular historical periods and reproduced by actors of power who have miscellaneous interests and goals. In Western societies, societal decision-making takes place on three levels of abstraction:

- **Level one:** Part of the general citizens have rights to elect representatives that have the role to translate the citizens' interests into policies and practices that would reflect their interests;

- **Level two:** The policies and practices are then supposed to realize the citizen's interests in the various formal institutions of power belonging to the state, regional organizations, municipalities, etc. Hence, the citizens take part in a system of indirect democracy in which elected representatives have the role to formulate the abstract interests of the miscellaneous individuals and groups of "the people" and realize them in laws and regulations;
- **Level three:** The "will of the people" is then brought further from the people to the executive branch of the state and other public institutions. On the way, democracy becomes even more indirect and detached from the people;
- **Level four:** The cleavage between the people, the representatives, and the public executive branch in this system becomes further amplified and extended in scale as populations grow depending on the size of jurisdictions.

Indeed, in recent decades, policy formation developed from being predominantly limited to regions and the interests of nation-states to international organizations such as the UN and military defense pacts. Moreover, the concept "security" in international politics is not anymore limited to military defense but is increasingly understood in terms of socio-ecological threats on global scale. The SDGs are a case at hand that demonstrate growing importance of internationalization in policy formation today. Later, we will discuss this issue in view of the rise of post-liberal oligarchic capitalism.

The levels of complication of decision-making escalate to the extent that particular type of knowledge and practices have to be formulated in ever more abstract way. This is necessary to reflect the miscellaneous definitions of interests, goals, and means the estranged actors in levels one to three have, so that they do not collide in disastrous chaos. The need for societal planning intensifies, that is, the type of planning that can be presented as acceptable. It has to be based on high level of abstraction, if the main colliding actors are to accept them and use them as a substance for long-term coalitions of collaboration. The bottom line is that insofar as science and research are to constitute this substance, science itself has to take on board ideological simplifications and starts functioning as ideology.[10]

Collaboration involves legitimizing[11] societal planning. If societal planning is to be accepted as legitimate, it must be based on these principles of "objectivity", "neutrality", and "science". However, if planned outputs fail to generate public policies that facilitate economic growth, a stage of rationality crisis threatens the legitimacy of the executive system. In addition to hinder output crises, societal planning needs to secure loyalty of the masses. Otherwise, the existing social formation faces an input crisis, namely a legitimation crisis. At this point, planning functionality starts to lose its rationality as loyalty stands and falls with normative values and cultures of the masses. However, rationality tends not to be accepted at any cost of ethics and virtues. In the context of indirect democracy with its large distances between the various groups and their miscellaneous interests, legitimation crisis is always a risk that undermines success in social and economic planning.

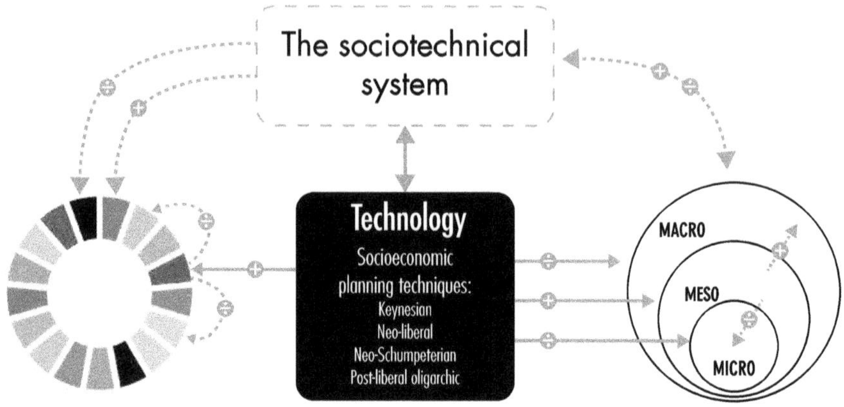

FIGURE 17.1 The analytical framework: Socioeconomic planning technique, developed from Sætra's Figure 2.5 in Chapter 2 of this book.

The history of capitalism has many stories of broken promises of social and economic planning and consequent legitimation crises. Let us look closer at this dimension of socioeconomic planning as a technique that is illustrated in Figure 17.1.

In the heydays of the competitive capitalist system, the virtues of possessive individualism found grounds in the expanding civil and family privatism. The ideology presumed that private initiatives and efforts by successful private competitive citizen would create his/her wealth and fortune. The needs of the family and the individuals were presumed to be relatively easily recognizable. However, gradually more basic needs were satisfied by the state. At the same time, it became more difficult for the private persons in the affluent society to define their needs. Hence, the principle of possessive individualism and civil privatism faltered. The wealth and fortune of the individual were decreasingly secured by private efforts and gradually replaced by the various forms of welfare states in capitalist societies (Esping-Andersen, 1989).

The virtue of possessive individualism was not only undermined by the necessary role of the state but even more by the rise of trusts and oligopoly and monopoly corporations (Sweezy, 1939).[12] The last decades of the 19th century and the first decades of the 21st century saw the rise of the virtue of what we prefer to call "possessive oligarchism", following unprecedented increase in worldwide wealth concentration and inequality. In the 1930s, the Great Depression sets in with worldwide economic depression and escalated rise of fascist and Nazi movements. This trend intensified again following the development of neoliberalist regimes since the 1980s and the rise of financial and digital oligarchy in the present era (Amin, 2014).

Since the 1930s, capitalism and its socioeconomic techniques have developed through four main alternatives, that is, Keynesianism, neoliberal, neo-Schumpeterian, and the contemporary post-liberal oligarchism which still is in its early stage of development. We have elsewhere analyzed the first three mentioned alternatives.[13] In the following, we will discuss the fourth alternative of post-liberal oligarchic capitalism.

17.3.3 Planning Post-liberal Oligarchic Capitalism

The new contemporary oligarchic era required new ideology to legitimize a resetting of the planning principles for economic growth, that is, a new oligarchist innovation system that must secure the interests of the rising oligarchist elite. The task at hand is to legitimize economic growth that would revive "possessive individualism" at the same time as "possessive oligarchism" must embellish. Simultaneously, the former Keynesian and neoliberalist regimes of capital accumulation had delivered leftovers of waste and climate change from their particular versions of economic growth policies. These leftovers were increasingly met with rejection and protests by the general public. Hence, in the aftermath of the financial crisis in 2007–2008, the contradictions escalated on national and global levels, not only in terms of economic inequality and polarization of power between the masses and the elites but also climate change that has turned into an urgent unsolvable problem due to economic growth (Fominaya & Cox, 2013). As climate change is a global problem, the economic policies/ideologies of Keynesianism, neoliberalist untamed liberalization of markets and national systems of innovation turned out to be rather useless techno-solutionist instruments to resolve the problem of escalating climate change. Indeed, the political regimes could not transform or generate new policy solutions accepted as "scientific economic planning".

However, the crises could not go on for too long time and the search for new policies continued. This search was intensified and made possible in the oligarchic dominated by the WEF with its annual conference in Davos, other regional conferences in locations across Africa, East Asia, Latin America, and India and additional annual meetings in oligarchized communist China and the United Arab Emirates (World Economic Forum, 2022). The engineer K. Schwab established the WEF in 1971, which allegedly has become the main ideological think-tank forum organized for and by multinational capital and oligarchs.[14]

The main thrust of the WEF is that it exploits the crises in the present balance of power between sovereign states. The WEF invites selected academic researchers, opinion givers, hegemonic ideologists, and members of the global political elite to join their political efforts. These individuals play the role of coalition builders for a new alternative developed by the WEF which we would like to call the "post-liberal alternative" (PLA). So, what does the PLA stand for?

The PLA is an ideology that is supported by willing economists and social scientists who aim at building a new era based on economic policy that focuses on satisfying predefined and prechosen multi-stakeholder interests. It aims at saving capitalism from socioeconomic crash that would threaten its essential need for economic (profit) growth as well as from letting humanity suffer in an environmental existential disaster.

The PLA emphasizes the view that technological development, particularly megatrends, generates long-term economic growth and prosperity. Much like what neo-Schumpeterians presume, technological megatrends generate major changes in social structures, the economy, and political life (Freeman, 1977 & 1987), the WEF presumes that industrial capitalism is presently going through the so-called fourth industrial revolution (4IR) that knits together AI, gene editing and advanced robotics blurring the lines between the physical, digital, and biological worlds. Innovations and inventions in these fields generate megatrends of societal transformation (Jonsson & Mósesdóttir, 2018).

Unlike neoliberalists but much like neo-Schumpeterians, the PLA acknowledges that inequality represents the greatest societal concern associated with the 4IR (Schwab, 2015 & 2016). What explains the rising gap in wealth between those dependent on capital as opposed to labor is that the largest beneficiaries of innovation tend to be the innovators, shareholders, and investors. Moreover, machines will also displace many low-skilled workers, while the diffusion of new technologies may result in a net increase in jobs for the highly skilled. According to Schwab (2015) the WEF's founder, growing polarization in income and wealth will in turn lead to an increase in social tensions. However, Schwab's analysis suffers from the same shortcoming as that of the neo-Schumpeterian. Both approaches lack an account of how social and political actors resist and/or seek to shape the diffusion of new technologies.[15] Instead, they are preoccupied with taxonomical categorization of the main features of technical change.

In the 1970s and 1980s, neo-Schumpeterian economists emphasized that developing strong national systems of innovation would be the right instruments to deliver future economic growth (Freeman, 1977, 1987). In recent years, the neo-Schumpeterian Mazzucato (2018) has followed suit with her "mission-orientated innovation policies". We claim that their policy initiative came too late as capital accumulation was already far too internationalized. The PLA of the WEF adheres to a quite different and much more one-sided policy alternative in line with the elitism of oligarchic capitalism. It rejects the idea that the sovereign state and the UN must play a leading role in solving the crises-ridden capitalist societies.

According to Schwab (2008), states and related international organizations are far too slow to implement the measures for societal transformation that would save capitalism. He argued that the global corporate elite has to engage in "Global Corporate Citizenship" and gain hegemonic role to rule over the outdated international system of states and global elites (Schwab, 2008).

Following suit, the WEF has since 2009 been working on its Global Redesign Initiative (GRI), which effectively argues for a transition away from intergovernmental decision-making toward a system of multi-stakeholder governance. The essential function of this initiative is to marginalize democratic governments by limiting their role to the negotiation of treaties which are then given the green light by democratically elected representatives. Thereafter, the treaties are delivered to a self-selected group of multi-"stakeholders" who make decisions about their execution on behalf of the general public. This WEF initiative of multi-stakeholder governance (MSG) is worrisome as it may undermine Western democracy.

As Gleckman claims: "What is ingenious and disturbing is that the WEF multi-stakeholder governance proposal does not require approval or disapproval by any intergovernmental body. Absent any intergovernmental action the informal transition to MSG as a partial replacement of multilateralism can just happen" (2016, p. 92).

The term itself, "stakeholder", has rather doubtful connotation and may be designed for framing proposes to confuse the general public. It conceals the immense differences in interests, role, power, and legitimacy that exist among the various actors that participate in multi-stakeholder initiatives. In practice, the multi-stakeholders are not equal and there is no distinction made between alleged "rights holders". Hence, no community is considered more affected than another by environmental destruction or human rights violations depriving individuals of their legitimate right to participate in decisions-making affecting their lives. Moreover, private corporation who are only accountable to their stakeholders and elected governments ("duty bearers") who have an obligation to act in the public interest are treated as partners with common interests (Manahan & Kumar, 2022, p. 7).

Following this train of thought, the WEF runs a network of "Global Future Councils" that are "dedicated to promoting innovative thinking to shape a more resilient, inclusive and sustainable future" (World Economic Forum, 2022). According to the WEF, "it assembles more than 1000 of the most relevant and knowledgeable thought leaders from academia, government, international organizations, business, and civil society, grouped in expertise-based thematic councils. It is an invitation-only community and members are nominated for a one-year period" (World Economic Forum, 2022).

In addition to the WEF's activities, the "Sustainable Development Solutions Network. A Global Initiative for the United Nations" (SDSN) works on initiating global coordination and standardization of statistics for the SDGs. As the homepage of the SDSN informs, it is not a part of the UN but a private sector network financed by oligarchs and private corporations. The aim is to influence international organizations and national statistical bureaus to fasten the process of developing and implementing the SDGs-based policies at national and international levels. The SDSN publishes annual reports on how the coordination work progresses.

Furthermore, the SDSN has moved the SDG initiative to a stage that may be called "stage 2". While stage 1 refers to UN's formulation and acceptance of the SDGs, stage 2 refers to implementing action agenda at the national level for the "multi-stakeholders" in the form of six transformations: The core of the six Transformations is the recognition that all 17 SDGs can be achieved through six major societal transformations, focused on (1) education and skills, (2) health and well-being, (3) clean energy and industry, (4) sustainable land use, (5) sustainable cities, and (6) digital technologies. All are guided by the twin principles to "leave no one behind" and "ensure circularity and decoupling . . . The six Transformations provide an action agenda for government ministries, businesses, and civil society" (Sachs et al., 2021, pp. 44–45).

As the SDSN acknowledges, the progress of instigating the six transformations is disappointing. The chief threshold is that the financing needs of low-income developing countries (LIDCs) are far greater than their governments can provide. The LIDCs will need a significant increase in fiscal space, which will require a combination of domestic and global fiscal policies (Sachs et al., 2021, p. 21).

The SDSN has no solutions to this problem, presumably because that would require a fundamental system shift and disruptive transition of capitalism. The system change would call for massive increase in state intervention, not only at national but at international level as well. It appears that a stage 3 is needed for the realization of the SDG involving initiatives at the international and national levels facilitating structural transition of "creative destruction".[16]

Moreover, in terms of developing effective governance strategies to transform political cultures and establish stable pathways for transition, the task ahead is enormous. Indeed, global differences in World Bank's Worldwide Governance Indicators (WGI) are huge between high-income OECD countries, low-income countries, and middle-income countries (World Bank, 2022). WGI refers to percentile rank (0–100) of indicators such as Voice and Accountability, Political Stability and Absence Violence/Terrorism, Government Effectiveness, Regulatory Quality, Rule of Law, and Control of Corruption. In 2020, the average score of high-income OECD countries was 82, low-income countries was 21, and middle-income countries was 37. As we have discussed earlier, the low-income countries are those that have the largest difficulties in implementing SDG and they score lowest in WGI.

17.4 DISCUSSION AND CONCLUSION: FUTURE TRENDS TOWARD POST-LIBERAL CORPORATIST OLIGARCHY

Our study of the PLA economic planning alternative is grounded in the main historical versions of the ideology of innovation and technology for economic growth. These alternatives involving Keynesianism, neoliberal, neo-Schumpeterian, and the contemporary post-liberal oligarchism appear to have resulted in a long-term crisis of liberal democracy and the crisis of its mode of legitimation of the capitalist system, that is, possessive individualism and oligarchism.

The core of the legitimation of the capitalist system is an ideology presuming societal planning based on "scientific economic theories" and socioeconomic planning technique that suffices to satisfy both possessive individualism and oligarchism by the way of securing permanent economic growth. However, there are some unfortunate foresights that worry us:

1. It is very likely that possessive individualism will be satisfied and in the short-term continue to keep the citizens of formally democratic societies apathetic.
2. The rise of multi-stakeholder governance of the WEF's oligarchic capitalism is directly aimed at undermining representative democracy by deliberately organizing the escalation of the power of oligarchic transnational corporations in planning the future policy formation.
3. As Stiglitz (2016) argues, economic policies based on orthodox economic theories have for decades advocated and generated extreme inequalities of income in most of the rich economies of the world. The experience of this development shows that increased inequality leads to reduced economic performance. Indeed, we would add that with post-liberal oligarchism, orthodox economic policies are likely to continue to generate growing inequality. Consequently, the legitimation crisis of capitalism is likely to deepen in the long term insofar as possessive individualism will be undermined and lead to decreasing apathy. Inequality and decreasing apathy may lead to mobilization of extremist political groups such as far-right fascist groups that increasingly threaten Western democracy.

4. Economic growth is essential for the survival of the capitalist system. However, economic growth is a major hindrance to the haltering climate change and the prospect of Green Growth is bleak (Hickel & Kallis, 2019). Research evidence do not support the assumption of the SDGs that economic growth and environmental sustainability can go hand in hand given the current state of technology (Hickel & Kallis, 2019).
5. Moreover, the reification of science, their increasingly external steering, and their bad faith due to their naïve objectivism and positivism are unlikely to counteract the rising era of the post-liberal alternative of "multi-stakeholder" governance under the hegemony of TNCs. In so far as the SDGs are grounded in profit-based economic growth, their attainment remains a utopia. The path dependency of the present system undermines Western democracy and the SDGs' participatory strategy. While the UN's strategy lacks sound policy and means for global action, the WEF elite at Davos and expert networks like the SDSN are giving birth to their strategy of post-liberal oligarchic corporatism.
6. The UN has not yet managed to develop a plan of actions to transform the SDGs into necessary policies that would lay the grounds for sustainable development regimes. The reason being the need for structural transition of capitalism and its "creative destruction" that requires extensive state interventionism on national and global scales to bridge the gap between low- and high-income countries. Moreover, as the WGIs indicate, the task ahead is enormous as effective governance strategies must be developed to transform political cultures and establish stable pathways in the low- and lower-middle-income economies of the world.

Having the previous prognosis in mind, the request for sustainable development appears to contradict in a fundamental way the capitalist order and its economic policy regimes. Hence, we conclude that the SDGs focusing on securing human well-being and social welfare (e.g., the SDGs 1, 2, 3, 4, 5, 6.7, 9, 10, 16) while at the same time fetishizing naïve techno-solutionism are rather unrealistic goals. In other words, capitalist sustainable transition appears to be a mission impossible.

NOTES

1 In this chapter, we use the term "techno-solutionism" rather than "techno-fetishism" that neo-Schumpeterians used in their book Cole et al. (1973) *Models of Doom. A Critique of the Limits to Growth.*

2 In terms of our present-day situation, we are challenged by constructing technological futures, that is, alternatives in molding nature–humans–society relations in the field of determining the aims and uses of technology such as AI. To what extent this challenge optimizes democratic participation of the general public or serves the interests of reified technocracy, capital accumulation and dominant elites, is among essential issues in contemporary politics. See, e.g., Sætra (2020).

3 For an early meta-critique of science of this kind, see Marx who wrote in 1844: "History itself is a real part of natural history and of nature's becoming man. Natural science will in time subsume the science of man just as the science of man will subsume natural science: there will be one science" (Marx, 1975, p. 355). For a comprehensive analysis, see Jonsson (2008).

4 As Foster (2015, p. 18) points at, the Forbes 2015 list of the top 500 largest companies in the world have revenues that mount to around 30% of 500 global private of world revenue.
5 See Sætra, 2018.
6 For further discussion, see Fosch-Villaronga, et al.'s chapter in the present book. See also Sætra (2021).
7 For a critical scrutiny of the concept of "structuration", see Stones (2001).
8 See Hacker (1964) for early analysis of the concept of "oligarchic capitalism".
9 According to more recent source, the global management consulting services market is expected to grow from \$819.79 billion in 2020 to \$895.46 billion in 2021 at a compound annual growth rate (CAGR) of 9.2%. The market is expected to reach \$1201.06 billion in 2025 at a CAGR of 8% (BusinessWire, 2021).
10 See Stiglitz (2002).
11 See Offe (2013).
12 The Marxist economist P. Sweezy (1939) was one of the first economists to put forward a theory of oligopolist competition.
13 See Jonsson (1991, 2015), and Mósesdóttir and Jonsson (2020).
14 With around 3000 paying members and selected participants – among which are investors, business leaders, political leaders, economists, celebrities, and journalists – for up to 5 days to discuss global issues across 500 sessions. WEF is funded by its 1,000 member companies, typically global enterprises with over five billion dollars in turnover (varying by industry and region). These enterprises are among the largest companies within their industry and/or country. In 2011, an annual membership costs \$52,000 for an individual member, \$263,000 for "Industry Partner", and \$527,000 for "Strategic Partner". An admission fee costs \$19000 per person. By 2014, annual fees for "Strategic Partner" had increased to \$628,000. However, these figures are not official WEF information as WEF keeps its finances secret (based on information from Wikipedia ("World Economic Forum," 2023) that may not be accurate, but gives rough indication of the subject matter).
15 See I., A. Ibrahim's Chapter 14 on the importance of politics and multi-stakeholders concerning implementation of SDGs.
16 The concept of "creative destruction" was coined by J. Schumpeter (see, e.g., H. Borgebund's Chapter 12). As J. Elliott (1978, pp. 148–169) has argued, K. Marx had already worked on a similar concept in his work *Grundrisse* in 1857–1858. Schumpeter recognized Marx' contribution in this field of study. While Schumpeter referred to "perennial gale of creative destruction", in which capitalism, through its creative success, leads to its own destruction, Marx highlights capitalism's "revolutionizing properties" in terms of technological change and mode of production that generates the surplus production, that is, surplus value upon which socialist society can be built as well as what today is referred to as de-growth and ecological socialism. See also Foster, 2022.

17.5 REFERENCES

Amin, S. (2014). *Capitalism in the age of globalization, the management of contemporary society.* Zed Books.
Archer, M. (1995). *Realist social theory: The morphogenetic approach.* Cambridge University Press.
Boudette, N. E., & Chokshi, N. (2021, October 21). U.S. will investigate Tesla's autopilot system over crashes with emergency vehicles. *The New York Times.* www.nytimes.com/2021/08/16/business/tesla-autopilot-nhtsa.html
Broecker, W. S. (2010). *The great ocean conveyor, discovering the trigger for abrupt climate change.* Princeton University Press.
BusinessWire. (2021). *Global management consulting services market report (2021 to 2030) – COVID-19 impact and recovery.* www.businesswire.com

Cole, H. S. D., Freeman, C., Jahoda, M., & Pavitt, K. (1973). *Models of doom: A critique of the limits to growth*. Universe Books.
Diaz, S. et al. (2019). *Global assessment report on biodiversity and ecosystem services of the Intergovernmental Science-Policy Platform on Biodiversity and Ecosystem Services*. Bonn IPBES Secretariat. https://doi.org/10.5281/zenodo.3831673
Elliott, J. E. (1978). Marx's "Grundrisse": Vision of capitalism's creative destruction. *Journal of Post Keynesian Economics, 1*(2). Taylor & Francis. www.jstor.org/stable/i407603.
Esping-Andersen, G. (1989). *The three worlds of welfare capitalism*. Polity Press.
Fominaya, F. C., & Cox, L. (2013). *Understanding European movements: New social movements, global justice struggles, anti-austerity protest*. Routledge.
Fosch-Villaronga, E., & Sætra, H. S. (2021). *Research in AI has Implications for Society: How do we Respond? The balance of power between science, ethics, and politics*. www.nomos-elibrary.de/10.5771/2747-5174-2021-1-60.pdf
Foster, J. B. (2015). The new imperialism of globalized monopoly-finance capital. An introduction. *Monthly Review, 67*(3). https://doi.org/10.14452/MR-074-03-2022-07_1
Foster, J. B. (2022). *Capitalism in the Anthropocene: Ecological ruin or ecological revolution*. Monthly Review Press.
Foster, J. B., & Clark, B. (2022). Socialism and ecological survival, an introduction. *Monthly Review, 74*(3). https://monthlyreview.org/2022/07/01/socialism-and-ecological-survival-an-introduction/
Freeman, C. (1977). *The Kondratiev long waves, technical change and unemployment*. Expert Meeting on Structural Determinants of Employment and unemployment. OECD.
Freeman, C. (1987). *Technology policy and economic performance; lessons from Japan*. Pinter Publishers.
Gadamer, H-G. (1977). *Philosophical hermeneutics*. University of California Press.
Gatehouse, G. (2022). *The coming storm: Q drops*. BBC Podcast. www.bbc.co.uk/sounds/play/p0bchpyg
Giddens, A. (1984). *The constitution of society*. Polity.
Gleckman, H. (2016). *Multi-stakeholder governance: A corporate push for a new form of global governance. Amsterdam*. The Transnational Institute. www.tni.org/en/publication/multi-stakeholderism-a-corporate-push-for-a-new-form-of-global-governance
Habermas, J. (1987). *Knowledge and human interests*. Polity Press.
Hacker, A. (Ed.). (1964). *The corporation takeover*. Harper & Row.
Hickel, J., & Kallis, G. (2020). Is Green growth possible? *New Political Economy, 25*(4), 469–486. https://doi.org/10.1080/13563467.2019.1598964
House of Lords and The House of Commons. (2021). *Draft online safety bill, report of session 2021–22*. British Parliament. https://committees.parliament.uk/committee/534/draft-online-safety-bill-joint-committee/
Jonsson, I. (1991). Keynes's general theory and structural competitiveness. In *Thjodmal: Yearbook of political economy and social sciences* (pp. 87–117). Félags-og hagvísindastofnun Íslands. www.researchgate.net/publication/338656621_Keynes's_General_Theory_and_Structural_Competitiveness
Jonsson, I. (2008). *Marx's Metascience – a dialogical approach to his thought*. Félags-og hagvísindastofnun Íslands. www.academia.edu/26299936/Marxs_Metascience_A_Dialogical_Approach_to_his_Thought
Jonsson, I. (2015). *The political economy of innovation and entrepreneurship: From theories to practice*. Routledge. https://ebookcentral.proquest.com/lib/hiof-ebooks/detail.action?docID=4442572
Jonsson, I., & Mósesdóttir, L. (2018). Innovation systems at a crossroads: Implementing systems of transformative innovation. In I. Bernhard (Ed.), *Uddevalla symposium 2018: Diversity, innovation, entrepreneurship – regional, urban, national and international perspectives*. University West.
Kolakowski, L. (1972). *Positivist philosophy, from Hume to the Vienna circle*. Pelican Books.
Lukács, G. (1971 [1923]). *History and class consciousness: Studies in Marxist dialectics*. The MIT Press.

Manahan, M. A., & Kumar, M. (2022). *The great takeover; mapping of multistakeholderism in global governance*. People's Working Group on Multistakeholderism. https://publicservices.international/resources/publications/mapping-of-multistakeholderism-in-global-governance?id=12623&lang=en

Marx, K. (1975). *Early writings*. Penguin.

Mazzucato, M. (2018). *Mission-oriented research & innovation in the European Union: A problem-solving approach to fuel innovation-led growth*. European Union.

Mósesdóttir, L., & Jonsson, I. (2020). Theorizing transformative innovation: The role of agency in real critical junctures. In I. Bernhard, U. Gråsjö, & C. Karlsson (Eds.), *Diversity, innovation and clusters: Spatial perspectives* (pp. 104–124). Edward Elgar Publishing. https://doi.org/10.4337/9781789902587.00010

Offe, C. (2013). Europe entrapped does the EU have the political capacity to overcome its current crisis? *European Law Journal, 19*(5), 595–611. https://doi.org/10.1111/eulj.12071

Rothchild, M. (2021). *The storm is upon us: How QAnon became a movement, cult, and conspiracy theory of everything*. Melville House Publishing.

Røvik, K. A. (2007). *Trender og translasjoner: ideer som former det 21. århundrets organisasjon*. Universitetsforlaget.

Sachs, J. et al. (2021). *Sustainable development report 2021; the decade of action for the sustainable development goals; includes the SDG index and dashboards*. Cambridge University Press. https://s3.amazonaws.com/sustainabledevelopment.report/2021/2021-sustainable-development-report.pdf

Sætra, H. S. (2018). Science as a vocation in the era of big data: The philosophy of science behind big data and humanity's continued part in science. *Integrative Psychological and Behavioral Science, 52*, 508–522. https://doi.org/10.1007/s12124-018-9447-5

Sætra, H. S. (2020, August) A shallow defence of a technocracy of artificial intelligence: Examining the political harms of algorithmic governance in the domain of government. *Technology in Society, 62*. https://doi.org/10.1016/j.techsoc.2020.101283

Sætra, H. S. (2021). Confounding complexity of machine action: A Hobbesian account of machine responsibility. *International Journal of Technoethics (IJT), 12*(1). www.igi-global.com/pdf.aspx?tid=269437&ptid=254203&ctid=4&oa=true&isxn=9781799861485

Sartre, J-P. (1972 [1943]). *Being and nothingness: An essay on phenomenological ontology*. Methuen &. Co LTD.

Schroyer, T. (1984). On finalization in science. *Theory and Society, 13*(5), 715–723. www.jstor.org/stable/657248

Schwab, K. (2008, January). Global corporate citizenship; working with governments and civil society. *Foreign Affairs, 87*, 107–118. www.proquest.com/magazines/global-corporate-citizenship-working-with/docview/214299073/se-2?accountid=43205

Schwab, K. (2015, December). The fourth industrial revolution: What it means and how to respond. *Foreign Affairs*. www.foreignaffairs.com/articles/2015-12-12/fourth-industrial-revolution

Schwab, K. (2016). *The fourth industrial revolution*. World Economic Forum.

Starr, A., Hall, I. R., Barker, S. et al. (2021, January). Antarctic icebergs reorganize ocean circulation during Pleistocene glacials. *Nature, 589*, 236–241. https://doi.org/10.1038/s41586-020-03094-7

Stiglitz, J. E. (2002). There is no invisible hand. *The Guardian*. www.theguardian.com/education/2002/dec/20/highereducation.uk1

Stiglitz, J. E. (2016). Inequality and economic growth. In M. Jacobs & M. Mazzucato (Eds.), *Rethinking capitalism: Economics and policy for sustainable and inclusive growth*. John Wiley & Sons Ltd.

Stones, R. (2001). Refusing the realism-structuration divide. *European Journal of Social Theory, 4*(2). SAGE.

Sustainable Development Solutions Network (SDSN). *A global initiative for the United Nations*. www.unsdsn.org/

Sweezy, P. (1939). Demand under conditions of oligopoly. *Journal of Political Economy, 47*(4), 568–573. www.journals.uchicago.edu/doi/abs/10.1086/255420.

United Nations. (2019). *The Future is now: Science for achieving sustainable development: Global sustainable development report 2019.* United Nations. https://sustainabledevelopment.un.org/content/documents/24797GSDR_report_2019.pdf

United Nations. (2022). *SDG progress report (2022).* https://unstats.un.org/sdgs/report/2022/

World Bank. (2022). *Worldwide governance indicators.* The World Bank. http://info.worldbank.org/governance/wgi/

World Economic Forum. (2022). *Global future councils: 2020–2022 term.*

World Economic Forum. World Economic Forum. (2023, February 24). In Wikipedia. https://en.wikipedia.org/wiki/World_Economic_Forum, last edited March 12, 2023.

CHAPTER 18

The Role of Technology in Alternatives to Growth-Based Sustainable Development

Henrik Skaug Sætra

CONTENTS

18.1 INTRODUCTION

> *Change in technology implies change in culture.*
>
> (Næss, 1989, p. 102)

The implications of technological development and change are vast, but how these implications are evaluated varies drastically. While some – the techno-optimists and Prometheans stand out – see humanity's savior in technology (Danaher, 2022; Farley & Smith, 2020), others argue that technology – at least in its current form – is to blame for a situation perceived to be undesirable and detrimental to our prospects of sustainable development (Watson, 2020). Some believe that growth and innovation have created societies in which human life has never been better (Pinker, 2011), while others have long argued that the downsides to economic growth, greater scale, and more complexity have negative consequences of great importance. Examples of the latter camp are those who call for various

DOI: 10.1201/9781003325086-18

versions of degrowth and decentralization as radical solutions to our current ailments, arguing that endless growth in a finite world is bound to be disastrous (Kerschner et al., 2018; Latouche, 2009; Næss, 1999).

Kerschner et al. (2018) provided the first review of the role of technology in *degrowth* research. In the research agenda developed through this review, they argue that exploring "new classical authors" is important, and this is a call here heeded. While authors such as Illich, Schumacher, Ellul, and others are heavily cited in the literature, the review also shows that, for example, Arne Næss and deep ecology and the alternative technology movement were not mentioned. This chapter argues that Næss' works and the sources that inspired him provide useful conceptualizations of different types of technologies and their role in achieving more sustainable outcomes across the three sustainability dimensions (Næss, 1989, 1999; Smith, 2005).

The deep tension between different approaches to technology becomes particularly clear when seen in light of the concept of sustainable development. While sustainable development establishes a set of goals that are in large part shared by all, great controversies surround *how* these goals are to be achieved. Of particular importance in this chapter is the role of technology and the need for growth discussed in the United Nations' conceptualizations of sustainability through the concept of sustainable development (Brundtland et al., 1987) and the Sustainable Development Goals (SDGs) (United Nations, 2015).

In this chapter, I explore these tensions through an exploration of the role of technology and growth as described in the 17 SDGs, the 169 targets, and the broader Agenda 2030 which contain them. I then proceed to present some radical alternative perspectives on growth and technology, which break fundamentally from the perspective derived from the SDGs. These alternatives are then used to explore the relationship between technology and growth in general. The key question asked is how technology can also play a role in non-growth-based approaches to sustainable development, thus departing from the technological determinism often found in common approaches to techno-solutionism as discussed throughout this book.

18.2 GROWTH AND THE ROLE OF TECHNOLOGY IN THE SUSTAINABLE DEVELOPMENT GOALS

Before examining the SDGs, it is worth having a brief look at its foundations in the concept of sustainable development, developed in the Brundtland commission's 1987 report *Our Common Future* (Brundtland et al., 1987). The basic aspects of this concept are described in Chapter 2, and I focus on the role of *growth* and *technology* in this chapter. The Brundtland report clearly argues that economic growth is required and that "[w]hat is needed now is a new era of economic growth – growth that is forceful and at the same time socially and environmentally sustainable" (Brundtland et al., 1987). This is further developed in chapter IV.3 of the report:

> We see instead the possibility for a new era of economic growth, one that must be based on policies that sustain and expand the environmental resource base. And we believe such growth to be absolutely essential to relieve the great poverty that is deepening in much of the developing world.
>
> (Brundtland et al., 1987)

The report is deeply pragmatic and sees the decoupling of economic growth and environmental impact as the *only* path that simultaneously allows us to address environmental *and* social sustainability – a position that is challenged later. Nevertheless, this perspective on the need for growth is largely accepted in mainstream corporate and intergovernmental forums on sustainable development today, as exemplified through, for example, efforts to achieve *green growth* (Jacobs, 2013). The report was, and still is, criticized for the emphasis on economic growth, and even those who seek to defend the report from the most severe criticisms tend to agree that economic growth is central to the notion of sustainable development there developed. Langhelle (1999), for example, attempts to show that while "economic growth is an important part of the message in *Our Common Future*, it is definitely not the entire message" – and the latter part is certainly not challenged here.

Another central message in *Our Common Future* is the role of new technology as a "mainspring of economic growth" (Brundtland et al., 1987). Technology is argued to be capable of "slowing the dangerously rapid consumption of finite resources", but the authors also acknowledge the dangers of "new forms of pollution and the introduction to the planet of new variations of life forms that could change evolutionary pathways" (Brundtland et al., 1987). Exacerbating these problems is the fact that polluting industries highly reliant on natural resources tend to grow rapidly in the developing world, which both need growth and have limited means of addressing environmental harms (Brundtland et al., 1987). One of the central features of the report is arguably the belief that deep pragmatism is required when attempting to address environmental challenges while simultaneously upholding social and environmental justice.

While the role of technology is emphasized, it is also explicitly accepted that new technologies introduce new risks. Arguing in favor of a precautionary principle of technology application focused on environmental sustainability, the authors argue that "national and international institutional mechanisms are needed to assess potential impacts of new technologies before they are widely used" (Brundtland et al., 1987). Furthermore, technologies of great potential also introduce challenges related to distribution, as developed nations have better access to new technologies, which might exacerbate – rather than alleviate – inequality (see Chapter 8). Technology transfer is thus essential (Brundtland et al., 1987), something also highlighted in the SDGs.

In Agenda 2030, it is stated that our time is one of "immense opportunity", and that information and communication technology (ICT) and increased connectedness on a global scale can "accelerate human progress" and "develop knowledge societies" (United Nations, 2015). Scientific and technological innovation are seen as key to both understanding and exploiting these immense opportunities (United Nations, 2015).

Two goals of obvious relevance are SDG 8 and SDG 9. Goal 8 is to "Promote sustained, inclusive and sustainable economic growth, full and productive employment and decent work for all", while SDG 9 is to "Build resilient infrastructure, promote inclusive and sustainable industrialization and foster innovation". Starting with growth, the targets of SDG 8 point to sustaining per capita economic growth "in accordance with national circumstances" with a specific goal of at least 7% gross domestic product (GDP) growth per annum for the least developed nations (8.1). Sustained growth and increased productivity are to be achieved through "technological upgrading and innovation" (8.2), and resource efficiency must also be improved in order to "decouple economic growth from environmental degradation" (8.4). Finally, target 8.a mentions the need for trade-related technical assistance to least developed countries.

SDG 9 is a compound goal, and in this context it is useful to distinguish between targets aimed at infrastructure, industry, and innovation (Sætra, 2022). In terms of infrastructure, 9.1 points toward "quality, reliable, sustainable and resilient infrastructure" to support economic development, human well-being, while also being affordable and providing "equitable access for all". 9.2 refers to increasing industry's role in GDP and employment significantly, while 9.4 states that both infrastructure and industry should increasingly adopt "clean and environmentally sound technologies and industrial processes". 9.5 proceeds in the same vein with an emphasis on the role of research in upgrading technological capabilities. 9.a refers to the need for "enhanced financial, technological and technical support" to those most in need, while 9.b targets the support of "domestic technology development, research and innovation in developing countries". Finally, 9.c points to the vital role of ICT and the need to increase access to such technologies, with an emphasis on "universal and affordable access to the Internet".

The broader agenda also encourages movement toward a world in which "development and the application of technology are climate-sensitive, respect biodiversity and are resilient" (United Nations, 2015), but further explanations of what this entails are omitted. While AI for climate change mitigation and adaption is currently a hot topic (Clutton-Brock et al., 2021; Rolnick et al., 2022), technology is not explicitly mentioned in SDG 13 on climate action. Neither is it mentioned in SDG 15 (life on land), while SDG 14 does aim to increase research capacity and transfer of marine technology (14.a).

Throughout the other goals, technology is presented as important for reducing poverty (SDG 1), ending hunger (SDG 2), as an area of importance related to education for those in least developed countries (SDG 4), empowering women through enabling technology (SDG 5), access to clean water and sanitation (SDG 6), and ensuring access to clean energy (SDG 7). In addition, a subsection of SDG 17 – partnership for reaching the goals – contains three goals focused on technology. 17.6 emphasizes the need for cooperation for promoting access to technology and a "global technology facilitation mechanism". 17.7 aims to promote technology transfer to developing countries on "favorable terms", while 17.8 focuses on operationalizing the Technology Bank for the Least Developed Countries[1] and using more enabling technology.

18.3 THE ALTERNATIVES TO GROWTH AND CENTRALIZATION

The SDGs and the foundational concept sustainable development accept growth as inevitable and give technology a major role in achieving this. While the SDGs almost exclusively discuss the positive potential of technology, we did see a discussion of the accompanying dangers in *Our Common Future* (Brundtland et al., 1987). Nevertheless, both the SDGs and sustainable development are based on a foundational techno-optimism, indicating that it is assumed that technology will, in balance, have a positive effect in terms of bringing about more of what is considered good. This aligns with what Danaher (2022) refers to as the preponderance view of techno-optimism. Seeing the qualifications made in the sustainable development framework, it does not represent techno-utopianism, but it is even further away from representing a variety of techno-pessimism or technophobia (Dinello, 2005).

But what are the alternatives to the techno-growth paradigm enshrined in sustainable development and the SDGs? I will here focus on attempts that challenge the need for growth, with an emphasis on how degrowth can in theory be technology based *or* technology limiting. This demonstrates that while technology is often tied to the growth imperative, it need not be.

18.3.1 The Critique of Growth

The scale of the challenges we now face across all sustainability dimensions is unprecedented, and the environmental challenges have even given rise to the concept of the *Anthropocene* – a new era of Earth's history "in which humankind had emerged as the most powerful influence upon global ecology" (McNeill & Engelke, 2014). Jørgen Randers, one of the authors of the famous Club of Rome report *Limits to growth* (Meadows et al., 1972), 40 years after the original report describes the world as small and fragile, populated by a "huge, dangerous and powerful" collection of human beings (Randers, 2012).

We might be dangerous, but we are also the ones in danger. One of the major figures in the "degrowth" community, Serge Latouche (2009), describes our current situation as follows:

> We are heading for a crash. We are in a performance car that has no driver, no reverse gear and no brakes and it is going to slam into the limitations of the planet.
>
> (Latouche, 2009, p. 2)

The feeling of impending catastrophe has triggered a wide range of responses, and the ones discussed as "radical" in this chapter focus on the need to stop senseless and damaging growth. This was a crucial issue for Arne Næss and deep ecology (Næss, 1999), but today the mainstream movement often connected with this fight is the *degrowth* community. However, despite its unfortunate consequences, growth tends to be pursued for seemingly very good reasons. Randers (2012), for example, sees great challenges associated with the projected plateauing, and then decline, of nations' GDP. Slower economic growth, or even *no growth* or a contraction, is a problem that will require drastic changes in our societies:

> It's not only the City analyst who will worry about my forecast of slowing economic growth in the rich world over the coming decades; most people feel that growth is desirable. The fundamental reason why most people favour growth is that it is the only way modern society has found to solve three problems effectively: poverty, unemployment, and pensions.
>
> (Randers, 2012, p. 5)

Latouche and others *have* discussed the end of economic growth at length, and having a plan for Randers' scenario seems indispensable (*even if* we should believe that continued growth is most likely):

> We know that simply contracting the economy plunges our societies into disarray, increases the rate of unemployment and hastens the demise of the health, social, educational, cultural and environmental projects that provide us with an indispensable minimal quality of life. It is not difficult to imagine the catastrophes that negative growth would bring about.
>
> (Latouche, 2009, p. 8)

For Latouche, "degrowth" is not an ideology, but a *necessity.* We must evaluate forms of society that don't require perpetual growth, he argues. Theoretically, he prefers the term "a-growth" (as in atheism), for approaches that "reject the irrational and quasi-idolatrous cult of growth for growth's sake" (Latouche, 2009). This – a move away from focusing on constantly expanding GDPs – is also a key issue for Næss and deep ecology (Næss, 1999). Næss says that we need to find the difference between *great* and *big*, and both Latouche and Næss seek to evaluate quality of life by criteria far removed from the ones based on consumption, etc. (Næss, 1999).

Latouche also discusses Kenneth Boulding's "spaceship Earth" metaphor and names him one of very few economists (at least in his time), to see the problems with maximizing consumption (Latouche, 2009). The "cowboy economy" is based on "predation and the pillaging of natural resources", while the "spaceman economy" is Boulding's suggestion for a sustainable way of using the resources we have available (Boulding, 1966). According to Latouche, he "concludes that anyone who believes that exponential growth can go on forever in a finite world is either a madman or an economist" (Latouche, 2009).

The critique of the "growth imperative" is nothing new, but it comes in various guises, such as today's degrowth (Kerschner et al., 2018). However, other varieties of "post-development" present somewhat similar critiques of mainstream notions of "development" and growth (Dunlap et al., 2021). The origins of both post-development and degrowth are often traced to the same people, and Ivan Illich is particularly often named a key source of inspiration (Dunlap et al., 2021; Kerschner et al., 2018).

Illich argued, among other things, that growth results from a mindset in which technology and the tools we use are not treated not as *means* but rather as *ends* (Samerski, 2018). Limiting our reliance on and use of technology was important for Illich (Samerski, 2018), and he was also the originator of the concept of *convivial* technology (Kerschner et al., 2018). Convivial technology enables us to "reestablish the 'autonomy' of humans from large hierarchical nondemocratic techno-structures powered by fossil fuels" (Kerschner et al., 2018) and can be analyzed through the dimensions *relatedness, access, adaptability, bio-interaction*, and *appropriateness* (Vetter, 2018). It is contrasted with *manipulative* technology which is both hard for users to understand and control, and which promotes heteronomous action instead of *autonomous action* (Samerski, 2018).

In their review of who the contributors to a special issue on degrowth and technology cite, Kerschner et al. (2018) find that Illich is a clear number one. Other frequently cited authors include, for example, Latouche, Georgescu-Roegen, Schumaher, and Ellul. In the technology for degrowth research agenda they develop, their first point refers to "New classical authors" and the need to broaden the literature on the role of technology in degrowth.

18.3.2 Næss and Deep Ecology

This chapter focuses on an author largely omitted from the citations of the articles just mentioned (Kerschner et al., 2018), namely Arne Næss whose writings are largely in line with post-development and degrowth (Dunlap et al., 2021). Næss' main work on deep ecology, *Ecology, community and Lifestyle* (Næss, 1989, 1999)[2] contains detailed discussions about the limitations of GDP-focused growth and what he refers to as soft, hard, and remote technologies. Næss' ideas are built upon, among other sources, Ellul's *The Technological Society* (1964). Readers of Ellul will see this in, for example, Næss' foregrounding of the politics of technology. As he relates this directly to environmental and economic concerns, his work is of particular relevance to understanding the radical potential of technology in the context of sustainable development.

My main purpose is not to present or develop Næss' full philosophy but to explore his concepts of different kinds of technologies and the general role of technology in achieving a more sustainable future. In doing so, we might also note that Næss himself cites many of the same authors as mentioned in Kerschner et al. (2018). In the 1999 version of the book, he cited, among others, Illich, Gerorgescu-Roegen, Schumacher, and Ellul (Næss, 1999). However, the first version of the book was named *Økologi og filosofi* (Ecology and Philosophy) and was released in 1971, which indicates that his philosophy in many ways preempts and was developed alongside the canon of degrowth.

In order to situate Næss' philosophy alongside this book's project, and also degrowth, certain basic aspects of his philosophy should be established. First of all, Næss distinguishes between *shallow* and *deep* ecology (Næss, 1973, 1999), where the shallow ecology movement is presented as follows:

> Fight against pollution and resource depletion. Central objective: the health and affluence of people in the developed countries.
>
> (Næss, 1973)

Deep ecology, on the other hand, relies on a perspective highlighting (a) humans as not existing *in* the environment but as part of it, (b) bisopherical egalitarianism in principle, (c) diversity and symbiosis, (d) anti-class posture, (e) the fight against pollution and resource depletion, (f) complexity over complication, and (g) local autonomy and decentralization (Næss, 1973). Deep ecology is an expressly normative system, and Næss developed his own version of it – referred to as Ecosophy T – described in the following eight points:

1. The flourishing of human and nonhuman life on Earth has intrinsic value. The value of nonhuman life forms is independent of the usefulness these may have for narrow human purposes.

2. Richness and diversity of life forms are values in themselves and contribute to the flourishing of human and nonhuman life on Earth.
3. Humans have no right to reduce the richness and diversity except to satisfy vital needs.
4. Present human interference with the nonhuman world is excessive, and the situation is rapidly worsening.
5. The flourishing of human life and cultures is compatible with a substantial decrease in the human population. The flourishing of nonhuman life requires such a decrease.
6. Significant change in life conditions for the better requires a change in policies. These affect basic economic, technological, and ideological structures.
7. The ideological change is mainly that of appreciating *life quality* (dwelling in situations with intrinsic value) rather than adhering to a high standard of living. There will be a profound awareness of the difference between big and great.
8. Those who subscribe to the foregoing points have an obligation directly or indirectly to participate in the attempt to implement the necessary changes (Næss, 1989, p. 29).

What is referred to by Næss as biospherical egalitarianism, or *ecocentrism*, can be contrasted with *technocentrism* or *technological environmentalism*. O'Riordan (1981) contrasts the two approaches and argues that the former lacks faith in large-scale technology, which is seen as requiring elitist expertise, being dependent on a central state, and anti-democratic. Materialism "for its own sake" is perceived as wrong, and the main reason for promoting growth would for the ecocentrists be to fulfill the basic needs of people living below subsistence levels, as emphasized by Næss (1999).

O'Riordan (1981) also describes the latter group – the technocentrists – composed of "environmental managers" and "cornucopians". The former highlight the need to create the right incentives (through taxes, fees, laws) to allow for the continuation of growth and resource exploitation. The latter group displays an even greater faith in humanity's ability to "find a way out of any difficulties either political, scientific or technological". Cornucopians emphasize human will and ingenuity, and believe that science and technology is how we can and will improve "the lot of the world's people" (O'Riordan, 1981), and they are thus representatives of the prometheanism described in Chapters 1 and 2.

18.3.3 Soft Technology in Deep Ecology

Næss shares with the technocentrists the opinion that we need to take technology more seriously. Not because it will necessarily solve all our challenges through its continued development and increased complexity, but because technology is instrumental for either *barring* or *opening* the paths to reaching our "ultimate ends" (Næss, 1989).

Næss discusses the need to invent and develop "ecologically satisfactory" technology and to progress from the study of technologies in isolation. Preempting the embedded analysis of Barley (2020) (see Chapter 2), Næss argues that all technologies must be analyzed as *systems* of technologies connected to both up- and downstream processes and technologies. Industrialized agriculture, for example, is a "higher order technology" composed of a vast myriad of technologies and material production processes (Næss, 1999, p. 123).

Soft technology is the label used to describe technologies developed with ecological responsibility in mind, and these are contrasted with *hard technologies*. Næss refers to ongoing – in his time – research, for example, by Robin Clarke and David Dickson, using such terms when discussing ecologically responsible technology, and the labels of soft and hard technology are not presented as his own. Of particular importance for understanding Næss' perspective is the *alternative technology* movement (Smith, 2005), which foreshadows much of what is written about the role of technology in degrowth today. This movement highlighted how technology mediates our relations with one another and nature and how technological solutions tend to only temporarily alleviate the problems we seek to solve through technology. *Unless*, that is, we realize that the problems discussed in this book are "about prioritising multiple social values that are always shifting and developing" (Smith, 2005) – to some degree as a direct result of our use and application of technology (Næss, 1999). The alternative technology movement was particularly interested in renewable energy, organic food and production, cooperative forms of living and producing, and small-scale and local infrastructures (Smith, 2005), all of which fits well with deep ecology and the call for smaller and more decentralized forms of living.

Næss relays Robin Clarke's comparison between soft and hard technology with a translated reference from Dickson (1974). The list is long, and Table 18.1 shows some selected items for comparison.

Næss argues that while instructive, the key distinction between *standardization* and *diversity* must be further emphasized. Of crucial importance is how the localized approach to technology generates great variation in technology and the production of these technologies (Næss, 1999). The benefits from such variation encompass not only the enhanced ability to adapt technology to local needs but also how increased diversity in technologies might generate fertile ground for innovation and continuous development. One example of a modern approach to technology based on such ideas is the "radical indigenous" *Lo-TEK*, described as a design movement aimed "at rebuilding an understanding of indigenous philosophy and vernacular architecture" for sustainable infrastructures (Watson, 2020). Such approaches are less *complicated* and of less *scale* than modern "hi-tech" solutions, but they

TABLE 18.1 Hard versus soft technology, translated excerpt from list reproduced in Næss (1999)

Hard technology	**Soft technology**
Use of materials and energy	Reversible use, recycling
Limited lifespan	Unlimited lifespan
Mass society oriented	Local society oriented
Alienated from nature	Integrated in nature
Growth-oriented	Stable
Alienates young and old	Integrates young and old
Too complicated to be understandable by the public	Understandable by the public

also highlight how traditional and indigenous solutions are also, in essence, *technological*. Rejecting technology is not a solution, but understanding what sort of technology is conducive to the future we desire is crucial.

While all might in principle agree that the goals of soft technology are commendable, the practical difficulties of implementing them are substantial, and often perceived as insurmountable. Næss refers to three key difficulties when political realities are brought to bear on the situation: The fear of decreased business profitability, decreased material standard of living, and the fear of unemployment (Næss, 1999). These take us straight back to the problem of conceiving an alternative to growth and the problems related to an end of growth discussed earlier (Latouche, 2009; Randers, 2012).

Another key point Næss makes is that hard technology is never purely instrumental. It cannot simply be appropriated, borrowed, or transferred between different societies without significant implications for the society into which it is introduced. Næss (1999) describes the impossible position of developing nations when they consider how much, or if at all, they can introduce hard technologies from industrialized countries without also having to introduce *undesirable* aspects of these societies. He describes initial optimism – gaining technology and preserving existing society – but argues that this "best of both worlds" approach has proven to be seriously flawed. Of crucial importance is how the preservation of existing *value systems* and *ideology* has proven to be exceedingly hard, because hard technology is embedded with its own values and ideology (Næss, 1999).

Technologies cannot successfully be imported and implemented in isolation; they require systems of technologies and, crucially, other non-material technologies related to how humans relate to each other, our work culture, etc.

> Stated briefly: one unleashes cultural invasion and increases ones dependence on others. Ones own culture dissolves. This can be referred to as the "domino hypothesis".
>
> (Næss, 1999, p. 128)

These are old debates and relate tightly to the questions of the values and politics inherent in technology (Ellul, 1964; Winner, 1977). Næss here refers to Farvar and Milton (1972), who stressed the importance of refraining from thoughtlessly transferring technologies to "developing countries" because culture is *a whole* in which technologies cannot be selectively introduced without drastically shifting broad sets of relations and institutions (Næss, 1999). Such considerations are just as important today – 50 years later – as technology transfer to developing and least developed nations is now so heavily emphasized (United Nations, 2015).

Before moving on, I close this section with Næss' own nine-point summary of the relationship between technology and "ecosophy T" (his deep ecology):

1. Objects produced by labor of a technical nature are in intimate interaction, not only with the means and the mode of production but also with all essential aspects of cultural activity.

2. Therefore, technology is intimately related directly or indirectly to other social institutions, for example, the sciences, the degree of centralized government, and beliefs about what is reasonable. *Change in technology implies change in culture.*

3. The height of technical development is primarily judged by the leading industrial states in terms of how the techniques can be assimilated into the economies of these states. The more advanced Western science, for example, quantum physics or electronics, a technique presupposes, the higher it is regarded. This untenable criterion of progressiveness is applied not only to our own technology but also to the technology of other cultures. This in turn leads to the general depreciation of the viability of foreign cultures.

4. The ecosophical criteria for progressiveness in technology are relative to ultimate normative objectives. Therefore, culture-neutral statements of the degree of advancement cannot be formulated.

5. The ecosophical basis for an appraisal of technique is the satisfaction of vital needs in the diverse local communities.

6. The objectives of the deep ecological movement do not imply any depreciation of technology or industry, but they imply general cultural control of developments.

7. Technocracies – societies to an overwhelming degree determined by technique and technology – can arise as a consequence of extreme division of labor and intimate merging of technologies of a higher order, combined with extremely specialized, centralized, and exclusive education of technologists. Although neither politicians, nor clergy, nor other groups with authority in the culture can test the explanations granted to the public, they can to some extent determine the political development. The extent of this influence is dependent upon many things: How much technical counter-expertise can be mobilized, and how willing the mass media are to present these counter-reports in a generally understandable form.

8. When a technique is replaced by another which requires more attention, education, and is otherwise more self-engaging and detached, the contact with the medium or milieu in which the technique acts is diminished. To the extent that this medium is *nature*, the engagement in nature is reduced in favor of engagement in the technology. The degree of inattentiveness or apathy increases and thus our awareness of the changes in nature caused by the technique decreases.

9. The degree of self-reliance for individuals and local communities diminishes in proportion to the extent a technique or technology transcends the abilities and resources of the particular individuals or local communities. Passivity, helplessness, and dependence upon "megasociety" and the world market increase (Næss, 1989, pp. 102–103).

18.3.4 Technological Determinism and Radical Sustainable Technology

> It would also be dangerous to suppose that any one group has full insight into and power over the techno-economic systems. The profundity of the crisis is due in part to its largely uncontrolled character: developments proceed at an accelerating pace even though no group, class, or nature has necessarily determined, planned, or accepted the next phase. Built-in mechanisms see to it that the tempo does not slacken. The cog-wheels have drawn us into the very machinery we thought was our slave. Reaching new objectives for progress necessitates greater insight into this machinery, not only within the elites of power, but also within the populace at large. The latter should participate as much as possible both in the formulation of new goals and in suggesting means to reach them.
>
> (Næss, 1989, p. 24)

Both proponents and opponents of the need for continued economic growth agree that technology has played a crucial role in the growth human societies have experienced thus far. However, saying that it has been instrumental in enabling growth need not by necessity imply that technology is the *cause* of such growth nor that technology cannot play an equally important role in the construction of more sustainable societies based on different economic and political foundations.

As discussed in Chapter 2, technology cannot be presumed to be completely neutral (Winner, 1980). Even the hammer, which is at times used as the example of a simple and straightforward contrast to modern complex technologies, is imbued with a certain logic. The old saying referring to how every problem is a nail to a person with a hammer demonstrates that this is not some mystical insight, but both an old and well-understood aspect of *all* technology. Even a hammer has its own politics of sorts, and while more complicated technologies could have political implications that are more difficult to unpack, I consider the fundamental question to be the same.

The main question, then, is whether the values and logic imbued in technologies determine historical development, or if we might either shape and change or control the political tendencies inherent in technologies in order to make sure that they are used in ways that promote our fundamental values (Næss, 1999). I have already emphasized how many philosophers of technology, like Winner (1977), highlight the autonomous force and inherent values of technology. This tends to lead to what some refer to as unintended broader consequences of technology (Collingridge, 1980), as we have seen represented in Næss' writings as well.

However, in order to have any hope of achieving a form of sustainable development that has some potential to address both social and environmental challenges without, for example, the need for rapid and drastic reductions of the human population, we must reject defeatist technological determinism (Barley, 2020; Heilbroner, 1994; Marx & Smith, 1994; Wyatt, 2008), and focus on how to assert social control of technology (Collingridge, 1980). This does not mean that the social power of technology is ignored or denied. It simply entails that we focus our efforts on how to evaluate technologies in terms of their potential implications (Kerschner et al., 2018), that the crucial role of *politics* in determining how to develop and apply technology is foregrounded, and that the competencies and institutions required for effective political control are strengthened (Næss, 1999).

As highlighted by Jasanoff (2016), the term "unanticipated" consequences is problematic, because we *know* that technology has broader social implications than those discovered through simplistic and isolationist analyses. Excusing designers and producers by referring to problems of anticipation is thus deeply troubling. Moreover, providing engineers and those in control with such "outs" could even enable them to weaponized technology – such as Winner's (1977) example of how bridges might be built with too little clearance for buses to pass, effectively keeping those parts of the population dependent on public transportation out of certain areas, etc.

By drawing not only on the philosophy of technology and knowledge of technology's autonomy, we must also draw upon the *sociology* of technology, which highlights how technologies' inherent qualities *underdetermine* historical development, and that social processes are integral to understanding the promotion, selection, and development of technologies (Smith, 2005). However, gaining control over technology is difficult, as Collingridge's dilemma illustrates; when technologies are new, we *can* relatively easily control them, but their consequences are also uncertain. When technologies mature and disseminate, however, we know their consequences but tend to have a hard time regaining control over them (Collingridge, 1980).

18.4 THE ROLE OF TECHNOLOGY IN RADICAL ALTERNATIVE FUTURES

This chapter has focused on the crucial role of technology in achieving *any* kind of future, and in particular how a clear form of techno-optimism – even solutionism – permeates Agenda 2030 and sustainable development in general. By exploring radical alternatives to techno-solutionism, we have seen that even among the "radicals" there are very different perceptions of what the role of technology is or should be. Some argue in favor of deep restraint (Samerski, 2018), while others highlight how other forms of technology – exemplified by Næss' notion of "soft technology" or Lo-TEK – are instrumental in achieving the alternative futures envisioned by those who do not see endless growth as the final solution to human and environmental ills.

The autonomous force of technology has also been highlighted. Not to quash hopes of controlling it, but rather to highlight the need to fight the aimless adoption of new technologies and the values and ideologies they permeate. Technocracy was for Næss the result of such an approach to technology, and combatting such technocracies is one of his key goals. This, he argues, must be done through political mobilization and the purposive pursuit of a deeper understanding of technological values and implications and the competencies required to gain control of it. He argued for mobilizing "technical counter-expertise" to wrest control from the technocrats, and also to examine how to make, for example, the media available for disseminating such expertise (Næss, 1999, p. 131). Academic institutions appear to be crucial for both promoting and building such expertise (Smith, 2005). When academia is connected to social movements, the alternative technology movement and Næss saw some hope for building post-growth societies not premised on the annihilation of technology (Næss, 1999; Smith, 2005).

For technology to have any chance of contributing to sustainable development, Næss (1999) argues that deep reforms of technical education on all levels are required, as engineers are no longer solely responsible for solving technical–economical problems – *humans* will also have to be considered. Furthermore, and crucially, deep reform of the relationships between democratic institutions and public institutions with technical goals is required (Næss, 1999). We thus return to the problem of the social control of technology (Collingridge, 1980), a problem that is, according to Næss, further problematized by how a technological logic inevitably wins out by way of *economic* arguments. Therefore, he also strongly emphasizes the need to resist the urge to turn politics into questions of economics and stresses the need to mobilize not only technical counter-expertise but also economic counter-expertise (Næss, 1999).

One approach to radical technology is the "methodological luddism" of Langdon Winner (1977). This does not entail rejecting all technology, but to realistically appraise the often inflated expectations of technology and to re-politicize technology and its control (Garcia et al., 2018), very much in line with the ecosophy of Arne Næss (Næss, 1999).

Radical approaches to technology can in theory be either techno-optimist or techno-pessimist (Danaher, 2022). However, the main perspective developed in this chapter, derived from a combination of an acceptance of the political nature of technology and an emphasis on the social control of technology to achieve a future clearly distinct from what we are currently headed for, could more plausibly be referred to as techno*realistic*. Technorealism is an existing concept (Bennahum et al., 1998), but one originally tied quite closely to ICT and not technology in general. The concept can, and perhaps should, be further developed to describe a control-focused and skeptical approach to technology.

Achieving sustainable development – broadly understood – might indeed require technology, but it cannot be based on a blind faith in technology-based growth. In fact, we might need "degrowth in the technological sphere" in order to achieve the necessary changes in the economic sphere (Samerski, 2018). However, the perspective here presented does not suggest that it makes sense to think we'll get by without technology at all. It is, instead, a question of making sure that we introduce and promote the right kinds of technology – soft technology – and that we always make sure to analyze technology as a part of broader socio-technical systems and never as isolated instruments divorced from values and ideology (Barley, 2020; Næss, 1999).

NOTES

1 www.un.org/technologybank/mandate
2 The latest edition of the Norwegian language version of this book (Økologi, samfunn og livsstil) is more comprehensive than the English language version, and I mainly rely on the Norwegian version in this discussion of Næss' views.

18.5 REFERENCES

Barley, S. R. (2020). *Work and technological change*. Oxford University Press.

Bennahum, D., Biggs, B. S., Borsook, P., Bowe, M., Garfinkel, S., Johnson, S., . . . Syman, S. (1998). *The principles of technorealism*. www.technorealism.org

Boulding, K. E. (1966). The economics of the coming spaceship earth. In H. Jarret (Ed.), *Environmental quality in a growing economy* (pp. 3–14). The Johns Hopkins University Press.

Brundtland, G. H., Khalid, M., Agnelli, S., Al-Athel, S., & Chidzero, B. (1987). *Our Common Future*: Report of the World Commission on Environment and Development. United Nations General Assembly document A/42/427.

Clutton-Brock, P., Rolnick, D., Donti, P. L., & Kaack, L. (2021). *Climate change and AI. recommendations for government action*. Global Partnership on AI.

Collingridge, D. (1980). *The social control of technology*. Frances Pinter.

Danaher, J. (2022). Techno-optimism: An analysis, an evaluation and a modest defence. *Philosophy & Technology, 35*(54). https://doi.org/10.1007/s13347-022-00550-2

Dickson, D. (1974). *Alternative technology and politics of technical change*. Fontana Books.

Dinello, D. (2005). *Technophobia! Science fiction visions of posthuman technology*. University of Texas Press.

Dunlap, A., Søyland, L. H., & Shokrgozar, S. (2021). Editorial introduction: Situating debates in post-development and degrowth. In A. Dunlap, L. H. Søyland, & S. Shokrgozar (Eds.), *Debates in post-development and degrowth: Volume 1*. Tvergastein.

Ellul, J. (1964). *The technological society*. Vintage Books.

Farley, H. M., & Smith, Z. A. (2020). *Sustainability: If it's everything, is it nothing?* Routledge.

Farvar, M. T., & Milton, J. P. (1972). *Careless technology*. Paper presented at the Conference on the Ecological Aspects of International Development (1968: Warrenton, VA).

Garcia, J. L., Jerónimo, H. M., & Carvalho, T. M. (2018). Methodological Luddism: A concept for tying degrowth to the assessment and regulation of technologies. *Journal of Cleaner Production, 197*, 1647–1653. https://doi.org/10.1016/j.jclepro.2017.03.184

Heilbroner, R. (1994). Technological determinism revisited. In L. Marx & M. R. Smith (Eds.), *Does technology drive history* (pp. 67–78). MIT Press.

Jacobs, M. (2013). Green growth. In R. Falkner (Ed.), *The handbook of global climate and environment policy* (pp. 197–214). Wiley Blackwell.

Jasanoff, S. (2016). *The ethics of invention: Technology and the human future*. W.W. Norton & Company.

Kerschner, C., Wächter, P., Nierling, L., & Ehlers, M.-H. (2018). Degrowth and technology: Towards feasible, viable, appropriate and convivial imaginaries. *Journal of Cleaner Production, 197*, 1619–1636. https://doi.org/10.1016/j.jclepro.2018.07.147

Langhelle, O. (1999). Sustainable development: Exploring the ethics of our common future. *International Political Science Review, 20*(2), 129–149. https://doi.org/10.1177/0192512199202002

Latouche, S. (2009). *Farewell to growth*. Polity.

Marx, L., & Smith, M. R. (1994). *Does technology drive history? The dilemma of technological determinism*. MIT Press.

McNeill, J. R., & Engelke, P. (2014). Into the Anthropocene: People and their planet. In A. Iriye (Ed.), *Global interdependence: The world after 1945* (pp. 365–533). Harvard University Press.

Meadows, D. H., Meadows, D. H., Randers, J., & Behrens III, W. W. (1972). *The limits to growth: Report for the Club of Rome's Project on the Predicament of Mankind*. Universe Books.

Næss, A. (1973). The shallow and the deep, long-range ecology movement. A summary. *inquiry, 16*(1–4), 95–100.

Næss, A. (1989). *Ecology, community and lifestyle: Outline of an ecosophy*. Cambridge university press.

Næss, A. (1999). *Økologi, Samfunn, Livisstil*. Bokklubben dagens bøker.
O'Riordan, T. (1981). *Environmentalism*. Pion.
Pinker, S. (2011). *The better angels of our nature: Why violence has declined*. Viking.
Randers, J. (2012). *2052: A global forecast for the next forty years*. Chelsea Green Publishing.
Rolnick, D., Donti, P. L., Kaack, L. H., Kochanski, K., Lacoste, A., Sankaran, K., . . . Waldman-Brown, A. (2022). Tackling climate change with machine learning. *ACM Computing Surveys (CSUR)*, *55*(2), 1–96.
Sætra, H. S. (2022). *AI for the sustainable development goals*. CRC Press.
Samerski, S. (2018). Tools for degrowth? Ivan Illich's critique of technology revisited. *Journal of Cleaner Production*, *197*, 1637–1646. https://doi.org/10.1016/j.jclepro.2016.10.039
Smith, A. (2005). The alternative technology movement: An analysis of its framing and negotiation of technology development. *Human Ecology Review*, *12*(2), 106.
United Nations. (2015). *Transforming our world: The 2030 agenda for sustainable development*. Division for Sustainable Development Goals.
Vetter, A. (2018). The matrix of convivial technology–assessing technologies for degrowth. *Journal of Cleaner Production*, *197*, 1778–1786. http://dx.doi.org/10.1016/j.jclepro.2017.02.195
Watson, J. (2020). *Lo-TEK: Design by radical indigenism*. Taschen.
Winner, L. (1977). *Autonomous technology: Technics-out-of-control as a theme in political thought*. MIT Press.
Winner, L. (1980). Do artifacts have politics? *Daedalus*, 121–136.
Wyatt, S. (2008). Technological determinism is dead; long live technological determinism. *The Handbook of Science and Technology Studies*, *3*, 165–180.

CHAPTER 19

Conclusion

The Promise and Pitfalls of Techno-solutionism for Sustainable Development

Henrik Skaug Sætra

CONTENTS

19.1 INTRODUCTION

Throughout the preceding chapters, a wide range of aspects related to technology and sustainability have been explored. Issues as varied as the use of technology in education, the engineering of our climate systems, and the use of technology for influencing human behavior have been analyzed. The perspectives and approaches have been as varied as the issues, and it is time to consider what the various findings indicate. What is the promise and what are the pitfalls associated with a techno-solutionist approach to sustainable development?

This chapter contains a synthesis of the key results presented in this book. I begin by presenting the findings related to the positive potential for sustainable development. Then, the limitations and problematic aspects of seeking technological solutions for sustainability-related challenges are discussed and seen in light of the positive potential. Finally, the concluding section considers the overall implications of this book, both in terms of the promise and pitfalls of techno-solutionism for sustainable development, but also with the goal of highlighting the problematic assumptions in and limitations associated with the concept of sustainable development and the Sustainable Development Goals (SDGs).

DOI: 10.1201/9781003325086-19

19.2 WHAT TECHNOLOGY *CAN* DO

The benefits of technology have been shown to be potentially significant and varied. Continuing with the framework presented in Chapter 2, where impacts are analyzed on the micro-, meso-, and macro-levels, we have seen impacts across all levels presented.

On the individual – or micro – level, we have seen how technology might be used to improve the situations of vulnerable individuals. Two examples are the use of VR-based interventions for people with autism and the use of filtering technology by trans persons. Both these cases demonstrate how technologies might be able to improve the situation of vulnerable groups. The former case shows how enabling technologies might be used to overcome individual challenges associated with, for example, disabilities and disorders (Chapter 7) – from serious to less severe. Such benefits of technology also extend to the use of various forms of assistive technologies. For example, technology that helps individuals see what their eyes cannot through computer vision applications or hear what their ears cannot through audio-to-text applications. While companies such as Google are frequently under scrutiny for purported privacy-unfriendly practices, their assistive technology in various guises can be quite useful for individuals with special needs (McNicholl et al., 2021), and such potential benefits must also be factored into the overall picture of the impacts of technology.

The latter case – the use of Shinigami Eyes discussed in Chapter 5 – shows how individuals might use technology to control their connections with the world around them. In the case discussed, the purpose is to avoid harmful individuals and information, and trans people exemplify a particularly vulnerable group for which such technological filtering is useful. Similar technologies could also be used to control who gets access to oneself, and thus serve as a two-way filter, as demonstrated by various privacy-preserving technologies (Brunton & Nissenbaum, 2015).

On the group (meso) level, the use of technology to combat discrimination is one example of how particular groups can benefit from new technologies. However, as discussed both in Chapter 9 and later, the overall potential might not be positive, even if there *are* aspects of the technology conducive to contributing positively to combatting discrimination. Other positive meso-level benefits have not been particularly heavily emphasized in this book, but we have seen how sustainable development and the SDGs expect technology to bring beneficial effects to various groups (see Chapters 8 and 18, and Sætra (2022)).

The positive impacts most emphasized by the contributors seem to arise at the societal (macro) level, and these relate to all dimensions of sustainable development. In the environmental dimension, geoengineering (Chapter 4) and the use of green nudges (Chapter 13) exemplify how technology might be used to both mitigate and adapt to climate and nature-related challenges. In the social dimension, the positive potential related to smart cities properly governed has been presented (Chapter 16), and also potential benefits related to the governance and legal frameworks capable of guiding digital innovation toward good social outcomes (Chapter 10). The latter chapter also touches upon the economic dimension of sustainable development and how technology relates to innovation. Central to this dimension is the idea that innovation and growth are based on technological development, as highlighted by Borgebund in Chapter 12.

In sum, the positive potential of technology has certainly not been dismissed and impacts across all levels and sustainability dimensions have been acknowledged. However, there is a clear discrepancy between such acknowledgment and the techno-optimism manifested in the sustainable development framework and in Agenda 2030 and the SDGs. The main discrepancy stems from how most of the contributors who have acknowledged positive impacts have also thoroughly stressed how technology simultaneously has significant negative impacts – at times *as* or *more* significant than the positive impacts. Not least, many of the positive impacts discussed seem to stem from how technology can be used to find solutions to challenges generated by existing technological solutions. Filtering social media users and content is one example of how the very challenges solved arise because of how technology platforms work. It is time, then, to consider the pitfalls of techno-solutionism and the more general limitations of technology for sustainable development.

19.3 WHAT TECHNOLOGY *CANNOT* DO

While we have seen that technology has positive macro- and micro-level potential, the negative impacts discussed by the contributors have tended to emphasize macro- and meso-level impacts. However, negative impacts related to individuals targeted by filtering technology and others harmed by hateful content online were discussed alongside positive benefits in Chapter 5. Furthermore, other general concerns related to new technologies, such as issues related to privacy infringements and manipulation, were addressed in Chapters 13 and 16.

The contributors have directed much attention toward negative meso-level impacts, and in particular Chapters 8 and 17 have dealt with fundamental consequences related to increased inequality generated by or reinforced by technology in combination with market forces and political systems. This is paradoxical and highly important when we consider how the SDGs promote technology as an inequality inhibitor. A concrete example in the domain of education is presented in Chapter 6. The tight link between technology and *power* presented in Chapter 15 serves to highlight how technology, when not properly controlled, might easily lead to outcomes quite the opposite of those envisioned for sustainable development and in the SDGs.

Finally, negative macro-level effects in the environmental and social dimensions have been emphasized. In Chapter 3, the materiality of technology, and artificial intelligence in particular, was connected with environmental challenges and the climate crisis. One of the major topics discussed in this book has been the challenges related to our political systems and the lack of social and political control of technology. Chapter 11 discusses fundamental governance challenges, while Chapter 14 highlighted the importance of politics through concrete cases of failures of technology for sustainable development. Finally, Chapters 17 and 18 have focused on fundamental challenges and more radical challenges related to how market forces and techno-optimism fuses together in a form of growth fetishism presented as fundamentally opposed to environmental, social, and economic sustainability.

While technology has a positive potential, we might conclude that technological solutions simultaneously tend to introduce new layers of complication in our social structures and consequently new sources of novel challenges. These might, of course, then be sought solved through ever-new technological solutions. The alternative discussed in Chapter 18, however, points toward the potential of technological degrowth and a form of methodological luddism. One key takeaway from the preceding discussion of the promise and pitfalls of techno-solution seems to be that technology tends to address symptoms and rarely the root causes of the problems we want to address. Natural, perhaps, as technology itself is often one of these very root causes.

19.4 TECHNO-SOLUTIONISM, SUSTAINABLE DEVELOPMENT, AND THE SUSTAINABLE DEVELOPMENT GOALS

What remains is to consider the overall implications for the sustainable development framework and the SDGs. Both were presented as clearly techno-optimist in Chapter 18, while Agenda 2030 and the SDGs contained fewer and less obvious caveats related to the potential negative impact of technology than the sustainable development framework. Natural, perhaps, as the SDGs are political goals based on negotiating a very wide array of national and regional concerns – ending up with a common denominator of goals that rarely fundamentally challenges the core values of powerful actors, be they private or public. Regarding the 17 goals, this book supports other work on the SDGs and technology in the overall conclusion that there is both positive and negative potentials of technology, and the limitations are strongest with regard to enabling political change (Sætra, 2022). This is an important conclusion, because without political change, within nations and globally, it seems unlikely that the overall effects of technology will be to promote the goals of, for example, reduced inequality (SDG 10), and *sustainable* and *inclusive* economic growth (SDG 8). The interconnectedness between the various goals has been stressed throughout this book, and Chapter 8 in particular explored the potential inconsistencies between SDG 9 and SDG 10. Despite this, technology has the potential to enable the reaching of most SDGs, but the main conclusion derived from the contributions in this book is that this will not happen without more social and political control of technology.

Sustainable development has received criticism both in this book and in general (Dunlap et al., 2021), and particularly the emphasis on *growth* and *development* has been problematized. Nevertheless, the pragmatism undergirding *Our Common Future* (Brundtland et al., 1987) can be argued to be both necessary and beneficial. In the short and intermediate term, the needs of those least well-off likely cannot be addressed by immediately ending growth and radically changing all political, social, and economic institutions. At least not without severe consequences for a wide range of individual and groups – and most likely the natural world. This is partly why Serge Latouche (2009) admits that degrowth is not a *humanist* philosophy:

> [B]ecause it is based upon a critique of development, growth, progress, technology and, ultimately, modernity and because it implies a break with Western centralism. It is no coincidence that most of those who inspired de-growth (Illich, Ellul, but also Claude Lévi-Strauss, Robert Jaulin, Marshall Sahlins and many others) denounce Western humanism.

One fundamental cause of the difference of opinion with regard to the prospects of technology might consequently be differences in philosophical perspectives and moral values.

These are questions beyond the scope of this concluding chapter, but they have been approached to some degree in different chapters and seem crucial for continuing the debate about what *sustainable development* is, and what our goals related to such development should be. While some may have abandoned hope of salvaging the very concept of sustainability and sustainable development (Dunlap et al., 2021), others seek to re-imagine and fight for a change in what we mean by these concepts and to rearrange their foundations (Farley & Smith, 2020). The latter approach might carry more potential for achieving real-world change, but radical shifts toward completely new societies seem unlikely through this path. The choice, then, might be between likely incremental change and unlikely radical change – an age-old question that will not be settled here.

However, we have argued that technology will play an integral role in whatever future we end up with. Even radical futures where negative social and environmental impacts are decoupled from growth *will* depend on technology. This book will hopefully allow for debates about what sort of technologies we want, and how we might be able to control technologies and their autonomous force.

19.5 REFERENCES

Brundtland, G. H., Khalid, M., Agnelli, S., Al-Athel, S., & Chidzero, B. (1987). *Our Common Future*: Report of the World Commission on Environment and Development. United Nations General Assembly document A/42/427.

Brunton, F., & Nissenbaum, H. (2015). *Obfuscation: A user's guide for privacy and protest.* MIT Press.

Dunlap, A., Søyland, L. H., & Shokrgozar, S. (2021). Editorial introduction: Situating debates in post-development and degrowth. In A. Dunlap, L. H. Søyland, & S. Shokrgozar (Eds.), *Debates in post-development and degrowth: Volume 1.* Tvergastein.

Farley, H. M., & Smith, Z. A. (2020). *Sustainability: If it's everything, is it nothing?* Routledge.

Latouche, S. (2009). *Farewell to growth.* Polity.

McNicholl, A., Casey, H., Desmond, D., & Gallagher, P. (2021). The impact of assistive technology use for students with disabilities in higher education: A systematic review. *Disability and Rehabilitation: Assistive Technology, 16*(2), 130–143. https://doi.org/10.1080/17483107.2019.1642395

Sætra, H. S. (2022). *AI for the sustainable development goals.* CRC Press.

Index

Note: *Italicized* page numbers in this index indicate illustrative materials.

For Product Safety Concerns and Information please contact our EU representative GPSR@taylorandfrancis.com
Taylor & Francis Verlag GmbH, Kaufingerstraße 24, 80331 München, Germany

www.ingramcontent.com/pod-product-compliance
Ingram Content Group UK Ltd.
Pitfield, Milton Keynes, MK11 3LW, UK
UKHW052220190926
475702UK00006B/1112
* 9 7 8 1 0 3 2 3 5 0 5 9 2 *